T0140598

# Lecture Notes in Computational Science and Engineering

**117**

Editors:

More information about this series at http://www.springer.com/series/3527

Tetsuya Sakurai • Shao-Liang Zhang •
Toshiyuki Imamura • Yusaku Yamamoto •
Yoshinobu Kuramashi • Takeo Hoshi
Editors

# Eigenvalue Problems: Algorithms, Software and Applications in Petascale Computing

EPASA 2015, Tsukuba, Japan,
September 2015

*Editors*
Tetsuya Sakurai
University of Tsukuba
Department of Computer Science
Tsukuba
Ibaraki, Japan

Shao-Liang Zhang
Nagoya University
Department of Electrical Engineering
and Computer Science
Nagoya
Aichi, Japan

Toshiyuki Imamura
RIKEN Advanced Institute
for Computational Science
Kobe
Hyogo, Japan

Yusaku Yamamoto
University of Electro-Communications
Graduate School of Informatics
and Engineering
Department of Communication
Engineering and Informatics
Tokyo, Japan

Yoshinobu Kuramashi
University of Tsukuba
Department of Particle Physics
Tsukuba
Ibaraki, Japan

Takeo Hoshi
Tottori University
Department of Applied Mathematics
and Physics
Tottori, Japan

ISSN 1439-7358 ISSN 2197-7100 (electronic)
Lecture Notes in Computational Science and Engineering
ISBN 978-3-319-87309-1 ISBN 978-3-319-62426-6 (eBook)
https://doi.org/10.1007/978-3-319-62426-6

Mathematics Subject Classification (2010): 65F15, 68-06


Softcover re-print of the Hardcover 1st edition 2017


Cover illustration from "exterior of EPOCHAL TSUKUBA" with kind permission by Tsukuba International Congress Center

Printed on acid-free paper

This Springer imprint is published by Springer Nature
The registered company is Springer International Publishing AG
The registered company address is: Gewerbestrasse 11, 6330 Cham, Switzerland

# Preface

The 1st International Workshop on Eigenvalue Problems: Algorithms, Software and Applications in Petascale Computing (EPASA2014) was held in Tsukuba, Ibaraki, Japan on March 7–9, 2014. The 2nd International Workshop EPASA (EPASA2015) was also held in Tsukuba on September 14–16, 2015. This volume includes selected contributions presented at EPASA2014 and EPASA2015.

EPASA2014 and EPASA2015 were organized with the support of the JST-CREST project "Development of an Eigen-Supercomputing Engine using a Post-Petascale Hierarchical Model" (Research Director: Tetsuya Sakurai, University of Tsukuba) in the Research Area "Development of System Software Technologies for post-Peta Scale High Performance Computing". The goal of this CREST project is to develop a massively parallel Eigen-Supercomputing Engine for post-petascale systems. Through collaboration with researchers in applied mathematics, HPC and application fields, two eigen-engines, z-Pares and EigenExa, have been developed.

The goal of the EPASA workshop is to bring together leading researchers working on the numerical solution of matrix eigenvalue problems to discuss and exchange ideas on state-of-the-art algorithms, software and applications in petascale computing. The workshop also aims to create an international community of researchers for eigenvalue problems.

The invited speakers of EPASA2014 were as follows:

- Peter Arbenz (ETH Zürich, Switzerland)
- Anthony P. Austin (University of Oxford, UK)
- Zhaojun Bai (University of California, Davis, USA)
- James Charles (Purdue University, USA)
- Stefan Güttel (The University of Manchester, UK)
- Tsung-Ming Huang (National Taiwan Normal University, Taiwan)

- Jun-ichi Iwata (The University of Tokyo, Japan)
- Hiroshi Kawai (Tokyo University of Science, Suwa, Japan)
- Gerhard Klimeck (Purdue University, USA)
- Bruno Lang (Bergische Universität Wuppertal, Germany)
- Ren-Cang Li (University of Texas at Arlington, USA)
- Wen-Wei Lin (National Chiao Tung University, Taiwan)
- Karl Meerbergen (KU Leuven, Belgium)
- Tsuyoshi Miyazaki (National Institute for Materials Science, Japan)
- Yuji Nakatsukasa (The University of Tokyo, Japan)
- Jose E. Roman (Universitat Politècnica de València, Spain)
- Yousef Saad (University of Minnesota, USA)
- Shigenori Tanaka (Kobe University, Japan)
- Françoise Tisseur (The University of Manchester, UK)
- Lloyd N. Trefethen (Oxford University, UK)
- Marián Vajteršic (University of Salzburg, Slovakia)
- Weichung Wang (National Taiwan University, Taiwan)

The invited speakers of EPASA2015 were as follows:

- Grey Ballard (Sandia National Laboratories, USA)
- Akihiro Ida (Kyoto University, Japan)
- Hiroyuki Ishigami (Kyoto University, Japan)
- Bruno Lang (Bergische Universität Wuppertal, Germany)
- Julien Langou (University of Colorado Denver, USA)
- Daijiro Nozaki (Technische Universität Dresden, Germany)
- Shin'ichi Oishi (Waseda University, Japan)
- Sho Shohiro (Osaka University, Japan)
- Daisuke Takahashi (University of Tsukuba, Japan)
- Françoise Tisseur (The University of Manchester, UK)
- Weichung Wang (National Taiwan University, Taiwan)
- Chao Yang (Lawrence Berkeley National Laboratory, USA)
- Rio Yokota (Tokyo Institute of Technology, Japan)

EPASA2014 and 2015 featured a total of 46 poster presentations focusing on recent developments in parallel eigensolvers, their theoretical analysis, software, applications, etc. More than 140 participants from various fields attended the workshops. This volume gathers 18 selected papers they produced.

We would like to express our gratitude to the participants of both EPASA workshops, the authors that contributed to this volume and the reviewers, whose valuable work helped to improve the quality of these proceedings. Neither workshops would have been possible without all the hard work of the committee members and the

CREST project members. Last but not least we would like to acknowledge Dr. Martin Peters and Ruth Allewelt from Springer for their support.

The 3rd International Workshop EPASA2018 will be held in March 2018.

Tsukuba, Japan — Tetsuya Sakurai
Nagoya, Japan — Shao-Liang Zhang
Kobe, Japan — Toshiyuki Imamura
Tokyo, Japan — Yusaku Yamamoto
Tsukuba, Japan — Yoshinobu Kuramashi
Tottori, Japan — Takeo Hoshi
April 2017

# Contents

**An Error Resilience Strategy of a Complex Moment-Based Eigensolver** ........ 1
Akira Imakura, Yasunori Futamura, and Tetsuya Sakurai

**Numerical Integral Eigensolver for a Ring Region on the Complex Plane** ........ 19
Yasuyuki Maeda, Tetsuya Sakurai, James Charles, Michael Povolotskyi, Gerhard Klimeck, and Jose E. Roman

**A Parallel Bisection and Inverse Iteration Solver for a Subset of Eigenpairs of Symmetric Band Matrices** ........ 31
Hiroyuki Ishigami, Hidehiko Hasegawa, Kinji Kimura, and Yoshimasa Nakamura

**The Flexible ILU Preconditioning for Solving Large Nonsymmetric Linear Systems of Equations** ........ 51
Takatoshi Nakamura and Takashi Nodera

**Improved Coefficients for Polynomial Filtering in ESSEX** ........ 63
Martin Galgon, Lukas Krämer, Bruno Lang, Andreas Alvermann, Holger Fehske, Andreas Pieper, Georg Hager, Moritz Kreutzer, Faisal Shahzad, Gerhard Wellein, Achim Basermann, Melven Röhrig-Zöllner, and Jonas Thies

**Eigenspectrum Calculation of the $O(a)$-Improved Wilson-Dirac Operator in Lattice QCD Using the Sakurai-Sugiura Method** ........ 81
Hiroya Suno, Yoshifumi Nakamura, Ken-Ichi Ishikawa, Yoshinobu Kuramashi, Yasunori Futamura, Akira Imakura, and Tetsuya Sakurai

**Properties of Definite Bethe–Salpeter Eigenvalue Problems** ........ 91
Meiyue Shao and Chao Yang

**Preconditioned Iterative Methods for Eigenvalue Counts** .................. 107
Eugene Vecharynski and Chao Yang

**Comparison of Tridiagonalization Methods Using High-Precision Arithmetic with MuPAT** .................................................... 125
Ryoya Ino, Kohei Asami, Emiko Ishiwata, and Hidehiko Hasegawa

**Computation of Eigenvectors for a Specially Structured Banded Matrix** ................................................................ 143
Hiroshi Takeuchi, Kensuke Aihara, Akiko Fukuda, and Emiko Ishiwata

**Monotonic Convergence to Eigenvalues of Totally Nonnegative Matrices in an Integrable Variant of the Discrete Lotka-Volterra System** ................................................................ 157
Akihiko Tobita, Akiko Fukuda, Emiko Ishiwata, Masashi Iwasaki, and Yoshimasa Nakamura

**Accuracy Improvement of the Shifted Block BiCGGR Method for Linear Systems with Multiple Shifts and Multiple Right-Hand Sides** ................................................................ 171
Hiroto Tadano, Shusaku Saito, and Akira Imakura

**Memory-Saving Technique for the Sakurai–Sugiura Eigenvalue Solver Using the Shifted Block Conjugate Gradient Method** .............. 187
Yasunori Futamura and Tetsuya Sakurai

**Filter Diagonalization Method by Using a Polynomial of a Resolvent as the Filter for a Real Symmetric-Definite Generalized Eigenproblem** ... 205
Hiroshi Murakami

**Off-Diagonal Perturbation, First-Order Approximation and Quadratic Residual Bounds for Matrix Eigenvalue Problems** ......... 233
Yuji Nakatsukasa

**An Elementary Derivation of the Projection Method for Nonlinear Eigenvalue Problems Based on Complex Contour Integration** ............ 251
Yusaku Yamamoto

**Fast Multipole Method as a Matrix-Free Hierarchical Low-Rank Approximation** ................................................................ 267
Rio Yokota, Huda Ibeid, and David Keyes

**Recent Progress in Linear Response Eigenvalue Problems** ................ 287
Zhaojun Bai and Ren-Cang Li

# An Error Resilience Strategy of a Complex Moment-Based Eigensolver

**Akira Imakura, Yasunori Futamura, and Tetsuya Sakurai**

**Abstract** Recently, complex moment-based eigensolvers have been actively developed in highly parallel environments to solve large and sparse eigenvalue problems. In this paper, we provide an error resilience strategy of a Rayleigh–Ritz type complex moment-based parallel eigensolver for solving generalized eigenvalue problems. Our strategy is based on an error bound of the eigensolver in the case that soft-errors like bit-flip occur. Using the error bound, we achieve an inherent error resilience of the eigensolver that does not require standard checkpointing and replication techniques in the most time-consuming part.

## 1 Introduction

In this paper, we consider complex moment-based eigensolvers for computing all eigenvalues located in a certain region and their corresponding eigenvectors for a generalized eigenvalue problem of the following form

$$A\boldsymbol{x}_i = \lambda_i B\boldsymbol{x}_i, \quad \boldsymbol{x}_i \in \mathbb{C}^n \setminus \{\boldsymbol{0}\}, \quad \lambda_i \in \Omega \subset \mathbb{C}, \tag{1}$$

where $A, B \in \mathbb{C}^{n\times n}$ and the matrix pencil $zB - A$ are assumed to be diagonalizable and nonsingular for any $z$ on the boundary of $\Omega$. Let $m$ be the number of target eigenpairs and $X_\Omega$ be an $n \times m$ matrix, whose columns are the target eigenvectors, i.e., $X_\Omega := [\boldsymbol{x}_i | \lambda_i \in \Omega]$.

For solving the generalized eigenvalue problem (1), Sakurai and Sugiura have proposed a projection type method that uses certain complex moment matrices constructed by a contour integral in 2003 [13]. Thereafter, several researchers have actively studied improvements and related eigensolvers based on the complex moment-based eigensolver [5–8, 12, 14, 17]. The concepts of Sakurai and Sugiura have also been extended to solve nonlinear eigenvalue problems [1–3, 18].

---

A. Imakura (✉) • Y. Futamura • T. Sakurai
University of Tsukuba, 1-1-1 Tennodai, Tsukuba, Ibaraki 305-8573, Japan
e-mail: imakura@cs.tsukuba.ac.jp; futamura@cs.tsukuba.ac.jp; sakurai@cs.tsukuba.ac.jp





T. Sakurai et al. (eds.), *Eigenvalue Problems: Algorithms, Software and Applications in Petascale Computing*, Lecture Notes in Computational Science and Engineering 117, https://doi.org/10.1007/978-3-319-62426-6_1

Recently, we analyzed error bounds of the Rayleigh–Ritz type complex moment-based eigensolver called the block SS–RR method [9]. In this paper, we apply the results of the analyses to the case that soft-errors like bit-flip occur. Using the error bound, we provide an error resilience strategy which does not require standard checkpointing and replication techniques in the most time-consuming part of the eigensolver.

The remainder of this paper is organized as follows. In Sect. 2, we briefly describe the basic concepts of the complex moment-based eigensolvers. In Sect. 3, we introduce the algorithm of the block SS–RR method and the results of its error bounds. We also introduce the parallel implementation of the block SS–RR method in Sect. 3. In Sect. 4, we propose an error resilience strategy for the block SS–RR method. In Sect. 5, we show some numerical results and we present conclusions in Sect. 6.

Throughout, the following notations are used. Let $V = [\boldsymbol{v}_1, \boldsymbol{v}_2, \ldots, \boldsymbol{v}_L] \in \mathbb{C}^{n \times L}$, then $\mathscr{R}(V)$ is the range space of the matrix $V$, and is defined by $\mathscr{R}(V) := \mathrm{span}\{\boldsymbol{v}_1, \boldsymbol{v}_2, \ldots, \boldsymbol{v}_L\}$. In addition, for $A \in \mathbb{C}^{n \times n}$, $\mathscr{K}_k^{\square}(A, V)$ are the block Krylov subspaces, $\mathscr{K}_k^{\square}(A, V) = \mathscr{R}([V, AV, \ldots, A^{k-1}V])$.

## 2 Complex Moment-Based Eigensolvers

As a powerful algorithm for solving the generalized eigenvalue problem (1), Sakurai and Sugiura have proposed the complex moment-based eigensolver in 2003 [13]. This is called the SS–Hankel method. To solve (1), they introduced the rational function

$$r(z) := \widetilde{\boldsymbol{v}}^{\mathrm{H}}(zB - A)^{-1}B\boldsymbol{v}, \quad \boldsymbol{v}, \widetilde{\boldsymbol{v}} \in \mathbb{C}^n \setminus \{\boldsymbol{0}\}, \tag{2}$$

whose poles are the eigenvalues $\lambda$ of the matrix pencil $zB - A$. They then considered computing all poles located in $\Omega$.

All poles located in a certain region of a meromorphic function can be computed by the algorithm in [11], which is based on Cauchy's integral formula,

$$r(a) = \frac{1}{2\pi \mathrm{i}} \oint_{\Gamma} \frac{r(z)}{z - a} \mathrm{d}z,$$

where $\Gamma$ is the positively oriented Jordan curve (i.e., the boundary of $\Omega$). By applying the algorithm in [11] to the rational function (2), the target eigenpairs $(\lambda_i, \boldsymbol{x}_i), \lambda_i \in \Omega$ of the generalized eigenvalue problem (1) are obtained by solving the generalized eigenvalue problem:

$$H_M^{<}\boldsymbol{u}_i = \theta_i H_M \boldsymbol{u}_i.$$

Here, $H_M$ and $H_M^<$ are small $M \times M$ Hankel matrices of the form

$$H_M := \begin{pmatrix} \mu_0 & \mu_1 & \cdots & \mu_{M-1} \\ \mu_1 & \mu_2 & \cdots & \mu_M \\ \vdots & \vdots & \ddots & \vdots \\ \mu_{M-1} & \mu_M & \cdots & \mu_{2M-2} \end{pmatrix}, \quad H_M^< := \begin{pmatrix} \mu_1 & \mu_2 & \cdots & \mu_M \\ \mu_2 & \mu_3 & \cdots & \mu_{M+1} \\ \vdots & \vdots & \ddots & \vdots \\ \mu_M & \mu_{M+1} & \cdots & \mu_{2M-1} \end{pmatrix},$$

whose entries consist of the following complex moments

$$\mu_k := \frac{1}{2\pi \mathrm{i}} \oint_\Gamma z^k r(z) \mathrm{d}z.$$

For details, refer to [13].

For more accurate eigenpairs, improvement of the SS–Hankel method has been proposed [14]. This improvement is based on the Rayleigh–Ritz procedure and is called the SS–RR method. Block variants of the SS–Hankel method and the SS–RR method have also been proposed [5, 6] for higher stability of the algorithms, specifically when multiple eigenvalues exist in $\Omega$. These are called the block SS–Hankel method and the block SS–RR method, respectively. An Arnoldi-based interpretation of the complex moment-based eigensolvers and the resulting algorithm have also been proposed [8]. The algorithm is named the block SS–Arnoldi method.

As another approach of the complex moment-based eigensolver, Polizzi has proposed the FEAST eigensolver for Hermitian generalized eigenvalue problems in 2009 [12] and then developed it further [17]. The FEAST eigensolver is an accelerated subspace iteration-type method, and a single iteration is closely connected to a special case of the block SS–RR method with $M = 1$.

The relationship among these complex moment-based eigensolvers was analyzed in [10].

## 3 The Block SS–RR Method

In this section, we introduce the algorithm of the block SS–RR method and the results of its error bounds.

### *3.1 Algorithm of the Block SS–RR Method*

Let $L, M \in \mathbb{N}$ be input parameters. Also let $V \in \mathbb{C}^{n \times L}$ be an input matrix, e.g., a random matrix. Then, we define an $n \times LM$ matrix

$$S := [S_0, S_1, \ldots, S_{M-1}],$$

where

$$S_k := \frac{1}{2\pi \mathrm{i}} \oint_{\Gamma} z^k (zB - A)^{-1} BV \mathrm{d}z. \tag{3}$$

Then, we have the following theorem; see e.g., [9].

**Theorem 1** *Let m be the number of eigenvalues of* (1) *and* rank$(S) = m$. *Then, we have*

$$\mathscr{R}(S) = \mathscr{R}(X_{\Omega}) = \mathrm{span}\{\boldsymbol{x}_i | \lambda_i \in \Omega\}.$$

Theorem 1 indicates that the target eigenpairs $(\lambda_i, \boldsymbol{x}_i), \lambda_i \in \Omega$ can be obtained by the Rayleigh–Ritz procedure with $\mathscr{R}(S)$. The above forms the basis of the block SS–RR method [5]. Continuous integration (3) is approximated by some numerical integration rule such as the $N$-point trapezoidal rule with $N \geq M - 1$. The approximated matrix $\widehat{S}_k$ is expressed as

$$S \approx \widehat{S}_k := \sum_{j=1}^{N} \omega_j z_j^k (z_j B - A)^{-1} BV, \tag{4}$$

where $z_j$ are the quadrature points, and $\omega_j$ are the corresponding weights. We also set

$$S \approx \widehat{S} := [\widehat{S}_0, \widehat{S}_1, \ldots, \widehat{S}_{M-1}]. \tag{5}$$

Here, $(z_j, \omega_j)$ are required to satisfy

$$\sum_{j=1}^{N} \omega_j z_j^k \begin{cases} \neq 0, & (k = -1) \\ = 0, & (k = 0, 1, \ldots, N-2) \end{cases}. \tag{6}$$

The algorithm of the block SS–RR method with numerical integration is consist of the following three steps:

Step 1. Solve $N$ linear systems with $L$ right-hand sides of the form:

$$(z_j B - A) W_j = BV, \quad j = 1, 2, \ldots, N. \tag{7}$$

Step 2. Construct the matrix $\widehat{S}$ by (4) and (5), where $\widehat{S}_k$ can be rewritten by using $W_j$ as follows:

$$\widehat{S}_k = \sum_{j=1}^{N} \omega_j z_j^k W_j, \quad k = 1, 2, \ldots, M-1. \tag{8}$$

**Algorithm 1** The block SS–RR method

**Input:** $L, M, N \in \mathbb{N}, V \in \mathbb{C}^{n \times L}, (z_j, \omega_j), j = 1, 2, \ldots, N$

**Output:** Approximate eigenpairs $(\widehat{\lambda}_i, \widehat{\boldsymbol{x}}_i)$ for $i = 1, 2, \ldots, LM$

1: Solve $W_j = (z_j B - A)^{-1} BV$ for $j = 1, 2, \ldots, N$

2: Compute $\widehat{S}_k = \sum_{j=1}^{N} \omega_j z_j^k W_j$ for $k = 0, 1, \ldots, M-1$ and set $\widehat{S} = [\widehat{S}_0, \widehat{S}_1, \ldots, \widehat{S}_{M-1}]$

3: Compute the orthogonalization of $\widehat{S} : Q = \mathrm{orth}(\widehat{S})$

4: Compute eigenpairs $(\theta_i, \boldsymbol{u}_i)$ of the generalized eigenvalue problem $Q^{\mathrm{H}} A Q \boldsymbol{u}_i = \theta_i Q^{\mathrm{H}} B Q \boldsymbol{u}_i$, and $(\widehat{\lambda}_i, \widehat{\boldsymbol{x}}_i) = (\theta_i, Q\boldsymbol{u}_i)$ for $i = 1, 2, \ldots, LM$

Step 3. Compute approximate eigenpairs by the Rayleigh–Ritz procedure as follows. Solve

$$Q^{\mathrm{H}} A Q \boldsymbol{u}_i = \theta_i Q^{\mathrm{H}} B Q \boldsymbol{u}_i,$$

and $(\widehat{\lambda}_i, \widehat{\boldsymbol{x}}_i) = (\theta_i, Q\boldsymbol{u}_i)$, where $Q = \mathrm{orth}(\widehat{S})$.

The algorithm of the block SS–RR method is summarized as Algorithm 1.

In practice, in order to reduce the computational costs and to improve accuracy, the matrix $\widehat{S}$ is replaced with a low-rank approximation obtained from the singular value decomposition. Moreover, $z^k$ is scaled for improving numerical stability. For details, refer to [5, 15].

The block SS–RR method has some parameters such as $L, M, N$, and these parameters strongly affect the performance of the method. In the current version of the software of the block SS–RR method, z-pares ver.0.9.6a [19], $N = 32, M = 16$ are used as the default parameters. The parameter $L$ is usually set such that $LM = 2m$, where $m$ is the number of the target eigenvalues in $\Omega$. The optimal parameters depend on the eigenvalue distribution, the required accuracy, computational environments and so on. For the details of how to set the parameters achieving good performance, refer to [15].

## 3.2 Error Bounds of the Block SS–RR Method

In [9], the error bounds of the block SS–RR method are analyzed. Here, we briefly introduce the results.

Let the matrix pencil $zB - A$ be diagonalizable, i.e.,

$$Y^{-1}(zB - A)X = z \begin{bmatrix} I_r & \\ & O_{n-r} \end{bmatrix} - \begin{bmatrix} \Lambda_r & \\ & I_{n-r} \end{bmatrix},$$

where $\Lambda_r := \mathrm{diag}(\lambda_1, \lambda_2, \ldots, \lambda_r)$ is a diagonal matrix, and $Y^{-1} := [\widetilde{\boldsymbol{y}}_1, \widetilde{\boldsymbol{y}}_2, \ldots, \widetilde{\boldsymbol{y}}_n]^{\mathrm{H}}$ and $X := [\boldsymbol{x}_1, \boldsymbol{x}_2, \ldots, \boldsymbol{x}_n]$ are nonsingular matrices. The generalized eigenvalue problem $A\boldsymbol{x}_i = \lambda_i B\boldsymbol{x}_i$ has $r := \mathrm{rank}(B)$ finite eigenvalues $\lambda_1, \lambda_2, \ldots, \lambda_r$ and

$n - r$ infinite eigenvalues. The vectors $\widetilde{\boldsymbol{y}}_i$ and $\boldsymbol{x}_i$ are the corresponding left and right eigenvectors, respectively. The filter function

$$f(\lambda_i) := \sum_{j=1}^{N} \frac{\omega_j}{z_j - \lambda_i}, \tag{9}$$

is commonly used for analysis of the complex moment-based eigensolvers [6, 16, 17]. Using this filter function, the matrix $\widehat{S}$ can be written as

$$\widehat{S} = \left(X_r f(\Lambda_r) \widetilde{X}_r^{\mathrm{H}}\right) [V, CV, \ldots, C^{M-1}V], \quad C := X_r \Lambda_r \widetilde{X}_r^{\mathrm{H}},$$

where $\Lambda_r := \mathrm{diag}(\lambda_1, \lambda_2, \ldots, \lambda_r), X_r := [\boldsymbol{x}_1, \boldsymbol{x}_2, \ldots, \boldsymbol{x}_r], \widetilde{X}_r := [\widetilde{\boldsymbol{x}}_1, \widetilde{\boldsymbol{x}}_2, \ldots, \widetilde{\boldsymbol{x}}_r]$ and $X^{-1} = \widetilde{X}^{\mathrm{H}} = [\widetilde{\boldsymbol{x}}_1, \widetilde{\boldsymbol{x}}_2, \ldots, \widetilde{\boldsymbol{x}}_n]^{\mathrm{H}}$. The error bound of the block SS–RR method in [9] can be simplified under some assumption on $V$ as follows.

**Theorem 2** *Let $(\lambda_i, \boldsymbol{x}_i)$ be the exact eigenpairs of the matrix pencil $zB - A$. Assume that $f(\lambda_i)$ are ordered in decreasing order of magnitude $|f(\lambda_i)| \geq |f(\lambda_{i+1})|$. Define $\mathscr{P}$ as the orthogonal projector onto the subspace $\mathscr{R}(\widehat{S})$. Then, we have*

$$\|(I - \mathscr{P})\boldsymbol{x}_i\|_2 \leq \alpha\beta_i \left| \frac{f(\lambda_{LM+1})}{f(\lambda_i)} \right|,$$

*where $\alpha = \|X\|_2 \|X^{-1}\|_2$, $\beta_i$ depends on the angle between the subspace $\mathscr{K}_M^{\square}(C, V)$ and each eigenvector $\boldsymbol{x}_i$.*

Moreover, in [9], the error bound has been proposed for the case in which the solution of the linear system for the $j'$-th quadrature point is contaminated as follows:

$$(z_{j'}B - A)^{-1}BV + E, \tag{10}$$

where $E \in \mathbb{C}^{n \times L}$ is an error matrix of $\mathrm{rank}(E) = L' \leq L$. Because of the contaminated solution (10), the matrix $\widehat{S}$ is also contaminated. We define the contaminated matrix as $\widehat{S}'$. The error bound of the block SS–RR method with the contaminated matrix in [9] can also be simplified under some assumption on $V$ as follows.

**Theorem 3** *Let $(\lambda_i, \boldsymbol{x}_i)$ be the exact eigenpairs of the matrix pencil $(A, B)$. Assume that $f(\lambda_i)$ are ordered in decreasing order of magnitude $|f(\lambda_i)| \geq |f(\lambda_{i+1})|$. Define $\mathscr{P}'$ as the orthogonal projector onto the subspace $\mathscr{R}(\widehat{S}')$. Then, we have*

$$\|(I - \mathscr{P}')\boldsymbol{x}_i\|_2 \leq \alpha\beta_i' \left| \frac{f(\lambda_{LM-L'+1})}{f(\lambda_i)} \right|,$$

*where $\alpha = \|X\|_2 \|X^{-1}\|_2$, $\beta_i'$ depends on the error matrix $E$ and the angle between the subspace $\mathscr{K}_M^{\square}(C, V)$ and each eigenvector $\boldsymbol{x}_i$.*

Here, we note that the values $\beta_i'$ is not equivalent to $\beta_i$, since $\beta_i'$ depends on error matrix $E$ and the contaminated quadrature point $j'$. $\beta_i'$ may become larger for $\lambda_i$ near $z_{j'}$ than others, specifically for the case where $L' = L$. For more details of these theorems, refer to [9].

## *3.3 Parallel Implementation of the Block SS–RR Method*

The most time-consuming part of the block SS–RR method is to solve $N$ linear systems with $L$ right-hand sides (7) in Step 1. For solving the linear systems, the block SS–RR method has hierarchical parallelism; see Fig. 1.

Layer 1. Contour paths can be performed independently.
Layer 2. The linear systems can be solved independently.
Layer 3. Each linear system can be solved in parallel.

By making the hierarchical structure of the algorithm responsive to the hierarchical structure of the architecture, the block SS–RR method is expected to achieve high scalability.

Because Layer 1 can be implemented completely without communications, here we describe a basic parallel implementation of the block SS–RR method for one contour path. Let $P$ be the number of MPI processes used for one contour path. Here

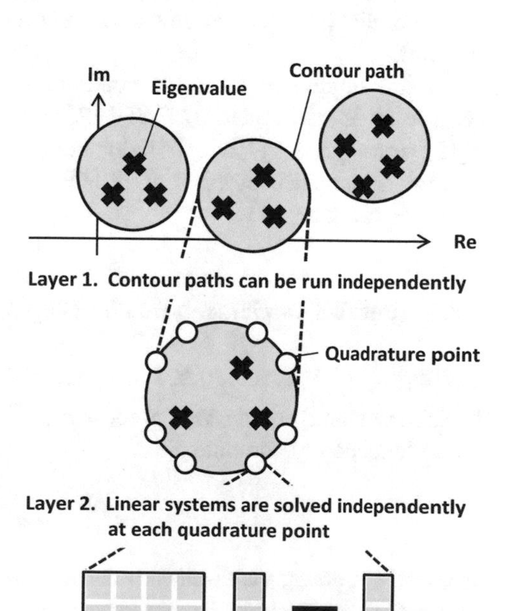


**Fig. 1** Hierarchical structure of the block SS–RR method

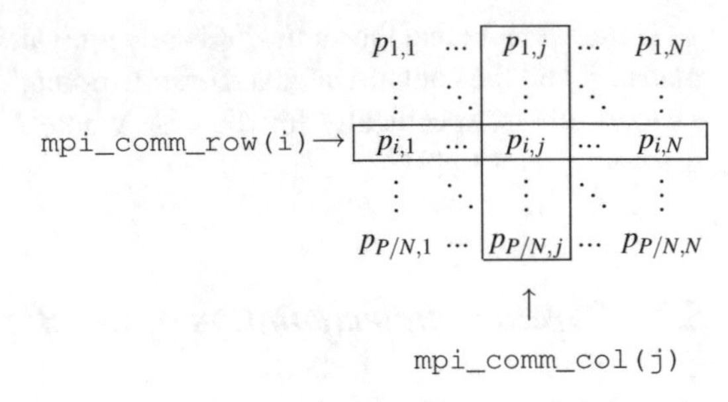


**Fig. 2** Processes grid and MPI sub-communicators

we assume $\mathrm{mod}(P, N) = 0$ for simplicity and consider two dimensional processes grid, i.e., $p_{i,j}, i = 1, 2, \ldots, P/N, j = 1, 2, \ldots, N$. Then, we also define MPI sub-communicators for $N$ MPI processes $p_{i,j}, j = 1, 2, \ldots, N$ as `mpi_comm_row(i)` ($i = 1, 2, \ldots, P/N$) and for $P/N$ MPI processes $p_{i,j}, i = 1, 2, \ldots, P/N$ as `mpi_comm_col(j)` ($j = 1, 2, \ldots, N$); see Fig. 2.

### 3.3.1 Parallel Implementation for Step 1

In Step 1, we need to solve $N$ linear systems (7). Because these linear systems are independent of $j$ (index of quadrature point), we can independently solve these $N$ linear systems in $N$ parallel. Each linear system $(z_jB - A)W_j = V$ is solved by some parallel linear solver on the MPI sub-communicator `mpi_comm_col(j)` in parallel.

In this implementation, the coefficient matrices $A, B$ and the input matrix $V$ require to be distributed to $P/N$ MPI processes in each MPI sub-communicator `mpi_comm_col(j)` with $N$ redundant. As a result, each solution $W_j$ of the linear system is also distributed to $P/N$ MPI processes in the MPI sub-communicator `mpi_comm_col(j)`.

### 3.3.2 Parallel Implementation for Step 2

Let $W_j^{(i)}, (i = 1, 2, \ldots, P/N, j = 1, 2, \ldots, N)$ be the distributed sub-matrix of $W_j$, which are stored by the MPI process $p_{i,j}$. Then, for constructing the matrix $\widehat{S}_k$ (8), we independently compute

$$W_{j,k}^{(i)} = \omega_j z_j^k W_j^{(i)}$$

in each MPI process without communication. Then, we perform `mpi_allreduce` on the MPI sub-communicator `mpi_comm_row(i)` with $N$ MPI processes in $P/N$

parallel as follows:

$$\widehat{S}_k^{(i)} = \sum_{j=1}^{N} W_{j,k}^{(i)}, \quad k = 0, 1, \ldots, M-1.$$

We also set

$$\widehat{S}^{(i)} = [\widehat{S}_0^{(i)}, \widehat{S}_1^{(i)}, \ldots, \widehat{S}_{M-1}^{(i)}],$$

where $\widehat{S}^{(i)}$ are the sub-matrix of $\widehat{S}$, which is redundantly stored by the MPI processes $p_{i,j}, j = 1, 2, \ldots, N$. In this implementation, the matrix $\widehat{S}$ are distributed in $P/N$ MPI processes in each MPI sub-communicator `mpi_comm_row(i)` with $N$ redundant.

### 3.3.3 Parallel Implementation for Step 3

We have two choices for parallel implementation for Step 3. The first choice is that all $P$ MPI processes perform the orthogonalization of $\widehat{S}$ and the Rayleigh–Ritz procedure. This choice makes it possible to work all MPI processes we can use; however, it needs to redistribution of the matrices $A, B$ and $\widehat{S}$.

The second choice is that only $P/N$ MPI processes in the MPI sub-communicator `mpi_comm_col(j)` perform this calculation. In this case, only $P/N$ MPI processes work and the others are just redundant; however, redistribution of the matrices $A, B$ and $\widehat{S}$ does not be required.

## 4 An Error Resilience Strategy of the Block SS–RR Method

With the recent development of high-performance computer, systems scale is drastically increasing. In such situation, fault management is considered to play an important role in large scale application. The fault can be classified to hardware fault and software fault. Here, we focus on software fault like bit-flip.

The most standard software fault tolerance techniques are checkpointing techniques. The checkpointing techniques save all correct data at some interval, and if some fault is detected then it restarts with the last correct data. These are efficient for the case that data size required to save is small and that interval between each checkpoint is small. On the other hand, large data size causes large I/O costs and large interval causes large recalculation costs when fault occurs.

The replication techniques are also very basic software fault tolerance techniques. Its basic idea is shown below. Let $P$ be the number of MPI processes we can use and $K$ be the number of redundancies. Firstly, we split MPI communicator into each $P/K$ MPI processes. The replication techniques restrict the parallelism to $P/K$, i.e., calculation is independently performed by $P/K$ MPI processes in each

MPI sub-communicator. Then, the correct solution is selected from $K$ solutions by e.g. a majority vote. These are efficient when the number of MPI processes is large such that target calculation does not show good scalability. However, if the target calculation shows good scalability, the replication techniques largely increase the execution time even if fault does not occur.

In this section, we consider an error resilience strategy of the block SS–RR method that can use all the MPI processes for the most time-consuming part, i.e., to solve the $N$ linear systems (7) in Step 1 and avoid resolving them even if fault occurs. Here, we assume the following software fault:

- Let $a \in \mathbb{F}$ be the correct value, where $\mathbb{F}$ is the set of floating point numbers. The fault occurs as the numerical error as follows:

$$a' \leftarrow a + e, \quad e \in \mathbb{F}, \tag{11}$$

  where $a' \in \mathbb{F}$ is the contaminated value. Here, $a, a', e$ are not "`Inf`" or "`Nan`".
- Unlike hardware faults, remaining calculation are correctly performed with the contaminated values.

## 4.1 Error Resilience Strategy

As shown in Sect. 3, the algorithm of the block SS–RR method and its parallel implementation can be divided into three steps: solving the linear systems, the numerical integration and the Rayleigh–Ritz procedure. Here, we consider error resilience of each step.

### 4.1.1 Error Resilience Strategy for Step 1

Step 1 is the most time-consuming part and also the most scalable part of the block SS–RR method. Therefore, standard checkpointing and replication techniques may not be efficient for computational costs. Hence, we introduce an alternative strategy to standard checkpointing and replication techniques for computational costs.

When fault occurs in Step 1, some kind of value(s) in calculation are replaced as (11) due to the fault. Then, the contamination is propagated to all MPI processes in the same MPI sub-communicator `mpi_comm_col(j)` via communication. As a result, the solution of the linear system is replaced as

$$W'_{j'} \leftarrow W_{j'} + E, \quad E \in \mathbb{F}^{n \times L}, \quad \mathrm{rank}(E) = L, \tag{12}$$

when fault occurs in the MPI process $p_{i,j'}$ associated with the $j'$-th linear system.

Here, we reconsider Theorems 2 and 3. Theorem 2 implies that the error bound of the block SS–RR method is evaluated by the ratio of the magnitude of the filter

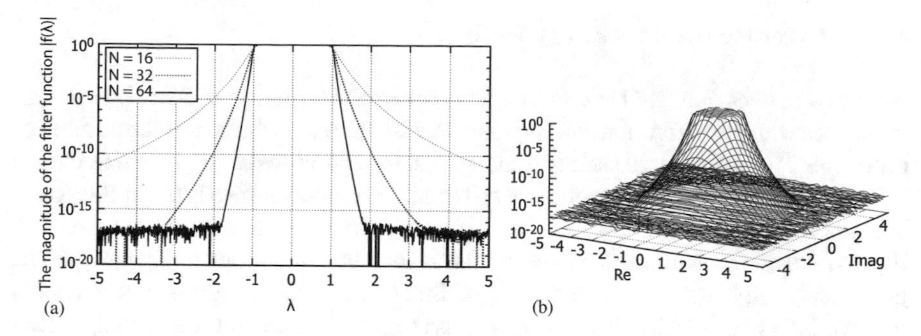


**Fig. 3** Magnitude of filter function $|f(\lambda)|$ of the $N$-point trapezoidal rule with $N = 16, 32, 64$ for the unit circle region $\Omega$. (**a**) On the real axis for $N = 16, 32, 64$. (**b**) On the complex plane for $N = 32$

function $|f(\lambda_i)|$ to the $(LM+1)$-th largest $|f(\lambda_{LM+1})|$. The magnitude of the filter function $|f(\lambda_i)|$ of the $N$-point trapezoidal rule with $N = 16, 32, 64$ for the unit circle region $\Omega$ is shown in Fig. 3. The filter function has $|f(\lambda)| \approx 1$ inside the region $\Omega$, $|f(\lambda))| \approx 0$ far from the region and $0 < |f(\lambda)| < 1$ outside but near the region. Because of Theorem 2 and the filter function, we usually set subspace size $LM$ such that $|f(\lambda_{LM+1})| \approx 0$ to compute the target eigenpairs $(\lambda_i, \boldsymbol{x}_i), \lambda_i \in \Omega$ with high accuracy.

Regarding the filter function, Theorem 3 implies that the accuracy of the block SS–RR method with the contaminated solution is evaluated by the ratio of the magnitude of the filter function $|f(\lambda_i)|$ to the $(LM-L+1)$-th largest $|f(\lambda_{LM-L+1})|$. Of course, Theorem 3 support the case when fault occurs in Step 1 like (12). Therefore, if we consider the case that fault occurs in Step 1, we just set subspace size $LM$ such that $|f(\lambda_{LM-L+1})| \approx 0$ in order to obtain the eigenpairs to high accuracy.

Here, we note that, when multiple faults occur in different quadrature points, i.e.,

$$W'_{j'_1} \leftarrow W_{j'_1} + E_1, \quad W'_{j'_2} \leftarrow W_{j'_2} + E_2, \quad E_1, E_2 \in \mathbb{F}^{n \times L}, \quad \mathrm{rank}(E_1) = \mathrm{rank}(E_2) = L,$$

then we can handle the fault in Step 1 by setting larger subspace $LM$ such that $|f(\lambda_{LM-2L+1})| \approx 0$.

This is an error resilience strategy for Step 1, which makes it possible to use all MPI processes for computing the $N$ linear systems (7) and to avoid resolving them even if fault occurs.

### 4.1.2 Error Resilience Strategy for Step 2

The computational cost for Step 2 is very small, and the data size is not exorbitant large. Therefore, we can apply checkpointing technique with small additional costs for Step 2.

#### 4.1.3 Error Resilience Strategy for Step 3

As noted in Sect. 3.3, we have two choices for implementation of Step 3: to use all processes with redistribution and to replicate without redistribution. If the number of processes $P$ is not so large such that this part shows good scalability, the first choice is better in terms of computational costs. If not, the second choice is better due to the costs of redistribution. In practice, we want to increase the number of processes $P$, if possible, during $N$ linear systems, which is the most time-consuming part, shows good scalability. And computation of $N$ independent linear systems is expected to have better scalability than one of the orthogonalization and the Rayleigh–Ritz procedure. Hence, we usually employ the second choice.

Therefore, we can apply replication technique without no additional costs for Step 3.

### 4.2 A Possibility of Development to Other Complex Moment-Based Eigensolvers

In Sect. 4, we proposed the error resilience strategy of the block SS–RR method which is based on the error analysis in [9]. Here, we consider a possibility of development of our strategy to other complex moment-based eigensolvers.

The proposed error resilient strategy is mainly based on Theorem 3 for the block SS–RR method. Similar theorems as Theorem 3 could be derived for other complex moment-based eigensolvers. One of the most important respects of Theorem 3 is that the subspace size $LM$ should be larger than the rank of error matrix $L'$, i.e., $LM > L'$. In the case of one linear solution is contaminated in the block SS–RR method with $M \geq 2$, the condition $LM > L \geq L'$ is always satisfied and this makes it possible to derive the proposed error resilient strategy.

For development of our strategy to other complex moment-based eigensolvers, we can expect that the proposed strategy is also utilized to other complex moment-based eigensolvers with high order complex moments such as the (block) SS–Hankel method and the block SS–Arnoldi method, although more detailed analyses and numerical experiments are required. Because these methods with $M > 2$ always satisfy the condition $LM > L \geq L'$ as well as the block SS–RR method.

On the other hand, the current proposed strategy may be difficult to recover the error of the complex moment-based eigensolvers only with low order complex moments such as the FEAST eigensolver [12, 17] and the Beyn method [3]. The subspace size of these methods is $L$ which is the same as the number of right-hand side of the linear systems. This indicates that the rank of the error matrix reaches the subspace size in the worst case. In this case, our strategy can not recover the error.

## 5 Numerical Experiments

In this section, we experimentally evaluate the results of the error resilience strategy specifically for Step 1.

### 5.1 Example I

For the first example, we apply the block SS–RR method with and without soft-error in Step 1 to the following model problem

$$A\boldsymbol{x}_i = \lambda \boldsymbol{x}_i,$$
$$A = \mathrm{diag}(0.01, 0.11, 0.21, \ldots, 9.91) \in \mathbb{R}^{100\times 100},$$
$$\lambda_i \in \Omega = [-1, 1],$$

and evaluate its accuracy.

We evaluate the relation between accuracy with the number of subspace size $LM$. To evaluate this relation, we fixed the parameters as $L = 10$ and $N = 32$, and tested four cases $M = 1, 2, 3, 4$ ($LM = 10, 20, 30, 40$). For this example, we set $\Gamma$ as the unit circle and the quadrature points as

$$z_j = \cos(\theta_j) + \mathrm{i}\sin(\theta_j), \quad \theta_j = \frac{2\pi}{N}\left(j - \frac{1}{2}\right)$$

for $j = 1, 2, \ldots, N$. We let fault occur at one of the following quadrature points,

$$z_{j'} = \begin{cases} z_1 = \cos\left(\frac{\pi}{32}\right) + \mathrm{i}\sin\left(\frac{\pi}{32}\right) \\ z_8 = \cos\left(\frac{15\pi}{32}\right) + \mathrm{i}\sin\left(\frac{15\pi}{32}\right) \\ z_{16} = \cos\left(\frac{31\pi}{32}\right) + \mathrm{i}\sin\left(\frac{31\pi}{32}\right) \end{cases}$$

The algorithm was implemented in MATLAB R2014a. The input matrix $V$ and the error matrix $E$ were set as different random matrices generated by the Mersenne Twister in MATLAB, and each linear system was solved by the MATLAB command "\".

We show in Table 1 the relation of the minimum and the maximum values of $\|\boldsymbol{r}_i\|_2$ in $\lambda_i \in \Omega$ with $LM$. Table 1(a) is for the case without fault and Table 1(b)–(d) are for the case when fault occurs in Step 1. We also show in Fig. 4 the residual 2-norm $\|\boldsymbol{r}_i\|_2 := \|A\boldsymbol{x}_i - \lambda_i B\boldsymbol{x}_i\|_2 / \|\boldsymbol{x}_i\|_2$ for the block SS–RR method with and without fault.

Table 1 shows that $\min_{\lambda_i\in\Omega} \|\boldsymbol{r}_i\|_2$ have approximately the same order as $|f(\lambda_{LM+1})|$ for the case without fault and as $|f(\lambda_{LM-L+1})|$ when fault occurs

**Table 1** Relation of accuracy of the block SS–RR method with *LM* when fault occurs in Step 1

| *(a) Without fault* | | | |
|---|---|---|---|
| *M (LM)* | $\lvert f(\lambda_{LM+1})\rvert$ | $min_{\lambda_i \in \Omega} \lVert \boldsymbol{r}_i \rVert_2$ | $max_{\lambda_i \in \Omega} \lVert \boldsymbol{r}_i \rVert_2$ |
| 1 (10) | $4.21 \times 10^{-1}$ | $1.76 \times 10^{-1}$ | $1.34 \times 10^{-1}$ |
| 2 (20) | $1.98 \times 10^{-10}$ | $2.29 \times 10^{-10}$ | $2.11 \times 10^{-9}$ |
| 3 (30) | $5.06 \times 10^{-16}$ | $1.44 \times 10^{-15}$ | $1.20 \times 10^{-14}$ |
| 4 (40) | $2.25 \times 10^{-17}$ | $2.03 \times 10^{-15}$ | $3.46 \times 10^{-15}$ |
| *(b) Fault occurs at* $z_1$ | | | |
| *M (LM)* | $\lvert f(\lambda_{LM-L+1})\rvert$ | $min_{\lambda_i \in \Omega} \lVert \boldsymbol{r}_i \rVert_2$ | $max_{\lambda_i \in \Omega} \lVert \boldsymbol{r}_i \rVert_2$ |
| 1 (10) | $1.00 \times 10^{0}$ | $5.23 \times 10^{-1}$ | $8.23 \times 10^{-1}$ |
| 2 (20) | $4.21 \times 10^{-1}$ | $1.84 \times 10^{-1}$ | $2.63 \times 10^{-1}$ |
| 3 (30) | $1.98 \times 10^{-10}$ | $2.43 \times 10^{-10}$ | $1.63 \times 10^{-8}$ |
| 4 (40) | $5.06 \times 10^{-16}$ | $6.57 \times 10^{-15}$ | $1.91 \times 10^{-13}$ |
| *(c) Fault occurs at* $z_8$ | | | |
| *M (LM)* | $\lvert f(\lambda_{LM-L+1})\rvert$ | $min_{\lambda_i \in \Omega} \lVert \boldsymbol{r}_i \rVert_2$ | $max_{\lambda_i \in \Omega} \lVert \boldsymbol{r}_i \rVert_2$ |
| 1 (10) | $1.00 \times 10^{0}$ | $5.42 \times 10^{-1}$ | $7.99 \times 10^{-1}$ |
| 2 (20) | $4.21 \times 10^{-1}$ | $1.04 \times 10^{-1}$ | $7.74 \times 10^{-1}$ |
| 3 (30) | $1.98 \times 10^{-10}$ | $5.11 \times 10^{-10}$ | $4.57 \times 10^{-9}$ |
| 4 (40) | $5.06 \times 10^{-16}$ | $5.16 \times 10^{-15}$ | $2.87 \times 10^{-14}$ |
| *(d) Fault occurs at* $z_{16}$ | | | |
| *M (LM)* | $\lvert f(\lambda_{LM-L+1})\rvert$ | $min_{\lambda_i \in \Omega} \lVert \boldsymbol{r}_i \rVert_2$ | $max_{\lambda_i \in \Omega} \lVert \boldsymbol{r}_i \rVert_2$ |
| 1 (10) | $1.00 \times 10^{0}$ | $5.54 \times 10^{-1}$ | $7.85 \times 10^{-1}$ |
| 2 (20) | $4.21 \times 10^{-1}$ | $4.11 \times 10^{-1}$ | $4.84 \times 10^{-1}$ |
| 3 (30) | $1.98 \times 10^{-10}$ | $7.05 \times 10^{-10}$ | $4.96 \times 10^{-9}$ |
| 4 (40) | $5.06 \times 10^{-16}$ | $3.71 \times 10^{-15}$ | $2.51 \times 10^{-14}$ |

in Step 1, respectively. Moreover, Fig. 4 shows that enough large subspace size ($LM = 40$ in this example) provides equally high accuracy independent of fault in Step 1.

## 5.2 Example II

For the second example, we apply the block SS–RR method with and without soft-error in Step 1 to the generalized eigenvalue problem AUNW9180 from ELSES matrix library [4]. The coefficient matrices $A, B$ are 9180 dimensional real sparse symmetric matrices and $B$ is also positive definite. We consider finding all eigenpairs $(\lambda_i, \boldsymbol{x}_i), \lambda_i \in \Omega = [0.119, 0.153]$. In this region, there exist 99 eigenvalues.

We set $\Gamma$ as the ellipse (center: 0.131, semi-major axis: 0.012 and semi-minor axis: 0.0012), and the quadrature points as

$$z_j = 0.131 + 0.012 \left(\cos(\theta_j) + 0.1\mathrm{i}\sin(\theta_j)\right),$$

$$\theta_j = \frac{2\pi}{N}\left(j - \frac{1}{2}\right)$$

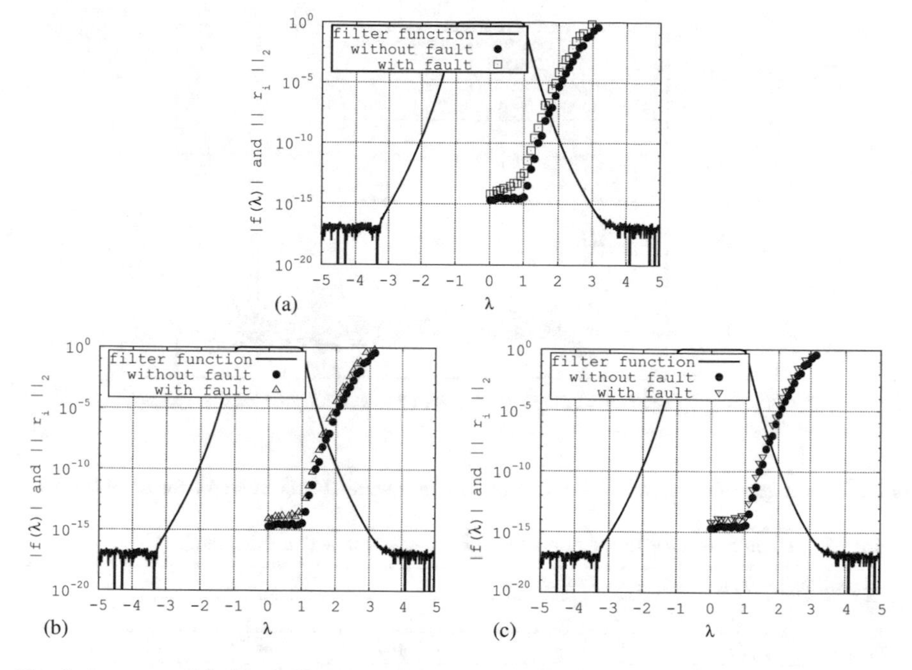


**Fig. 4** Accuracy of the block SS–RR method with $L = 10, M = 4, N = 32$ when fault occurs in Step 1. (**a**) Fault occurs at $z_1$. (**b**) Fault occurs at $z_8$. (**c**) Fault occurs at $z_{16}$

for $j = 1, 2, \dots, N$. We also set parameters as $L = 25, M = 8, N = 32$ for the case without fault and as $L = 25, M = 10, N = 32$ when fault occurs in Step 1.

The input matrix $V$ and the error matrix $E$ were set as different random matrices generated by the Mersenne Twister, and each linear system was solved by "cluster_sparse_solver" in Intel MKL. Here, we note that, in this numerical experiment, we solved only $N/2$ linear systems with multiple right-hand sides for $j = 1, 2, \dots, N/2$, because the linear solution $W_{N-j}$ can be constructed from $W_j$ using a symmetric property of the problem.

The numerical experiments were carried out in double precision arithmetic on 8 nodes of COMA at University of Tsukuba. COMA has two Intel Xeon E5-2670v2 (2.5 GHz) and two Intel Xeon Phi 7110P (61 cores) per node. In this numerical experiment, we use only CPU part. The algorithm was implemented in Fortran 90 and MPI, and was executed with 8 [node] × 2 [process/node] × 8 [thread/process].

We show in Fig. 5 the residual 2-norm $\|\boldsymbol{r}_i\|_2 := \|A\boldsymbol{x}_i - \lambda_i B\boldsymbol{x}_i\|_2 / \|\boldsymbol{x}_i\|_2$ for the block SS–RR method with and without fault. This shows that, by increasing subspace size $LM$, the block SS–RR method with fault can achieve approximately the same accuracy as the case without fault.

Table 2 shows that the computation time of the block SS–RR method without fault using 1–16 processes and the computation time of the block SS–RR method

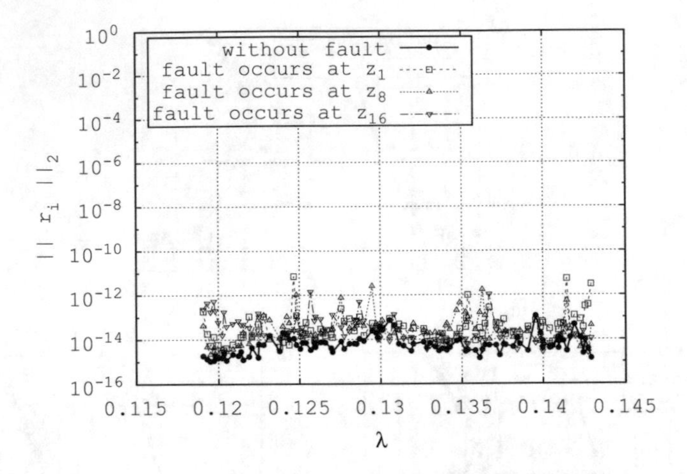


**Fig. 5** Accuracy of the block SS–RR method with and without fault in Step 1 for AUNW9180

**Table 2** Computation time of the block SS–RR method with and without fault

| *(a) without fault* ($L = 25, M = 8, N = 32$) | | | | | |
|---|---|---|---|---|---|
| | *Time [s]* | | | | |
| *#process* | *Step 1* | *Step 2* | *Step 3* | *MISC* | *Total* |
| 1 | 2.14E+02 | 4.20E−05 | 4.92E−01 | 1.18E−01 | 2.15E+02 |
| 2 | 1.06E+02 | 1.05E−02 | 4.76E−01 | 7.87E−02 | 1.06E+02 |
| 4 | 5.30E+01 | 1.49E−02 | 4.77E−01 | 6.34E−02 | 5.36E+01 |
| 8 | 2.66E+01 | 2.05E−02 | 4.79E−01 | 5.73E−02 | 2.72E+01 |
| 16 | 1.34E+01 | 1.56E−02 | 4.78E−01 | 5.33E−02 | 1.39E+01 |
| *(b) with fault* ($L = 25, M = 10, N = 32$) | | | | | |
| | *Time [s]* | | | | |
| *#process* | *Step 1* | *Step 2* | *Step 3* | *MISC* | *Total* |
| 16 | 1.37E+01 | 2.09E−02 | 6.44E−01 | 1.10E−02 | 1.44E+01 |

with fault using 16 processes. This result indicates that Step 1 of the SS–RR method is the most time-consuming. We can also observe from this result that the proposed strategy recovers software faults with very small additional computational costs.

## 6 Conclusion

In this paper, we investigated the error resilience strategy of the Rayleigh–Ritz type complex moment-based parallel eigensolver (the block SS–RR method) for solving generalized eigenvalue problems. Based on the analyses of the error bound of the method, we provided the error resilience strategy which does not require standard checkpointing and replication techniques in the most time-consuming and the most scalable part. From our numerical experiment, our strategy recovers software faults like bit-flip with small additional costs.

**Acknowledgements** The authors would like to thank the anonymous referees for their useful comments. This research was supported partly by University of Tsukuba Basic Research Support Program Type A, JST/CREST, JST/ACT-I (Grant No. JPMJPR16U6) and KAKENHI (Grant Nos. 25286097, 25870099, 17K12690). This research in part used computational resources of COMA provided by Interdisciplinary Computational Science Program in Center for Computational Sciences, University of Tsukuba.

## References

1. Asakura, J., Sakurai, T., Tadano, H., Ikegami, T., Kimura, K.: A numerical method for nonlinear eigenvalue problems using contour integrals. JSIAM Lett. **1**, 52–55(2009)
2. Asakura, J., Sakurai, T., Tadano, H., Ikegami, T., Kimura, K.: A numerical method for polynomial eigenvalue problems using contour integral. Jpn. J. Ind. Appl. Math. **27**, 73–90 (2010)
3. Beyn, W.-J.: An integral method for solving nonlinear eigenvalue problems. Lin. Algor. Appl. **436**, 3839–3863 (2012)
4. ELSES matrix library, http://www.elses.jp/matrix/
5. Ikegami, T., Sakurai, T.: Contour integral eigensolver for non-Hermitian systems: a Rayleigh-Ritz-type approach. Taiwan. J. Math. **14**, 825–837 (2010)
6. Ikegami, T., Sakurai, T., Nagashima, U.: A filter diagonalization for generalized eigenvalue problems based on the Sakurai-Sugiura projection method. J. Comput. Appl. Math. **233**, 1927–1936 (2010)
7. Imakura, A., Sakurai, T.: Block Krylov-type complex moment-based eigensolvers for solving generalized eigenvalue problems. Numer. Algor. **75**, 413–433 (2017)
8. Imakura, A., Du, L., Sakurai, T.: A block Arnoldi-type contour integral spectral projection method for solving generalized eigenvalue problems. Appl. Math. Lett. **32**, 22–27 (2014)
9. Imakura, A., Du, L., Sakurai, T.: Error bounds of Rayleigh–Ritz type contour integral-based eigensolver for solving generalized eigenvalue problems. Numer. Algor. **71**, 103–120 (2016)
10. Imakura, A., Du, L., Sakurai, T.: Relationships among contour integral-based methods for solving generalized eigenvalue problems. Jpn. J. Ind. Appl. Math. **33**, 721–750 (2016)
11. Kravanja, P., Sakurai, T., van Barel, M.: On locating clusters of zeros of analytic functions. BIT **39**, 646–682 (1999)
12. Polizzi, E.: A density matrix-based algorithm for solving eigenvalue problems. Phys. Rev. B **79**, 115112 (2009)
13. Sakurai, T., Sugiura, H.: A projection method for generalized eigenvalue problems using numerical integration. J. Comput. Appl. Math. **159**, 119–128 (2003)
14. Sakurai, T. Tadano, H.: CIRR: a Rayleigh-Ritz type method with counter integral for generalized eigenvalue problems. Hokkaido Math. J. **36**, 745–757 (2007)
15. Sakurai, T., Futamura, Y., Tadano, H.: Efficient parameter estimation and implementation of a contour integral-based eigensolver. J. Algor. Comput. Technol. **7**, 249–269 (2014)
16. Schofield, G., Chelikowsky, J.R., Saad, Y.: A spectrum slicing method for the Kohn-Sham problem. Comput. Phys. Commun. **183**, 497–505 (2012)

17. Tang, P.T.P., Polizzi, E.: FEAST as a subspace iteration eigensolver accelerated by approximate spectral projection. SIAM J. Matrix Anal. Appl. **35**, 354–390 (2014)
18. Yokota, S., Sakurai, T.: A projection method for nonlinear eigenvalue problems using contour integrals. JSIAM Lett. **5**, 41–44 (2013)
19. z-Pares, http://zpares.cs.tsukuba.ac.jp/

# Numerical Integral Eigensolver for a Ring Region on the Complex Plane

**Yasuyuki Maeda, Tetsuya Sakurai, James Charles, Michael Povolotskyi, Gerhard Klimeck, and Jose E. Roman**

**Abstract** In the present paper, we propose an extension of the Sakurai-Sugiura projection method (SSPM) for a circumference region on the complex plane. The SSPM finds eigenvalues in a specified region on the complex plane and the corresponding eigenvectors by using numerical quadrature. The original SSPM has also been extended to compute the eigenpairs near the circumference of a circle on the complex plane. However these extensions can result in division by zero, if the eigenvalues are located at the quadrature points set on the circumference. Here, we propose a new extension of the SSPM, in order to avoid a decrease in the computational accuracy of the eigenpairs resulting from locating the quadrature points near the eigenvalues. We implement the proposed method in the SLEPc library, and examine its performance on a supercomputer cluster with many-core architecture.

## 1 Introduction

We consider a generalized eigenvalue problem $A\boldsymbol{x} = \lambda B\boldsymbol{x}$, where $A, B \in \mathbb{C}^{n\times n}$, $\lambda \in \mathbb{C}$ is an eigenvalue, and $\boldsymbol{x} \in \mathbb{C}^n\backslash\{\boldsymbol{0}\}$ is an eigenvector. Eigenvalue problems arise in many scientific applications such as in quantum transport models, where the self-energy is required to describe the charge injection and extraction effect of the contact. To compute the self-energy exactly, one needs to compute all of the

Y. Maeda (✉) • T. Sakurai
Department of Computer Science, University of Tsukuba, 1-1-1 Tennodai, Tsukuba, Ibaraki 305-8573, Japan
e-mail: maeda@mma.cs.tsukuba.ac.jp; sakurai@cs.tsukuba.ac.jp

J. Charles • M. Povolotskyi • G. Klimeck
Network for Computational Nanotechnology Purdue University, West Lafayette, IN 47906, USA
e-mail: charlejs.james@gmail.com; mpovolot@purdue.edu; gekco@purdue.edu

J.E. Roman
D. Sistemes Informàtics i Computació, Universitat Politècnica de València, Camí de Vera s/n, Valencia 46022, Spain
e-mail: jroman@dsic.upv.es



T. Sakurai et al. (eds.), *Eigenvalue Problems: Algorithms, Software and Applications in Petascale Computing*, Lecture Notes in Computational Science and Engineering 117, https://doi.org/10.1007/978-3-319-62426-6_2

eigenpairs, however it is enough for practical applications to compute only some of the eigenpairs. In [9], it is necessary to obtain the eigenvalues $\lambda = e^{ik\Delta}$ near the circumference of a circle on the complex plane $|\lambda| = 1$ and corresponding eigenvectors, where $\Delta$ is the lattice period length, and $k$ is the wave number.

The shift-invert Arnoldi method is a widely used method for obtaining interior eigenpairs[12]. This method computes eigenvalues close to a shift point and the corresponding eigenvectors. It is hard to obtain eigenpairs near the circumference with the shift-invert Arnoldi method. The Sakurai-Sugiura projection method (SSPM)[7, 8, 13] has been proposed for computing eigenvalues in a given region, and the corresponding eigenvectors, with contour integration. The SSPM finds eigenvalues in a domain surrounded by an integration path, by solving linear systems of equations at the quadrature points with numerical quadrature. An extension of the SSPM for calculating eigenvalues in the arc-shaped region by dividing the circumference of a circle into several arcs, and computing the eigenpairs for each line was proposed in [10]. This extension allows effective parallel computing of the eigenpairs in each arc. However, the quadrature points are set on the arc, and when the eigenpairs are located at the quadrature points, division by zero arises in the calculations.

In this paper, we present an alternative extension of the SSPM by setting two arcs, which avoids a decrease in the computational accuracy of the eigenpairs resulting from locating the quadrature points near the eigenvalues, and allows parallel computation.

We test the proposed method in SLEPc (the Scalable Library for Eigenvalue Problem Computations) [5].

This paper is organized as follows. In Sect. 2, we review the SSPM and an extension of the method for arcs. In Sect. 3, we propose an extension of the SSPM for the partial ring region and implement it in SLEPc. In Sect. 4, we discuss the results of the numerical experiments, and our conclusions are presented in Sect. 5.

## 2 An Extension of the SSPM for Arcs

In this section, we introduce the SSPM for generalized eigenvalue problems[13] and show an extension of the SSPM for the ring region on the complex plane[10]. The extension divides the ring region into several arcs, and calculates the eigenpairs near each arc. In the extension, we construct a subspace that contains the eigenvectors associated with the eigenvalues near the arc.

First, we introduce the SSPM. Let $\Gamma$ be a positively oriented closed Jordan curve on the complex plane. The SSPM approximates eigenvalues inside of the closed Jordan curve $\Gamma$ and corresponding eigenvectors, using a two-step procedure. The first step is to construct the subspace with a filtering for eigenvectors, and the second step is to extract the eigenpairs inside the closed Jordan curve.

We now introduce the procedure for constructing the subspace. Suppose that $m$ eigenvalues are located inside $\Gamma$, let $V$ be a $n \times L$ matrix, the column vectors of which are linearly independent, and let $S = [S_0, S_1, \ldots, S_{M-1}]$ where $S_k$ are $n \times L$ matrices be $n \times LM$ matrices which are determined through contour integration,

$$S_k = \frac{1}{2\pi \mathrm{i}} \oint_{\Gamma} z^k (zB - A)^{-1} BV \mathrm{d}z, \text{ for } k = 0, 1, \ldots, M-1, \tag{1}$$

where $zB - A$ is a regular matrix pencil on $z \in \Gamma$, and $M$ is chosen such that $LM > m$.

We assume that the matrix pencil $\mu B - A$ is diagonalizable for any $\mu$; regular matrices $X = (\boldsymbol{x}_1, \boldsymbol{x}_2, \ldots, \boldsymbol{x}_n)$ and $Y = (\boldsymbol{y}_1, \boldsymbol{y}_2, \ldots, \boldsymbol{y}_n)$ that satisfy $Y^{\mathrm{H}}(\mu B - A)X = (\mu I - \Lambda)$ exist, where $\Lambda$ is the diagonal matrix with elements $\lambda_1, \lambda_2, \ldots, \lambda_n$ on the diagonal. From the residue theorem,

$$S_k = \sum_{i=1}^{n} f_k(\lambda_i) \boldsymbol{x}_i \boldsymbol{y}_i^{\mathrm{H}} BV,$$

where $\boldsymbol{y}_i$ and $\boldsymbol{x}_i$ are the left and right eigenvector of $\mu B - A$ respectively, and $f_k(\lambda_i)$ is a filter function that satisfies

$$f_k(x) = \frac{1}{2\pi \mathrm{i}} \oint_{\Gamma} \frac{z^k}{z - x} \mathrm{d}z = \begin{cases} x^k, & x \in G, \\ 0, & \text{otherwise}, \end{cases}$$

where $G$ is the interior region of $\Gamma$. Eigenvalues outside $\Gamma$ are filtered out with the filter function $f_k(\lambda_i)$. Thus the components of $S$ in the direction of eigenvectors with eigenvalues outside $\Gamma$ are reduced.

In the case that the Jordan curve $\Gamma$ is a circle with a center $\gamma$ and a radius $\rho$, an $N$-point trapezoidal rule can be applied to compute (1) numerically, that is

$$S_k \approx \hat{S}_k = \sum_{j=1}^{N} w_j \zeta_j^k X_j, \tag{2}$$

where

$$z_j = \gamma + \rho \mathrm{e}^{\frac{2\pi \mathrm{i}}{N}(j + \frac{1}{2})}, \quad w_j = \frac{z_j - \gamma}{\rho N}, \quad \zeta_j = \frac{z_j - \gamma}{\rho}, \quad j = 0, 1, \ldots, N-1,$$

are quadrature points, normalized quadrature points and corresponding weights, respectively, and $X_j$, $j = 0, 1, \ldots, N-1$ are the solutions of linear systems with multiple right-hand side vectors,

$$\left(z_j B - A\right) X_j = BV, \quad j = 0, 1, \ldots, N-1. \tag{3}$$

The filter function $f_k(x)$ is approximated by the $N$-point trapezoidal rule as

$$f_k(x) \approx \hat{f}(x) x^k = \sum_{j=1}^{N} \frac{w_j}{z_j - x} x^k, \quad 0 \leq k \leq N-1, \tag{4}$$

where $\hat{f}(x)$ is a rational function. The rational function $\hat{f}(x)$ and eigenvectors in $\hat{S}_k$ depend on $z_j, w_j, \zeta_j$ and $N$. In this case, the rational function $\hat{f}(x)$ decays outside the circle[7, 11]. Thus the components of $\hat{S}_k$ in the direction of eigenvectors with eigenvalues outside $\Gamma$ are small.

Next, we introduce the procedure for the approximation of eigenpairs using the Rayleigh-Ritz approach for the SSPM[7]. Let the singular value decomposition (SVD) of $\hat{S} = [\hat{S}_0, \hat{S}_1, \ldots, \hat{S}_{M-1}] \in \mathbb{C}^{n\times(LM)}$ be $\hat{S} = Q\Sigma W^{\mathrm{H}}$, where $Q = [\boldsymbol{q}_1, \boldsymbol{q}_2, \ldots, \boldsymbol{q}_{LM}] \in \mathbb{C}^{n\times LM}$, $\Sigma = \mathrm{diag}(\sigma_1, \sigma_2, \ldots, \sigma_{LM})$, $\sigma_1 \geq \sigma_2 \geq \ldots \geq \sigma_{LM}$ and $W \in \mathbb{C}^{LM\times LM}$. We omit singular values less than $\delta$, and construct $\hat{Q} = [\boldsymbol{q}_1, \boldsymbol{q}_2, \ldots, \boldsymbol{q}_K] \in \mathbb{C}^{n\times K}$, where $K > m$, and $\sigma_K \geq \delta \geq \sigma_{K+1}$. We solve the small eigenvalue problem

$$(\alpha_i \hat{Q}^{\mathrm{H}} B \hat{Q} - \hat{Q}^{\mathrm{H}} A \hat{Q})\boldsymbol{u}_i = \boldsymbol{0}, \quad \hat{Q}^{\mathrm{H}} A \hat{Q}, \hat{Q}^{\mathrm{H}} B \hat{Q} \in \mathbb{C}^{K\times K},$$

where $\alpha_i$ is the eigenvalue of the matrix pencil $\alpha_i \hat{Q}^{\mathrm{H}} B \hat{Q} - \hat{Q}^{\mathrm{H}} A \hat{Q}$ and $\boldsymbol{u}_i$ is the eigenvector corresponding to $\alpha_i$. Then the eigenvalues of the matrix pencil $A - \lambda B$ are approximated by $\lambda_i \approx \alpha_i$, and the corresponding approximate eigenvectors are given by $\boldsymbol{x}_i \approx \hat{Q}\boldsymbol{u}_i$ for $i = 1, 2, \ldots, K$. Some approximated eigenvalues may appear outside $\Gamma$. We keep eigenvalue $\lambda_i$ inside $\Gamma$ for $i = 1, 2, \ldots, \tilde{m}$, where $\tilde{m}$ is the number of approximated eigenvalues inside $\Gamma$, and discard the rest.

We can compute the eigenpairs in a specific circle by using the SSPM. When many eigenvalues exist in the circle, we have to set a large value for $LM$, and thus the computational cost for computing the eigenpairs is high. In some applications, the eigenpairs near the circumference of the circle are also required. When computing these eigenpairs, the computational cost can be reduced with an extension of the SSPM for the arc as follows[10]. In the extension, the procedure for constructing the subspace is different, but the procedure for extracting eigenpairs remains the same.

Let $\mathbb{L}$ be the arc with center $\gamma$, radius $\rho$, starting angle $\theta_{\mathrm{a}}$ and ending angle $\theta_{\mathrm{b}}$,

$$\mathbb{L} : z = \gamma + \rho \mathrm{e}^{\mathrm{i}\theta}, \quad \theta_{\mathrm{a}} \leq \theta \leq \theta_{\mathrm{b}},$$

where $0 \leq \theta_{\mathrm{a}} < \theta_{\mathrm{b}} \leq 2\pi$. Quadrature points $z_j$, normalized quadrature points $\zeta_j$ and corresponding weights $w_j$ are given by

$$z_j = \gamma + \rho \mathrm{e}^{\mathrm{i}\theta_j},\ \zeta_j = \cos\left(\frac{2j+1}{2N}\pi\right),\ w_j = \frac{T_{N-1}(\zeta_j)}{N}, \quad j = 0, 1, \ldots, N-1, \tag{5}$$

where, $T_k(x)$ is the Chebyshev polynomial of the first kind of degree $k$, and $\theta_j = \theta_{\mathrm{a}} + (\theta_{\mathrm{b}} - \theta_{\mathrm{a}})\frac{\zeta_j+1}{2}$. $\zeta_j$ are $N$ Chebyshev points in the interval $[-1, 1]$, and $z_j$ are points in $\mathbb{L}$. The matrices $\hat{S}_k$ in (2) are computed with $z_j, \zeta_j, w_j$ in (5). According to [10], the rational function $\hat{f}(x)$ in (4) with $z_j, \zeta_j, w_j$ in (5) decays outside the arc

$\mathbb{L}$. Thus the eigenvectors associated with the eigenvalues outside $\mathbb{L}$ are filtered out. Then we extract the eigenpairs using a Rayleigh-Ritz approach, and we can obtain the eigenpairs near $\mathbb{L}$.

For computing the eigenpairs in a ring region, we divide the circumference of the ring into $D$ arcs $\mathbb{L}_d,\ d = 1, 2, \ldots, D$ with $\theta_{\mathrm{a}}^{(d)}, \theta_{\mathrm{b}}^{(d)},\ d = 1, 2, \ldots, D$. Then we compute the eigenpairs on each arc.

## 3 Extension of the SSPM for the Partial Ring Region

In the extension of the SSPM, quadrature points may lie on the arc. Division by zero arises when eigenpairs are located at quadrature points. Therefore, to avoid division by zero, quadrature points should be located sufficiently far from the arc. The filter function, which is approximated by the rational function, is dependent on the quadrature points, and decays outside of the arc. Thus components of eigenvectors in $\hat{S}$ decrease when eigenvalues are farther from quadrature points. When the eigenpairs are located away from the quadrature points, the accuracy of the approximated eigenpairs is reduced. We propose an alternative extension of the SSPM, which avoids a decrease in the computational accuracy of the eigenpairs resulting from locating the quadrature points near the eigenvalues. The proposed method uses alternative formulations for $z_j, \zeta_j, w_j$, and derive the filter function which decays outside of the partial ring region.

Let $\mathbb{L}^{\pm}$ be two arcs such that

$$\mathbb{L}^{\pm} : z = \gamma + \rho^{\pm} \mathrm{e}^{\mathrm{i}\theta}, \quad \theta_{\mathrm{a}} \leq \theta \leq \theta_{\mathrm{b}},$$

where $\gamma$ is the center, and $\rho^{+},\ \rho^{-}$ are the outer and inner radii of the arcs that satisfy $\rho^{+} > \rho^{-}$, and $\theta_{\mathrm{a}}, \theta_{\mathrm{b}}$ are the starting and ending angles that satisfy $0 \leq \theta_{\mathrm{a}} < \theta_{\mathrm{b}} \leq 2\pi$.

Quadrature points $z_j,\ j = 0, 1, \ldots, N-1$ are Chebyshev points on $\mathbb{L}^{+}$ and $\mathbb{L}^{-}$,

$$z_j = \begin{cases} z_j^{+}, & (0 \leq j < N^{+}) \\ z_{j-N^{+}}^{-}, & (N^{+} \leq j < N^{+} + N^{-}) \end{cases},$$

where $z_j^{+}$ are $N^{+}$ Chebyshev points on $\mathbb{L}^{+}$, and $z_j^{-}$ are $N^{-}$ Chebyshev points on $\mathbb{L}^{-}$ defined by (5), and $N = N^{+} + N^{-}$. In the SSPM, a weight for the quadrature $\{w_0, w_1, \ldots, w_{N-1}\}$ is set to satisfy the following equation for computing the eigenpairs inside $\Gamma$ [14],

$$\sum_{j=0}^{N-1} w_j \zeta_j^k = \begin{cases} 1, & (k = -1) \\ 0, & (k = 0, 1, \ldots, N-2) \end{cases}. \tag{6}$$

In the proposed method, we compute the eigenpairs between two arcs. The weights for a quadrature $w_j$ are defined by barycentric weight [2],

$$w_j = (-1)^{N+1} \frac{\prod_{k=0}^{N-1}(\zeta_k)}{\prod_{k=0,k\neq i}^{N-1}(\zeta_k - \zeta_k)},$$

where

$$\zeta_j = \frac{2(z_j - \gamma)}{\rho_1 + \rho_2}, \quad j = 0, 1, \ldots, N-1.$$

The barycentric weight is used for computing weight for quadrature, which satisfies (6). Then, we construct the matrix $\hat{S}_k$ by (2). The procedure after constructing $\hat{S}_k$ is then the same as the SSPM in Sect. 2.

Figures 1 and 2 show schematics of the quadrature points in the SSPM for an arc and a partial ring region, and Figs. 3 and 4 show rational functions $\hat{f}(x)$ defined by (4) in each extension for $N = 32$ quadrature points. In Fig. 3, we set $\gamma = 0$, $\rho = 1$, $\theta_a = 0$ and $\theta_b = \pi$. In Fig. 4, we set $\gamma = 0$, $\rho^+ = 1.01$, $\rho^- = 0.99$, $\theta_a = 0$, $\theta_b = \pi$, $N^+ = 24$ and $N^- = 8$. In the extension for arcs as well as for the partial ring region, the rational function $\hat{f}(x)$ decays outside of the two arcs. Thus

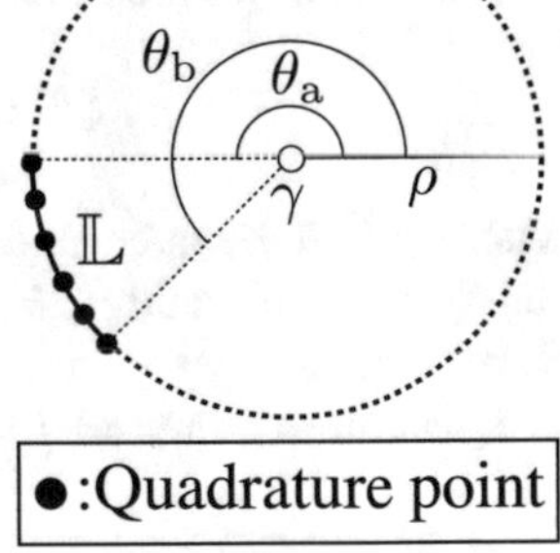


**Fig. 1** Quadrature points for the SSPM for the arc

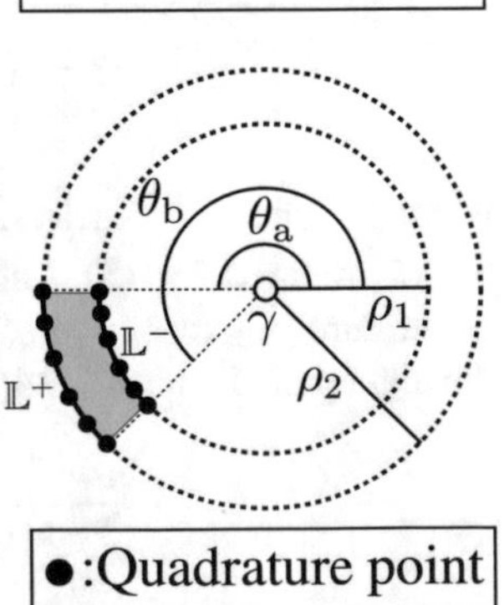


**Fig. 2** Quadrature points for the proposed method

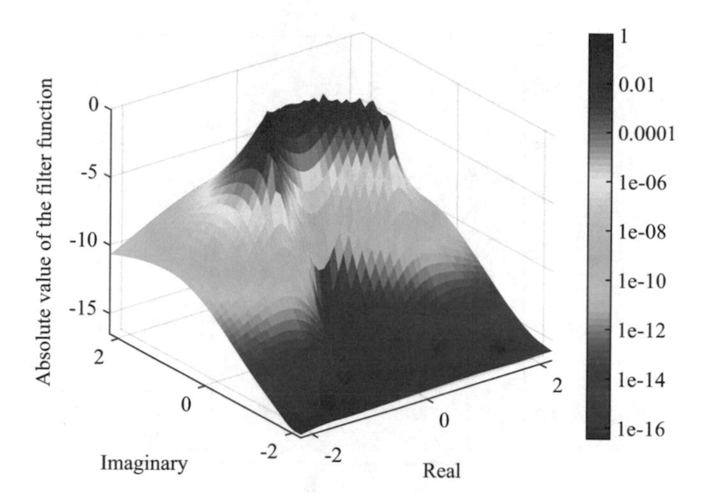


**Fig. 3** Filter function for the SSPM for the arc

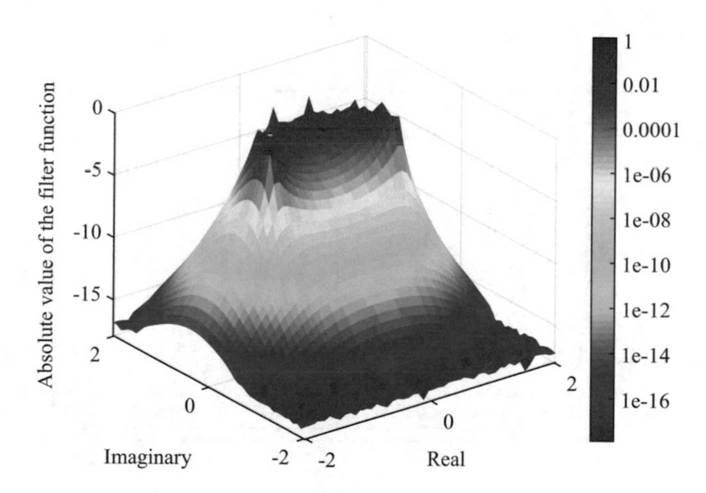


**Fig. 4** Filter function for the proposed method

the components of $\hat{S}_k$ in the direction of eigenvectors associated with eigenvalues outside $\Gamma$ are small.

Figures 6 and 7 show the rational functions $\hat{f}(x)$ on the two lines in Fig. 5. We compute absolute value of the rational function for the SSPM for the arc and the proposed method with $N^+ = 24, N^- = 8$ and $N^\pm = 16$. The parameters $\gamma, \rho, \rho^\pm, \theta_\mathrm{a}$ and $\theta_\mathrm{b}$ are the same as in Figs. 3 and 4. In Figs. 6 and 7, the horizontal axis indicates angle of the line1 and imaginary axis, respectively. In Fig. 6, we can see that the gap between maximum value and minimum value of the rational function for the proposed method is smaller than the gap for the SSPM for the arc. The rational function for the proposed method with $N^+ = 24, N^- = 8$ are similar

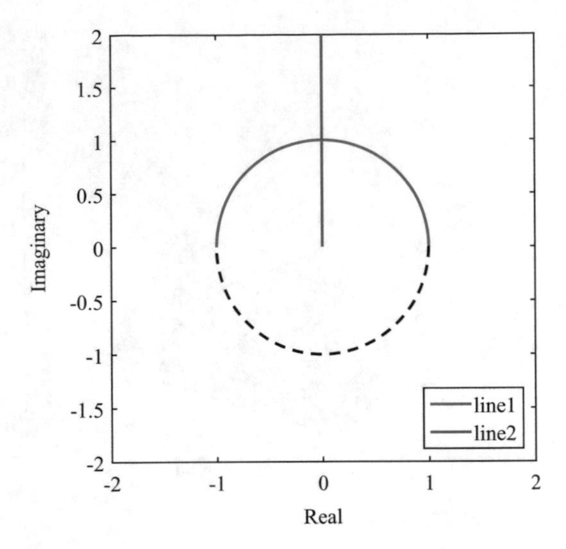


Fig. 5 Two lines for filter function

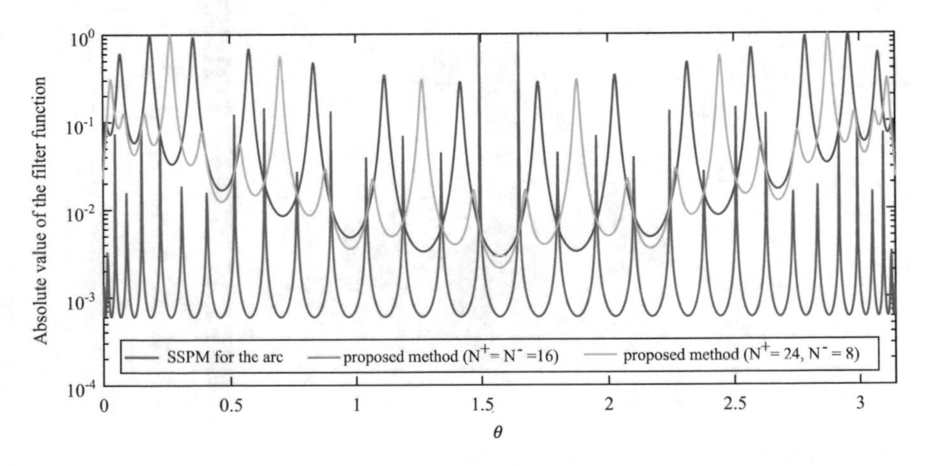


Fig. 6 Absolute value of the filter function on the line1

to that with $N^{\pm} = 16$. Thus the components of $\hat{S}_k$ in the direction of eigenvectors associated with eigenvalues near the arc for the proposed method is more equable than that for the SSPM for arc. In Fig. 7, we can see that the rational functions for the proposed method decays outside of the circle more rapidly than that for the SSPM for arc. However, the rational functions for the proposed method decays inside of the circle more slowly than that for the SSPM for arc. The rational function for the proposed method with $N^{+} = 24, N^{-} = 8$ are similar to that with $N^{\pm} = 16$. Thus the components of $\hat{S}_k$ in the direction of eigenvectors associated with eigenvalues outside the circle for the proposed method are smaller than that for the SSPM

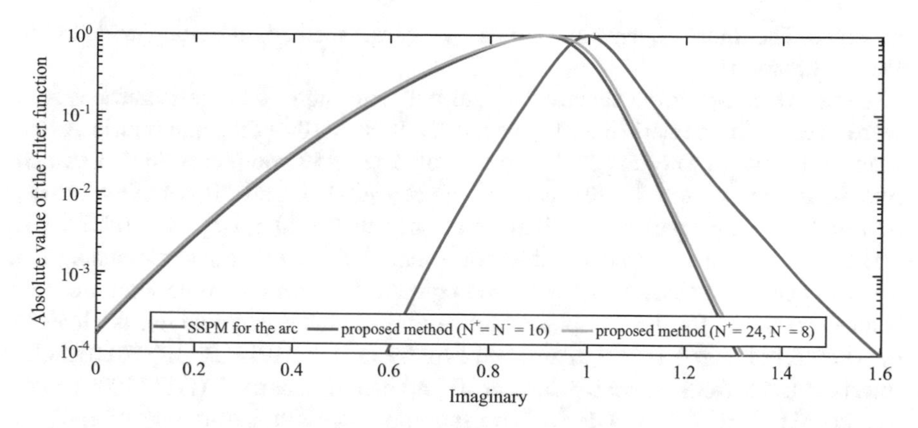


**Fig. 7** Absolute value of the filter function on the line2

for arc, and the components of $\hat{S}_k$ in the direction of eigenvectors associated with eigenvalues inside the circle for the proposed method are larger than that for the SSPM for arc.

In the proposed method, we compute the eigenpairs in a partial ring region. To do this, we divide the ring region into $D$ partial ring regions by $\theta_{\mathrm{a}}^{(d)}, \theta_{\mathrm{b}}^{(d)}$, $d = 1, 2, \ldots, D$. Then we compute the eigenpairs using the proposed method in each partial ring region.

The SSPM has potential for hierarchical parallelism: (I) each region can be computed independently, (II) linear systems at each quadrature point can be solved independently, (III) multiple right-hand sides of the linear systems can be solved simultaneously. Therefore we can assign different tasks for solving the linear systems to each parallel processor. Parallel implementations of the SSPM have been developed, such as Bloss[6], z-Pares[4] and CISS (Contour Integral Spectral Slicing). In CISS, the parallelism of (I) and (II) is implemented in SLEPc, along with an extension for the arc.

## 4 Numerical Example

In this section, we present numerical examples of the proposed method. We implement the proposed method in SLEPc, and compare the performance of the proposed method with that of the extension for the arc.

Experiments are performed on the supercomputer cluster of many-core architecture COMA (PACS-IX) at the Center for Computational Sciences, the University of Tsukuba. COMA has a total of 393 nodes providing 1.001 PFLOPS at optimum performance. Each node has dual CPU (Intel Xeon E5-2670v2), dual MIC (Intel Xeon Phi 7110P), and 64 GB memory, and the CPU has 10 cores and the MIC has

61 cores. The linear systems are solved by a direct method, in particular PCLU in PETSc library[1].

First, we compare the accuracy of the two extensions. Four test matrices $A, B$ were used in the numerical experiments: (I) $1000 \times 1000$ diagonal matrix $A$ and identity matrix $B$. The diagonal elements of $A$ are 950 complex values, real and imaginary part of which are random between $[0, 0.4]$, and 50 complex values, real and imaginary part of which are random between $[0.5, 1.1]$ (SAMPLE). (II) $501 \times 501$ diagonal matrix $A$ and identity matrix $B$. The diagonal elements of $A$ are 500 complex values, that are spaced equiangularly on the circle with a center 0 and a radius 1 on the complex plane, and 1 complex value which is close to quadrature point $(z_1 + 10^{-10})$ in the SSPM for the arc (SAMPLE2). (III) $5000 \times 5000$ matrix $A$ taken from the matrix market[3] and identity matrix $B$ (OLM5000). (IV) $11{,}520 \times 11{,}520$ matrix $A, B$ derived from computation of the self-energy of a silicon nanowire with a $6 \times 6\,\mathrm{nm}^2$ cross section[9] (SI11520). In both extensions, we divide the ring region into $D = 4$ partial ring regions, and we set $\rho^{\pm} = \rho \pm \beta$. Parameters for the ring region, starting angles and ending angles are given in Table 1. The remaining parameters for both extensions were $N^{+} = 24$, $N^{-} = 8$, $N = 32$, $L = 32$, $M = 8$ and $\delta = 10^{-12}$.

Table 2 shows the accuracy of the two extensions. max(res) and min(res) are the maximum and minimum values of the residuals $\|A\boldsymbol{x}_i - \lambda_i B\boldsymbol{x}_i\|_2 / (\|A\|_{\mathrm{F}} + |\lambda_i| \|B\|_{\mathrm{F}})$, respectively. The proposed method shows similar accuracy to the SSPM for the arc for the case (I), (III) and (IV). In the case (II), we can see that the maximum value of the residuals of the proposed method is smaller than that of the SSPM for the arc. Because 1 eigenvalue is very close to the quadrature point for the SSPM for the arc, a component of $\hat{S}_k$ in the direction of eigenvector associated with eigenvalue near the quadrature point becomes large, and other components become small relatively. These results indicate that the accuracy of residuals become low when the eigenpairs

**Table 1** Parameters for the proposed method and the extension for the arc

| | $\gamma$ | $\rho$ | $\beta$ | $[\theta_{\mathrm{a}}^{(1)}, \theta_{\mathrm{b}}^{(1)}]$ | $[\theta_{\mathrm{a}}^{(2)}, \theta_{\mathrm{b}}^{(2)}]$ | $[\theta_{\mathrm{a}}^{(3)}, \theta_{\mathrm{b}}^{(3)}]$ | $[\theta_{\mathrm{a}}^{(4)}, \theta_{\mathrm{b}}^{(4)}]$ |
|---|---|---|---|---|---|---|---|
| SAMPLE | 0 | 0.8 | 0.3 | $[0, 0.5\pi]$ | $[0.5\pi, \pi]$ | $[\pi, 1.5\pi]$ | $[1.5\pi, 2\pi]$ |
| SAMPLE2 | 0 | 1 | 0.01 | $[0, 0.5\pi]$ | $[0.5\pi, \pi]$ | $[\pi, 1.5\pi]$ | $[1.5\pi, 2\pi]$ |
| OLM5000[3] | −5 | 6.6 | 0.1 | $[0, 0.5\pi]$ | $[0.5\pi, \pi]$ | $[\pi, 1.5\pi]$ | $[1.5\pi, 2\pi]$ |
| SI11520[9] | −1 | 1 | 0.03 | $[0, 0.38\pi]$ | $[0.38\pi, \pi]$ | $[\pi, 1.62\pi]$ | $[1.62\pi, 2\pi]$ |

**Table 2** Accuracy of the two extensions

| | Arc | | | Partial ring region | | | |
|---|---|---|---|---|---|---|---|
| | Total of $\tilde{m}$ | Max (res) | Min (res) | Total of $\tilde{m}$ | Max (res) | Min (res) | Exact number of eigenpairs |
| SAMPLE | 50 | $5.5 \times 10^{-8}$ | $3.7 \times 10^{-15}$ | 50 | $3.2 \times 10^{-8}$ | $4.2 \times 10^{-13}$ | 50 |
| SAMPLE2 | 501 | $1.8 \times 10^{-6}$ | $1.5 \times 10^{-15}$ | 501 | $2.2 \times 10^{-10}$ | $8.6 \times 10^{-13}$ | 501 |
| OLM5000[3] | 26 | $1.2 \times 10^{-16}$ | $8.9 \times 10^{-18}$ | 26 | $1.1 \times 10^{-16}$ | $1.6 \times 10^{-17}$ | 26 |
| SI11520[9] | 332 | $3.9 \times 10^{-11}$ | $2.3 \times 10^{-14}$ | 332 | $7.2 \times 10^{-9}$ | $4.4 \times 10^{-13}$ | 332 |

**Table 3** Computational time for the proposed method for different numbers of processes

| # processes | 1 | 2 | 4 | 8 | 16 | 32 | 64 | 128 |
|---|---|---|---|---|---|---|---|---|
| Total | 308,800 | 153,170 | 76,623 | 40,601 | 23,439 | 15,268 | 10,463 | 8245 |
| Ideal | 308,800 | 151,300 | 75,650 | 37,825 | 18,913 | 9456 | 4728 | 2364 |

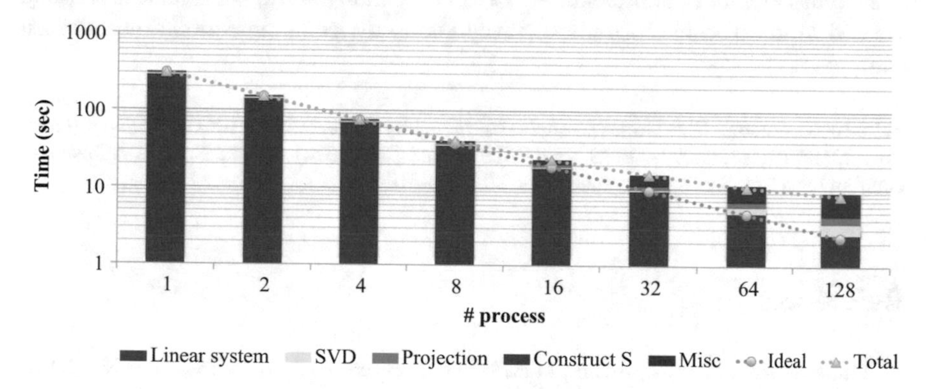


**Fig. 8** Details of the computational time for the proposed method with different numbers of processes (SI11520[9])

are closed at the quadrature points, and the proposed method improves the accuracy by locating the quadrature points sufficiently far from the arc.

Next, we evaluate the parallel performance of the proposed method. We investigate how the computational time varies as the number of processes increases. We use test matrix SI11520[9] in this experiment. We implement the proposed method in SLEPc. Here, the number of processes is set to $1, 2, 4, 8, 16, 32, 64$ and $128$, and other parameter values are the same as in the above experiment.

Table 3 shows the computational time for the proposed method with different numbers of processes. Total is the computational time for the proposed method, and Ideal is the ideal time (Total of 1 process)/(# processes). Figure 8 shows the details of the computational time. In Table 3 and Fig. 8, we can see that the proposed method has a good scaling for solving linear systems in (3), but is saturated for constructing $\hat{S}$ in (2) and computing the SVD due to the increase in communication time for each process. Thus the computational time for the proposed method is close to the ideal time for up to 8 processes but increases for 16 or more processes.

## 5 Conclusion

In the present paper, we presented an extension of the SSPM for a partial ring region. The filter function for the extension is similar to that for the existing SSPM for the arc. We implemented the SSPM for a partial ring region using SLEPc, and demonstrated that the method can be parallelized. The performance of the method was examined on a supercomputer cluster with many-core architecture. The results showed that the accuracy of the proposed method was similar to that of the SSPM for an arc, and the proposed method improves the accuracy by locating the quadrature points sufficiently far from the arc. We demonstrate that our implementation on SLEPc has efficient parallelism. As an area for future work, we intend to develop the SSPM to avoid the loss in efficiency due to the communication time between computers.

**Acknowledgements** This work was supported in part by JST/CREST, MEXT KAKENHI (Grant Nos. 25104701, 25286097), JSPS KAKENHI (Grant No. 26374) and the Spanish Ministry of Economy and Competitiveness under grant TIN2013-41049-P.

## References

1. Balay, S., Abhyankar, S., Adams, M.F., Brown, J., Brune, P., et al.: PETSc. Web page http://www.mcs.anl.gov/petsc.Cited2Sep2016
2. Berrut, J.-P., Trefethen, L.N.: Barycentric Lagrange interpolation. SIAM Rev. **159**(3), 501–517 (2004)
3. Boisvert, R., Pozo, R., Remington, K., Miller, B., Lipman, R.: NIST: MatrixMarket. http://math.nist.gov/MatrixMarket/.Cited24Dec2015
4. Futamura, Y., Sakurai, T.: z-Pares: Parallel Eigenvalue Solver. http://zpares.cs.tsukuba.ac.jp/. Cited24Dec2015
5. Hernandez, V., Roman J. E., Vidal V.: SLEPc: A scalable and flexible toolkit for the solution of eigenvalue problems. ACM Trans. Math. Softw. **31**(3), 351–362 (2005). https://doi.org/10.1145/1089014.1089019
6. Ikegami, T.: Bloss Eigensolver. https://staff.aist.go.jp/t-ikegami/Bloss/.Cited24Dec2015
7. Ikegami, T., Sakurai, T.: Contour integral eigensolver for non-Hermitian systems: a Rayleigh-Ritz-type approach. Taiwan. J. Math. **14**, 825–836 (2010)
8. Ikegami, T., Sakurai, T., Nagashima, U.: A filter diagonalization for generalized eigenvalue problems based on the Sakurai-Sugiura projection method. J. Comput. Appl. Math. **233**, 1927–1936 (2010)
9. Luisier, M., Schenk, A., Fichtner, W., Klimeck, G.:Atomistic simulation of nanowires in the sp3d5s* tight-binding formalism: from boundary conditions to strain calculations. Phys. Rev. B **74**, 205323 (2006). https://doi.org/10.1103/PhysRevB.74.205323
10. Maeda, Y., Sakurai, T.: A method for eigenvalue problem in arcuate region using contour integral. IPSJ Trans. Adv. Comput. Syst. **8**(4), 88–97 (2015) (Japanese)
11. Murakami, H.: filter diagonalization method by the linear combination of resolvents. IPSJ Trans. Adv. Comput. Syst. **49**(SIG2(ACS21)), 66–87 (2008) (Japanese)
12. Saad, Y.: Numerical Methods for Large Eigenvalue Problems, 2nd edn. SIAM, Philadelphia (2011)
13. Sakurai, T., Sugiura, H.: A projection method for generalized eigenvalue problems using numerical integration. J. Comput. Appl. Math. **159**, 119–128 (2003)
14. Sakurai, T., Futamura, Y., Tadano, H.: Efficient parameter estimation and implementation of a contour integral-based eigensolver. J. Algor. Comput. Technol. **7**, 249–269 (2013)

# A Parallel Bisection and Inverse Iteration Solver for a Subset of Eigenpairs of Symmetric Band Matrices

**Hiroyuki Ishigami, Hidehiko Hasegawa, Kinji Kimura, and Yoshimasa Nakamura**

**Abstract** The tridiagonalization and its back-transformation for computing eigenpairs of real symmetric dense matrices are known to be the bottleneck of the execution time in parallel processing owing to the communication cost and the number of floating-point operations. To overcome this problem, we focus on real symmetric band eigensolvers proposed by Gupta and Murata since their eigensolvers are composed of the bisection and inverse iteration algorithms and do not include neither the tridiagonalization of real symmetric band matrices nor its back-transformation. In this paper, the following parallel solver for computing a subset of eigenpairs of real symmetric band matrices is proposed on the basis of Murata's eigensolver: the desired eigenvalues of the target band matrices are computed directly by using parallel Murata's bisection algorithm. The corresponding eigenvectors are computed by using block inverse iteration algorithm with reorthogonalization, which can be parallelized with lower communication cost than the inverse iteration algorithm. Numerical experiments on shared-memory multi-core processors show that the proposed eigensolver is faster than the conventional solvers.

H. Ishigami (✉)
Graduate School of Informatics, Kyoto University, Yoshida-Honmachi, Sakyo-ku, Kyoto-shi, Kyoto, Japan

Yahoo Japan Corporation, Kioi Tower, 1-3 Kioicho, Chiyoda-ku, Tokyo, Japan
e-mail: hishigami@amp.i.kyoto-u.ac.jp; hishigam@yahoo-corp.jp

H. Hasegawa
Faculty of Library, Information and Media Science, University of Tsukuba, Kasuga 1-2, Tsukuba, Japan
e-mail: hasegawa@slis.tsukuba.ac.jp

K. Kimura • Y. Nakamura
Graduate School of Informatics, Kyoto University, Yoshida-Honmachi, Sakyo-ku, Kyoto-shi, Kyoto, Japan
e-mail: kkimur@amp.i.kyoto-u.ac.jp; ynaka@i.kyoto-u.ac.jp



T. Sakurai et al. (eds.), *Eigenvalue Problems: Algorithms, Software and Applications in Petascale Computing*, Lecture Notes in Computational Science and Engineering 117, https://doi.org/10.1007/978-3-319-62426-6_3

## 1 Introduction

A subset of eigenpairs of a real symmetric matrix, i.e. eigenvalues and the corresponding eigenvectors, is required in several applications, such as the vibration analysis, the molecular orbital computation, and the kernel principal component analysis. As the scale of the problems appearing in such applications becomes significantly larger, the parallel processing is crucial for reducing the overall execution time. As a result, there has been an increase in the demand for highly scalable algorithms for computing a subset of the eigenpairs in parallel processing.

Let us consider computing a subset of eigenpairs of a real symmetric dense matrix on the basis of the orthogonal similarity transformations. In general cases, the eigenpairs of the target symmetric matrix are computed through the following three phases: The first phase is *tridiagonalization*, which is to reduce the target matrix to symmetric tridiagonal form by orthogonal similarity transformations. The second phase is to compute eigenpairs of the symmetric tridiagonal matrix. The last phase is *back-transformation*, which is to compute the eigenvectors of the original matrix from the computed eigenvectors of the symmetric tridiagonal matrix by using the orthogonal transformations generated for the first phase.

The following two-step framework [7, 8] is widely used for the tridiagonalization: The first step is to reduce the symmetric dense matrix to symmetric band form and the second step is to reduce the symmetric band matrix to symmetric tridiagonal form. Several efficient parallel algorithms based on this framework has been proposed in [2–4, 7, 17], etc. However, it is pointed out in [4] that the band reduction in the second step and its corresponding back-transformation remains to be the bottleneck of the overall execution time in massively parallel processing if the eigenvectors are required.

To overcome the problem in the tridiagonalization and the corresponding back-transformation, the authors consider not reducing a real symmetric band matrix to the tridiagonal form, but directly computing the eigenpairs of the real symmetric band matrix. The bisection and inverse iteration algorithms [6] for real symmetric band matrices can be used for this purpose. It is to be noted that the bisection algorithm can compute the desired eigenvalues of the target matrices and the inverse iteration algorithm gives the corresponding eigenvectors. Their implementation for real symmetric band matrices is proposed in [12, 20]. As shown in numerical results on Sect. 4.1, the bisection algorithm proposed in [20] is faster than that proposed in [12]. Let us name the algorithm in [12] *Gupta's bisection algorithm* and that in [20] *Murata's bisection algorithm*, respectively. The inverse iteration algorithm with reorthogonalization is used for computing the corresponding eigenvectors both in [12, 20].

In this paper, the authors propose the following parallel algorithm for computing desired eigenpairs of real symmetric band matrices: the desired eigenvalues are computed by using the parallel implementation of Murata's bisection algorithm and the corresponding eigenvectors are computed by using *block inverse iteration algorithm with reorthogonalization (BIR algorithm)* [14], which is a variant of the inverse iteration algorithm with reorthogonalization. Then, the performance of the proposed methods is evaluated through numerical experiments on shared-memory multi-core processors.

The rest of this paper is organized as follows. Section 2 gives a review of the bisection algorithms for computing eigenvalues of real symmetric band matrices proposed in [12, 20], and then shows their parallel implementation for shared-memory multi-core processors. Section 3 shows the inverse iteration algorithm with reorthogonalization and the BIR algorithm for computing eigenvectors of real symmetric band matrices, and then shows their parallel implementation for shared-memory multi-core processors. Section 4 shows results of numerical experiments on shared-memory multi-core processors to evaluate the performance of the proposed parallel algorithm, which computes eigenvalues of target matrices by using parallel Murata's bisection algorithm and computes the corresponding eigenvectors by using the BIR algorithm. We end with conclusions and future work in Sect. 5.

## 2 Eigenvalue Computation of Symmetric Band Matrices

The bisection algorithm [6] computes the desired eigenvalues of a real symmetric matrix by updating repeatedly the half-open intervals $(\mu^L, \mu^R]$. The computation of $\nu(\mu)$ is required for updating the intervals, where $\mu$ is a value in the intervals $(\mu^L, \mu^R]$ and $\nu(\mu)$ is the number of eigenvalues of the target matrix that are less than $\mu$, and is the most time-consuming part of the bisection algorithm.

In this section, we introduce Gupta's bisection algorithm [12] and Murata's bisection algorithm [20] for computing the desired eigenvalues of a real $n \times n$ symmetric band matrix $B$ with half-bandwidth $w$ and then show a parallel implementation of them. These two algorithms differ in the computation of $\nu(\mu)$.

### 2.1 *Gupta's Bisection Algorithm*

Gupta's bisection algorithm employs *Martin-Wilkinson's Gaussian elimination* [19] for computing $\nu(\mu)$.

The computation of $\nu(\mu)$ by employing Martin-Wilkinson's Gaussian elimination is based on Sturm's theorem [21]. In this case, all the principal minor determinant of $B - \mu I$ is required. Martin-Wilkinson's Gaussian elimination is adequate for this purpose since economical partial pivoting strategy is implemented to it from the viewpoint of both the numerical stability and the number of floating-point operations. Note that the number of floating-point operations in Martin-Wilkinson's Gaussian elimination is $O(w^2 n)$.

Martin-Wilkinson's Gaussian elimination is mainly composed of the BLAS (Basic Linear Algebra Subprograms [9]) 1 routines such as vector operations. Whenever the partial pivoting occurs in Martin-Wilkinson's Gaussian elimination, the number of floating-point operations increases and, moreover, a pattern of the data access changes. As a result, Gupta's bisection algorithm is difficult to achieve a high performance from the viewpoint of data reusability.

## 2.2 Murata's Bisection Algorithm

Murata's bisection algorithm [20] employs not only Martin-Wilkinson's Gaussian elimination but also the *LU* factorization without pivoting for computing $\nu(\mu)$.

The computation of $\nu(\mu)$ by employing the *LU* factorization without pivoting is based on Sylvester's law of inertia [21]. In this case, we consider a *LU* factorization without pivoting $B - \mu I = LU$, where $L$ is an $n \times n$ lower triangular matrix with lower bandwidth $w$ and $U$ is an $n \times n$ unit upper triangular matrix with upper bandwidth $w$. On the basis of Sylvester's law of inertia, $\nu(\mu)$ is equivalent to the number of negative values in diagonal elements of $L$.

The number of floating-point operations in Martin-Wilkinson's Gaussian elimination is about 3 times higher than that in the *LU* factorization without pivoting. In addition, the cache hit ratio of the *LU* factorization without pivoting is higher than that of Martin-Wilkinson's Gaussian elimination owing to absence of any pivoting. However, owing to both the rounding errors and the absence of any pivoting, the *LU* factorization without pivoting sometimes fails or the resulting elements of this factorization may be not correct even if accomplished.

As a result, Murata's bisection algorithm computes $\nu(\mu)$ in the following way: In the early stage of computing eigenvalues of $B$, Murata's bisection algorithm employs the *LU* factorization without pivoting for computing rapidly $\nu(\mu)$. In addition, if $\mu$ is significantly close to a certain eigenvalue, Murata's bisection algorithm employs Martin-Wilkinson's Gaussian elimination for computing accurately $\nu(\mu)$. Consequently, Murata's bisection algorithm is expected to be faster than Gupta's algorithm for computing the desired eigenvalues of $B$.

The several acceleration techniques for the *LU* factorization algorithm has been proposed. As shown in [13], the optimal implementation for vector processors is implemented to the *LU* factorization without pivoting and the overall execution time for Murata's bisection algorithm becomes shorter. On recent scalar processors with the large cache, the *LU* factorization had better to be implemented using the block algorithm for enforcing higher cache hit ratio. Hence, in this paper, the block *LU* factorization of real symmetric band matrices is introduced into Murata's bisection algorithm for the purpose of improving further its performance on shared-memory multi-core processors.

**Algorithm 1** Parallel bisection algorithm for symmetric band matrices

```
 1: function ParallelBandBisection(B, ℓ)
 2:     Set μ_1^L, μ_1^R                              ▷ Use Gerschgorin theorem, etc.
 3:     k_b := 1, k_e := 1
 4:     repeat
 5:         !$omp parallel do private(μ_k)
 6:         do k := k_b to k_e                                              ▷ k_e ≤ ℓ
 7:             μ_k := (μ_k^L + μ_k^R)/2
 8:             Compute ν_B(μ_k)
 9:         end do
10:         Update the intervals μ_k^L, μ_k^R for k = k_b, ..., k_e and the indices k_b, k_e
11:     until All of the desired eigenvalues meets the stopping criteria
12:     return λ̃_k := (μ_k^L + μ_k^R)/2 for k = 1, ..., ℓ
13: end function
```

## 2.3 *Parallel Implementation of Bisection Algorithm*

In this paper, Murata's and Gupta's bisection algorithms are implemented on the basis of the `dstebz` routine [15], which is provided in LAPACK (Linear Algebra PACKage [1]) and computes the desired eigenvalues of real symmetric tridiagonal matrices. A pseudocode of their parallel implementation is shown in Algorithm 1, where $\ell$ is the number of the desired eigenvalues.

The computation of $\nu(\mu_k)$ on line 8 is applied different algorithms for Murata's and Gupta's bisection algorithm as mentioned in Sects. 2.1 and 2.2, respectively. In addition, the computation of $\nu(\mu_k)$ is performed in parallel with respect to each search point $\mu_k$ by employing the OpenMP directive shown in line 5.

Note that an initial interval $\mu_1^L$ and $\mu_1^R$ are set on line 2. $\mu_1^L$ and $\mu_1^R$ are, at first, set as the lower and upper bounds derived from Gerschgorin theorem [11] and then are refined by the iterative computation of $\nu(\mu_1^L)$ and $\nu(\mu_1^R)$ in the same way as shown on lines 7 and 8. Moreover, several criteria are designed for stopping the binary search (a **repeat-until** iteration on lines 4–11) in the `dstebz` routine and its subroutine `dlaebz` and we apply the modified criteria on line 11 for computing eigenvalues of $B$. For more details, see the `dstebz` and `dlaebz` routines [15].

Note that the above-mentioned parallel bisection algorithm requires the working memory for computing $\nu(\mu)$ independently on each computation thread. The amount of the working memory for Martin-Wilkinson's Gaussian elimination is $(3w + 1)n$ per a computation thread and is larger than that for the block $LU$ factorization. Thus, we have to spend about $t(3w + 1)n$ working memory for performing parallel Murata's and Gupta's bisection algorithm, where $t$ is the number of the computation threads.

# 3 Eigenvector Computation of Symmetric Band Matrices

The inverse iteration algorithm with reorthogonalization is used in [12, 20] for computing the eigenvectors of a real symmetric band matrix $B$. In this section, we consider applying *block inverse iteration algorithm with reorthogonalization (BIR algorithm)* [14] for computing the eigenvectors of $B$, which is a variant of the inverse iteration algorithm with reorthogonalization.

## 3.1 Inverse Iteration Algorithm with Reorthogonalization

We first introduce the inverse iteration with reorthogonalization for computing the eigenvectors of $B$. In the followings, let $\lambda_k$ be an eigenvalue of the target matrix such that $\lambda_1 \geq \lambda_2 \geq \cdots \geq \lambda_\ell$ ($\ell \leq n$) and $\boldsymbol{q}_k$ be the corresponding eigenvector to $\lambda_k$, respectively. Moreover, let $\tilde{\lambda}_k$ be an approximate value of $\lambda_k$, obtained by some eigenvalue computation algorithm such as the bisection algorithm, and $\boldsymbol{v}_k^{(0)}$ be a starting vector for $k = 1, \ldots, \ell$. Then the inverse iteration is to generate a sequence of vectors $\boldsymbol{v}_k^{(i)}$ by solving iteratively the following linear equation:

$$\left(B - \tilde{\lambda}_k I\right) \boldsymbol{v}_k^{(i)} = \boldsymbol{v}_k^{(i-1)}, \quad i = 1, 2, \ldots, \tag{1}$$

where $I$ is the $n \times n$ identity matrix. If $|\tilde{\lambda}_k - \lambda_k| \ll |\tilde{\lambda}_j - \lambda_k|$ for $j \neq k$ is satisfied, $\boldsymbol{v}_k^{(i)}$ converges to $\boldsymbol{q}_k$ as $i \to \infty$.

If some of the eigenvalues are very close to one another, we must reorthogonalize all the corresponding eigenvectors to these eigenvalues. Hereafter, such eigenvalues are referred to as *clustered eigenvalues*. *Peters-Wilkinson's criterion* [22] is applied in `dstein` [15], which is a LAPACK routine for computing eigenvectors of a real symmetric tridiagonal matrix $T$, as dividing eigenvalues of $T$ into clusters. In Peters-Wilkinson's criterion, $\lambda_{k-1}$ and $\lambda_k$ are regarded as belonging to the same cluster if $|\tilde{\lambda}_{k-1} - \tilde{\lambda}_k| \leq 10^{-3}\|T\|_1$ is satisfied ($2 \leq k \leq \ell$). In the followings, we also use Peters-Wilkinson's criterion for dividing eigenvalues of a real symmetric band matrix $B$ into clusters. However, $\|B\|_1 (= \|B\|_\infty)$ is not adequate to use in this criterion since $\|B\|_1$ becomes significantly large according to $w$, the half-bandwidth of $B$. To the contrary, since $\|B\|_2$ satisfies

$$\|B\|_2 = \sup_{\boldsymbol{x} \in \mathbb{R}^n} \frac{\|B\boldsymbol{x}\|_2}{\|\boldsymbol{x}\|_2} \geq \max_i |\lambda_i|, \tag{2}$$

a good lower bound of $\|B\|_2$ is obtained by $\max(\tilde{\lambda}_1, \tilde{\lambda}_n)$, where both $\tilde{\lambda}_1$ and $\tilde{\lambda}_n$ do not depend on $w$. Thus, in this paper, Peters-Wilkinson's criterion for computing the eigenvectors of $B$ is designed by using $\tilde{\lambda}_1$ and $\tilde{\lambda}_n$.

**Algorithm 2** Inverse iteration algorithm with reorthogonalization for symmetric band matrices

```
1: function BandInv(B, ℓ, λ̃_1, ..., λ̃_ℓ)
2:   do k := 1 to ℓ
3:     i := 0
4:     Generate an initial random vector: v_k^(0)
5:     LU factorization with partial pivoting: B − λ̃_k = P_k L_k U_k     ▷ Call dgbtrf
6:     repeat
7:       i := i + 1
8:       Normalize v_k^(i−1) to q_k^(i−1)
9:       Solve P_k L_k U_k v_k^(i) = q_k^(i−1)                            ▷ Call dgbtrs
10:      if k > 1 and |λ̃_{k−1} − λ̃_k| ≤ 10^−3 × max(|λ̃_1|, |λ̃_n|), then
11:        Reorthogonalize v_k^(i) to q_k^(i) by employing MGS algorithm
12:      else
13:        k_1 := k
14:      end if
15:    until some condition is met.
16:    Normalize v_k^(i) to q_k^(i)
17:    Q_k := [Q_{k−1} q_k^(i)]
18:  end do
19:  return Q_ℓ = [q_1 ··· q_ℓ]
20: end function
```

Algorithm 2 shows a pseudocode of the inverse iteration algorithm with reorthogonalization for computing the $\ell$ eigenvectors of $B$ and is designed on the basis of the `dstein` routine. As well as the `dstein`, the modified Gram-Schmidt (MGS) algorithm [11] is applied to the reorthogonalization part on line 11. On line 10, Peters-Wilkinson's criterion with the above-mentioned modification is applied for dividing eigenvalues of $B$ into clusters. For solving the linear equation (1), we once perform the *LU factorization with the partial pivoting (PLU factorization)* of $B - \tilde{\lambda}_k I$ by employing the `dgbtrf` routine (line 5), and then iteratively obtain $\boldsymbol{v}_k^{(i)}$ by employing the `dgbtrs` routine (line 9). Note that both `dgbtrf` and `dgbtrs` routines are provided in LAPACK. The `dgbtrf` routine is implemented on the basis of the block algorithm of the *PLU* factorization and is composed of the BLAS 2 and 3 routines. In addition, the `dgbtrs` routine is mainly composed of the BLAS 2 routines and requires $P_k$, $L_k$, and $U_k$, which are the resulting elements of the *PLU* factorization by the `dgbtrf` routine. For this purpose, we have to store $P_k$, $L_k$, and $U_k$ in the working memory and their amount is about $(3w + 1)n$.

In this paper, let us consider that the inverse iteration algorithm with reorthogonalization is parallelized by employing the parallel BLAS routines.

**Algorithm 3** Block inverse iteration algorithm with reorthogonalization for computing the corresponding eigenvectors to clustered eigenvalues of symmetric band matrices

```
1: function BandBIR(B, r, ℓ̂, λ̃_1, ..., λ̃_ℓ̂)
2:   Set an n × r matrix Q_0 be Q_0 := O
3:   do j := 1 to ℓ̂/r
4:     i := 0
5:     Generate Q_{j,r}^(0) := [q_{(j-1)r+1}^(0) ··· q_{jr}^(0)]
6:     !$omp parallel do
7:     do k =: (j − 1)r + 1 to jr
8:       LU factorization with partial pivoting: B − λ̃_k = P_k L_k U_k    ▷ Call dgbtrf
9:     end do
10:    repeat
11:      i := i + 1
12:      !$omp parallel do
13:      do k =: (j − 1)r + 1, ..., jr
14:        Solve P_k L_k U_k v_k^(i) = q_k^(i−1)    ▷ Call dgbtrs
15:      end do
16:      V_{j,r}^(i) := V_{j,r}^(i) − Q_{(j−1)r} Q_{(j−1)r}^T V_{j,r}    ▷ Call dgemm ×2
17:      QR factorization: V_{j,r}^(i) = Q̄_{j,r}^(i) R_{j,r}^(i)
18:      Q̄_{j,r}^(i) := Q̄_{j,r}^(i) − Q_{(j−1)r} Q_{(j−1)r}^T Q̄_{j,r}^(i)    ▷ Call dgemm ×2
19:      QR factorization: Q̄_{j,r}^(i) = Q_{j,r}^(i) R_{j,r}^(i)
20:    until converge
21:    Q_{jr} := [Q_{(j−1)r} Q_{j,r}^(i)]  (Q_r := [Q_{1,r}^(i)])
22:  end do
23:  return Q_ℓ̂ = [q_1 ··· q_ℓ̂]
24: end function
```

## 3.2 Block Inverse Iteration Algorithm with Reorthogonalization

A pseudocode of the block inverse iteration algorithm with reorthogonalization (BIR algorithm) for computing the corresponding eigenvectors to the clustered eigenvalues of $B$ is shown in Algorithm 3, where $\hat{\ell}$ is the number of eigenvalues belonging to a certain cluster and $r$ is a block parameter determined arbitrarily by users ($r \leq \hat{\ell}$). For convenience, we assume the $r$ is a divisor of $\hat{\ell}$. Note that the BIR algorithm corresponds to the inverse iteration algorithm with reorthogonalization in Algorithm 2 if $r = 1$.

The BIR algorithm is composed of two parts as well as the inverse iteration algorithm with reorthogonalization. The one is to solve $r$ linear equations simultaneously. For this part, the `dgbtrf` and `dgbtrs` routines are employed as well as the inverse iteration algorithm with reorthogonalization. Different from the inverse iteration algorithm with reorthogonalization, the computation of solving simultaneously $r$ linear equations can be parallelized in terms of $k$ since the linear equations are independent of each other. Thus, the computation of this part is parallelized by the OpenMP directives shown on lines 6–9 and lines 12–15. Note that we have to spend $r(3w + 1)n$ working memory to store $P_k$, $L_k$, and $U_k$ corresponded to the $r$ linear equations for the above purpose.

The other is the block reorthogonalization part as shown on lines 16 to 19. In Algorithm 3, the block Gram-Schmidt algorithm with reorthogonalization (BCGS2 algorithm) [5] is used for this part. The BCGS2 algorithm is mainly composed of the matrix multiplication, which is one of the BLAS 3 routines. Thus, the `dgemm` routines are employed to the computation on lines 16 and 18 in Algorithm 3. As well as the BIR algorithm proposed in [14], we consider that the recursive implementation of the classical Gram-Schmidt algorithm [24] is applied to the computation of the QR factorization on lines 17 and 19, which is also mainly composed of the matrix multiplications. In this paper, the block reorthogonalization part is parallelized by employing the parallel BLAS routines.

The BIR algorithm corresponds to the simultaneous inverse iteration algorithm [10] if $r = \hat{\ell}$. Thus, the simultaneous inverse iteration algorithm always spends the larger amount of memory than the BIR algorithm does. Note that the memory use for the BIR algorithm is almost equal to that for the parallel bisection algorithms mentioned in 2.3 if $r$ is set to the number of the computation threads.

### 3.3 Remark on Inverse Iteration Algorithms

The relationship between the inverse iteration algorithm with reorthogonalization and the BIR algorithm is analogous to that between the $LU$ factorization and the block $LU$ factorization. Thus, the number of floating-point operations in the BIR algorithm is almost equal to that in the inverse iteration algorithm with reorthogonalization.

As mentioned before, both the inverse iteration algorithm with reorthogonalization and the BIR algorithm are composed two parts: solving linear equation and the (block) reorthogonalization part. Assuming $\ell$ is the number of the desired eigenvectors of $B$, the number of floating-point operations is $O(\ell w^2 n)$ in solving linear equations and is $O(\hat{\ell}_{\max}^2 n)$ in the reorthogonalization part, where $\hat{\ell}_{\max}$ is the number of eigenvalues belonging to the largest eigenvalue cluster ($\hat{\ell}_{\max} \leq \ell$). As a result, solving linear equations is occupied with much of the execution time for computing eigenvectors of $B$ by the inverse iteration algorithm with reorthogonalization as long as it is not a case that $\ell$ is much larger than $w$. The same is true of the BIR algorithm.

## 4 Performance Evaluation

This section gives experimental results on shared-memory multi-core processors to evaluate the performance of the proposed eigensolver, which computes eigenvalues of real symmetric band matrices by employing parallel Murata's bisection algorithm mentioned in Sect. 2 and computes the corresponding eigenvectors by the BIR algorithm mentioned in Sect. 3.2.

Table 1 shows our experimental environment, which is one node of Appro Green Blade 8000 at ACCMS, Kyoto University. All the experiments were performed by `numactl --interleave=all` to control NUMA policy. In addition, each code was run with 16 threads on the condition that the `KMP_AFFINITY` was set to "none" in all the numerical experiments except for the performance evaluation of the parallel efficiency of the eigensolvers in Sect. 4.3. Note that the `KMP_AFFINITY` is an environment variable for controlling the OpenMP thread affinity. The Intel Math Kernel Library (MKL) was used for the parallel execution of the BLAS and LAPACK routines and the OpenMP directives are also used for the thread parallelization as mentioned in Sects. 2 and 3. The block size $r$ of the BIR algorithm in the proposed eigensolver was set to $r = 16$ in all the experiments, which is equal to the number of cores in the experimental environment shown in Table 1. Since the working memory for the BIR algorithm is almost equal to that for the parallel bisection algorithm, the total memory use can be easily estimated on this condition. Note that the maximum number of iterations in both the BIR algorithm and the inverse iteration algorithm with reorthogonalization is set to 5, as well as the `dstein` routine provided in LAPACK. In fact, the number of iterations in both of them is 3 in all the experiments.

The following $n \times n$ symmetric band matrices with half-bandwidth $w$ were used for test matrices in the performance evaluation, whose elements are set random numbers in the range $[0, 1)$: $B_1$ is set to $n = 20,000$ and $w = 64$; $B_2$ is set to $n = 40,000$ and $w = 256$. In the experiments, the largest $\ell$ eigenvalues of them and the corresponding eigenvectors are computed, where $\ell$ is set to $\ell = 250, 500, 1000$.

**Table 1** Specifications of the experimental environment

| One node of Appro Green Blade 8000 at ACCMS | |
|---|---|
| CPU | Intel Xeon E5-2670@2.6 GHz, 16 cores (8 cores × 2 sockets) |
| | L3 cache: 20 MB × 2 |
| RAM | DDR3-1600 64 GB, 136.4 GB/s |
| Compiler | Intel C++/Fortran Compiler 15.0.2 |
| Options | `-O3 -xHOST -ipo -no-prec-div` |
| | `-openmp -mcmodel=medium -shared-intel` |
| Run command | `numactl --interleave=all` |
| Software | Intel Math Kernel Library 11.2.2 |

### 4.1 Performance Evaluation of Murata's Bisection Algorithm

To evaluate the performance of parallel Murata's bisection algorithm in Sect. 2, we compared the execution time for computing the desired eigenvalues of real symmetric band matrices by using parallel Murata's bisection algorithm with that by using Gupta's bisection algorithm. Their codes are parallelized by employing the OpenMP directives as shown in Sect. 2.3.

Figure 1a and b show the execution times for computing the $\ell$ largest eigenvalues of $B_1$ and $B_2$, respectively. According to the expectation in Sect. 2.2, we observe that Murata's bisection algorithm is faster than Gupta's bisection algorithm in all the cases.

Table 2a and b show the number of computing $\nu(\mu)$ on the basis of the block *LU* factorization algorithm and Martin-Wilkinson's Gaussian elimination. These tables indicate that most of the computation of $\nu(\mu)$ is performed by the block *LU* factorization-based algorithm in Murata's bisection. In addition, the total number of computing $\nu(\mu)$ in Murata's bisection is almost the same as that in Gupta's bisection. We also observe that the total number of computing $\nu(\mu)$ in both bisection algorithms depends on $\ell$, the number of the desired eigenvalues.

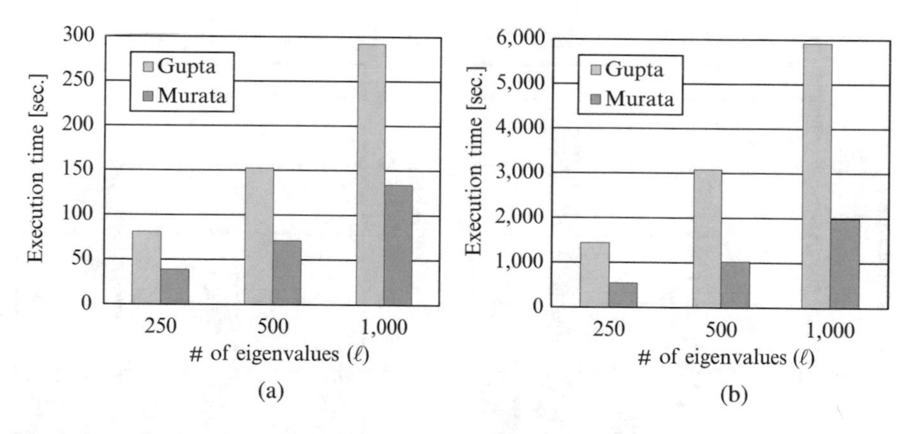


**Fig. 1** Execution times for computing the largest $\ell$ eigenvalues of real symmetric band matrices by using parallel Murata's bisection algorithm and parallel Gupta's bisection algorithm. (**a**) Cases of $B_1$. (**b**) Cases of $B_2$

**Table 2** The number of computing $\nu(\mu)$ on the basis of the block *LU* factorization algorithm and Martin-Wilkinson's Gaussian elimination when we compute the largest $\ell$ eigenvalues by employing parallel Murata's bisection algorithm and parallel Gupta's bisection algorithm

| (a) Cases of $B_1$ | | | | |
|---|---|---|---|---|
| # of eigenvalues ($\ell$) | | 250 | 500 | 1000 |
| Murata | Block *LU* | 9538 | 18,802 | 36,852 |
| | M-W | 108 | 209 | 692 |
| Gupta | M-W | 9896 | 19,511 | 38,544 |

| (b) Cases of $B_2$ | | | | |
|---|---|---|---|---|
| # of eigenvalues ($\ell$) | | 250 | 500 | 1000 |
| Murata | Block *LU* | 9030 | 17,728 | 34,719 |
| | M-W | 186 | 506 | 1449 |
| Gupta | M-W | 9216 | 18,234 | 36,168 |

## 4.2 Performance Evaluation of BIR Algorithm

In order to evaluate the performance of the proposed eigenvector computation algorithm in Sect. 3, we compared the execution time for computing of the eigenvectors corresponding to the $\ell$ largest eigenvalues of real symmetric band matrices by using the proposed algorithm ("BIR") with that by using the inverse iteration algorithm with reorthogonalization ("Inv"). Their codes are parallelized by employing the Intel MKL and the OpenMP directives as shown in Sect. 3. In addition, the $\ell$ largest eigenvalues of the target matrices are obtained by using parallel Murata's bisection algorithm.

Figure 2 shows the execution times for computing the corresponding eigenvectors to the $\ell$ largest eigenvalues of the target matrices and their details, where "Solving equation" denotes the execution time for solving linear equations (1) and "Reorthogonalization" denotes the execution time for the reorthogonalization part performed by the MGS algorithm in "Inv" or the BCGS2 algorithm in "BIR". Figure 2a and b correspond to the case of $B_1$ and $B_2$, respectively. These figures show that "BIR" is faster than "Inv" in all the cases. According to the discussion about the number of floating-point operations in each part mentioned in Sect. 3.3, the

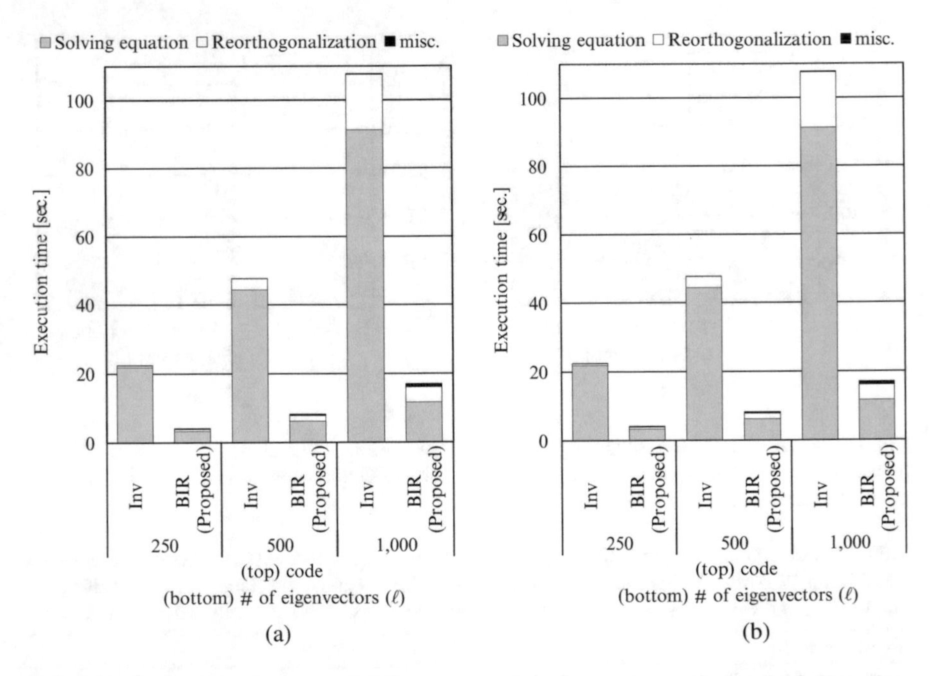


**Fig. 2** Execution times for computing the corresponding eigenvectors to the largest $\ell$ eigenvalues of symmetric band matrices by using the block inverse iteration algorithm with reorthogonalization ("BIR") and the inverse iteration algorithm with reorthogonalization ("Inv") and their details. (**a**) Cases of $B_1$. (**b**) Cases of $B_2$

"Solving equation" part occupies most parts of the execution time of the eigenvector computation in all the cases.

We also observe that the execution time for the "Solving equation" part of "BIR" is significantly shorter than that of "Inv" in all the cases. As mentioned in Sect. 3, the parallelization of the "Solving equation" part in "BIR" differs from that in "Inv". In the BIR algorithm, each linear equation is solved on each computation thread, and thus any barrier synchronization between the computation threads does not occur until all the computation assigned to each computation thread is finished. On the other hand, since the inverse iteration algorithm with reorthogonalization is parallelized by employing the parallel BLAS routines, the barrier synchronization between the computation threads occurs each time the BLAS routine is called. Moreover, the BLAS-based computations in the `dgbtrf` and `dgbtrs` routines are difficult to achieve good performance in parallel processing since the size of vectors and matrices appearing in these computations is too small. From the above reasons, the "Solving equation" part of "BIR" achieves the higher performance in parallel processing than that of "Inv".

Finally, we examined the effect of the block size $r$ on the performance of the BIR algorithm. As mentioned in Sect. 3.2, the block reorthogonalization part of the BIR algorithm includes many matrix multiplications, and thus, the performance of the BIR algorithm depends on that of the routine for the matrix multiplications `dgemm`. In addition, the `dgemm` routine is difficult to achieve the better performance if the size of the matrices appearing in the computation is sufficiently large. Figure 3 shows the execution times for computing the corresponding eigenvectors to the largest 1000 eigenvalues of $B_1$ by using the BIR algorithm with different block size $r$. From this figure, we observe that the BIR algorithm with $r = 128$ or 256 is somewhat faster than that with $r = 16$ for computing the 1000 eigenvectors of $B_1$. In spite of this tendency, we set $r$ as the number of cores in the proposed eigensolver. This is because $r$ must be smaller than the number of the desired eigenvectors as mentioned in Sect. 3.2.

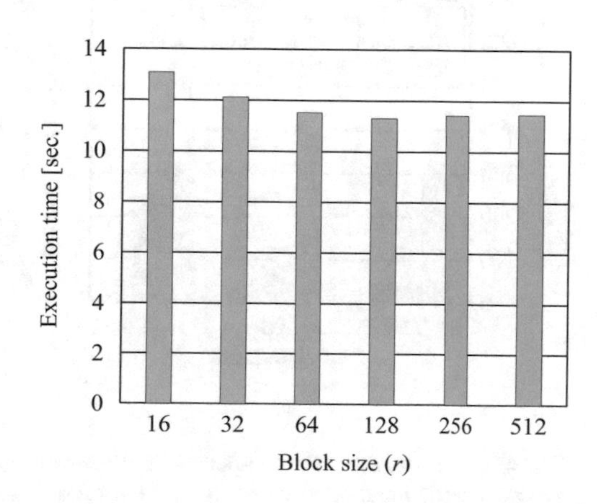


**Fig. 3** Execution times for computing the corresponding eigenvectors to the largest 1000 eigenvalues of $B_1$ by using the block inverse iteration algorithm with reorthogonalization

## 4.3 Performance Evaluation of Proposed Band Eigensolver

In order to evaluate the performance of the proposed symmetric band eigensolver, we compared the proposed solver ("Murata+BIR") with the two conventional eigensolvers. One of the conventional solvers is also to compute eigenvalues by using Murata's bisection algorithm and to compute the corresponding eigenvectors by using the inverse iteration algorithm with reorthogonalization shown in Algorithm 2 and is referred to as "Murata+Inv". The codes of "Murata+BIR" and "Murata+Inv" are also parallelized by employing the Intel MKL and the OpenMP directives as shown in Sects. 2.3 and 3. The other conventional solver is "dsbevx" provided in Intel MKL, which is a LAPACK routine for computing a subset of eigenpairs of real symmetric band matrices through the tridiagonalization. Note that "dsbevx" employs the `dsbtrd` routine to tridiagonalize the target band matrix in the way proposed in [17], the `dstebz` routine to compute the desired eigenvalues of the real symmetric tridiagonal matrix, and the `dstein` routine to compute the corresponding eigenvectors.

Figure 4a and b show the overall execution time for computing the eigenpairs corresponding to the $\ell$ largest eigenvalues of $B_1$ and $B_2$, respectively. We observe that the proposed eigensolver, "Murata+BIR", is faster than the conventional solvers in all the cases. Figure 5a and b show the details of the overall execution time for "Murata+BIR" and "Murata+Inv". We observe that most of the execution time in "Murata+BIR" remains to be occupied by that of the eigenvalue computation using parallel Murata's bisection algorithm in all the cases. One reason of this result is that the number of floating-point operations in "Murata", Murata's bisection algorithm, is much higher than that in "BIR". The other reason is that the execution time for eigenvector computation in "Murata+BIR" is significantly reduced from that in "Murata+Inv" as mentioned in Sect. 4.2.

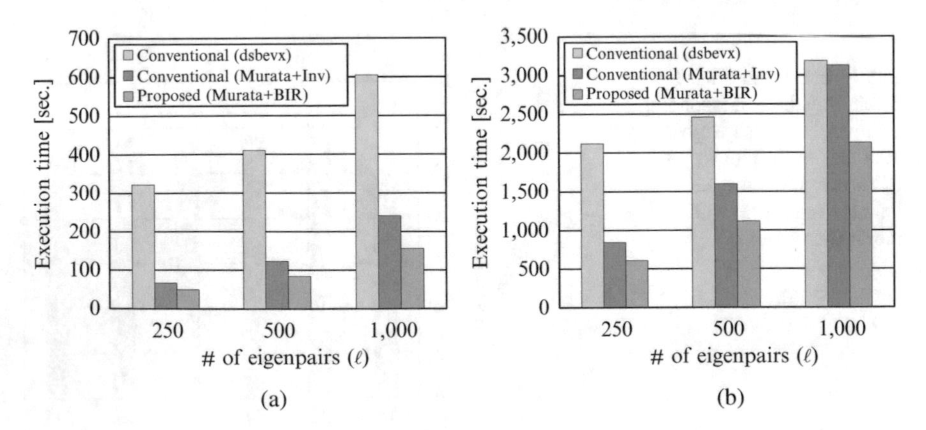


**Fig. 4** Execution times for computing the eigenpairs corresponding to the largest $\ell$ eigenvalues of real symmetric band matrices by using the proposed solver ("Murata+BIR") and the conventional solvers ("Murata+Inv" and "dsbevx"). (**a**) Cases of $B_1$. (**b**) Cases of $B_2$

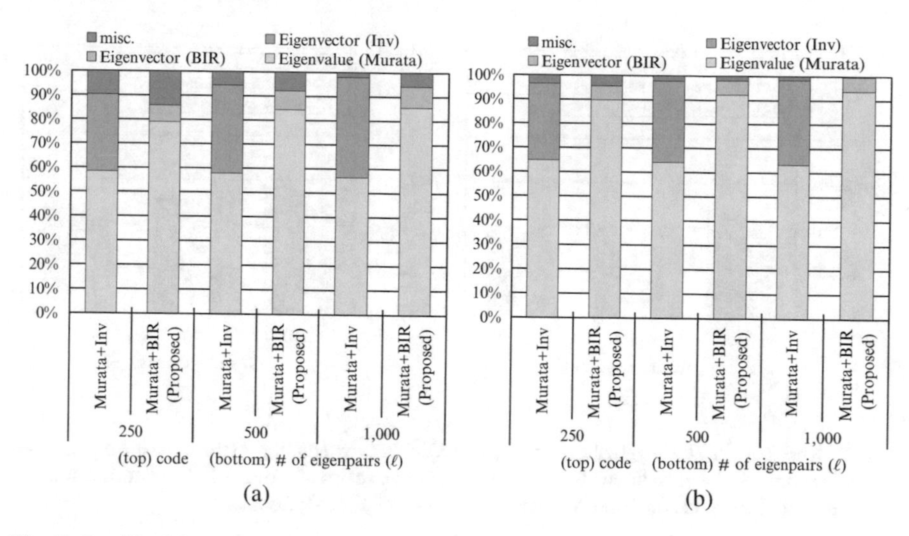


**Fig. 5** Details of the execution times for computing the eigenpairs corresponding to the largest $\ell$ eigenvalues of real symmetric band matrices by using "Murata+BIR" and "Murata+Inv". (**a**) Cases of $B_1$. (**b**) Cases of $B_2$

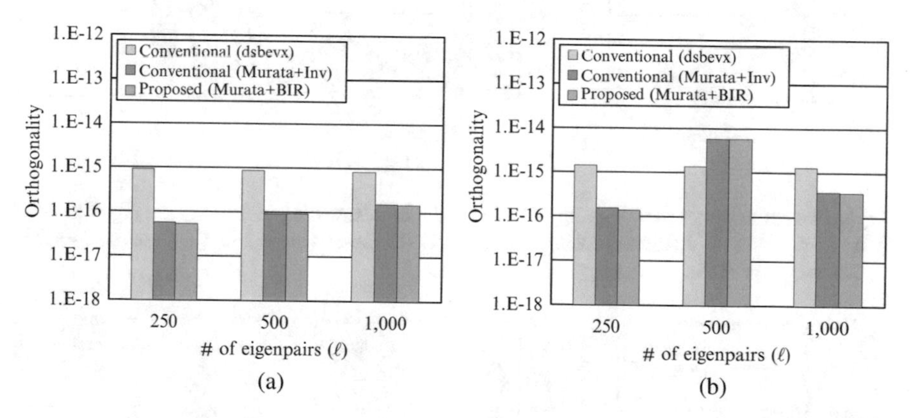


**Fig. 6** Orthogonality $\|Q_\ell^\top Q_\ell - I\|_\infty/\ell$ of the corresponding eigenvectors to the largest $\ell$ eigenvalues of real symmetric band matrices by using the proposed solver ("Murata+BIR") and the conventional solvers ("Murata+Inv" and "dsbevx"). (**a**) Cases of $B_1$. (**b**) Cases of $B_2$

The accuracy of the eigenpairs computed by using the proposed solver and the conventional solvers is shown as follows. Figure 6a and b show the orthogonality $\|Q_\ell^\top Q_\ell - I\|_\infty/\ell$ of $B_1$ and $B_2$, respectively. Similarly, Fig. 7a and b show the residual $\|B_i Q_\ell - Q_\ell D_\ell\|_\infty/\ell$ of $B_1$ and $B_2$, respectively. Note that $D_\ell = \mathrm{diag}(\tilde{\lambda}_1, \cdots, \tilde{\lambda}_\ell)$ and $Q_\ell = [\boldsymbol{q}_1 \cdots \boldsymbol{q}_\ell]$. These figures show that the proposed eigensolver computes the desired eigenpairs as accurately as the conventional solvers.

To evaluate the parallel efficiency, we compared the overall execution times with 1, 2, 4, 8, and 16 threads for computing the eigenpairs corresponding to the $\ell$

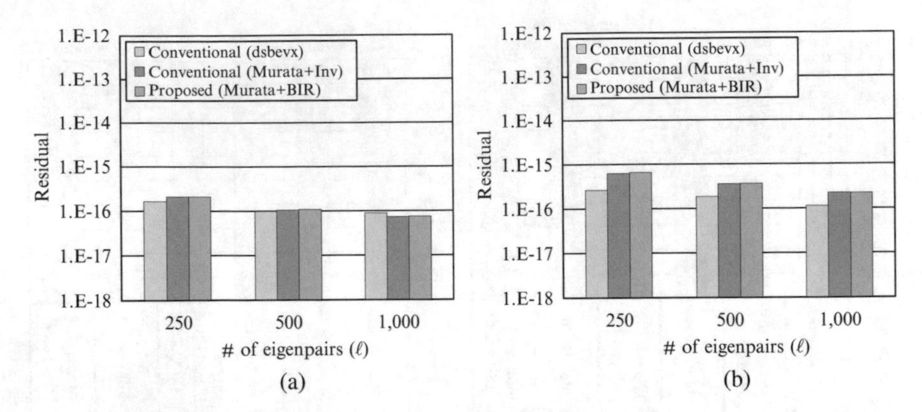


**Fig. 7** Residual $\|B_i Q_\ell - Q_\ell D_\ell\|_\infty/\ell$ of the eigenpairs corresponding to the largest $\ell$ eigenvalues of real symmetric band matrices by using the proposed solver ("Murata+BIR") and the conventional solvers ("Murata+Inv" and "dsbevx"). (**a**) Cases of $B_1$. (**b**) Cases of $B_2$

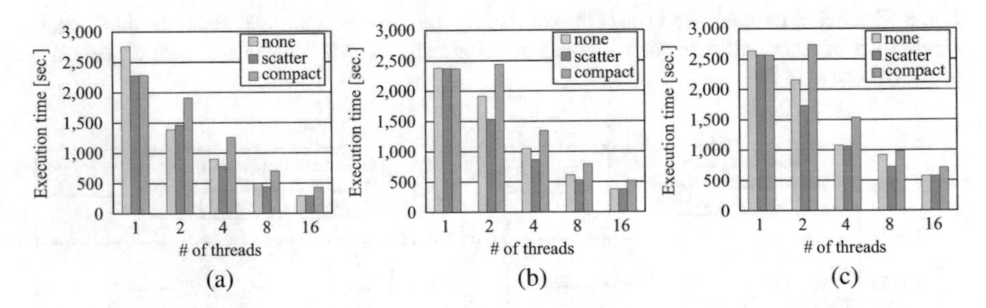


**Fig. 8** Execution times for computing the eigenpairs corresponding to the largest $\ell$ eigenvalues of $B_1$ by using "dsbevx" in different KMP_AFFINITY. (**a**) Cases of $\ell = 250$. (**b**) Cases of $\ell = 500$. (**c**) Cases of $\ell = 1000$

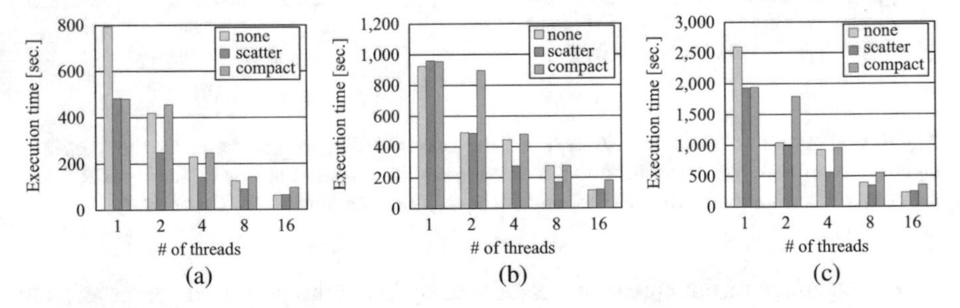


**Fig. 9** Execution times for computing the eigenpairs corresponding to the largest $\ell$ eigenvalues of $B_1$ by using "Murata+Inv" in different KMP_AFFINITY. (**a**) Cases of $\ell = 250$. (**b**) Cases of $\ell = 500$. (**c**) Cases of $\ell = 1000$

largest eigenvalues of the test matrices. Figures 8, 9, and 10 show the cases of $B_1$. Figures 11, 12, and 13 show the cases of $B_2$. In these figures, we also compared the execution times of each code run using the different KMP_AFFINITY environment variables: "none", "scatter", and "compact". From these figures, we observe that

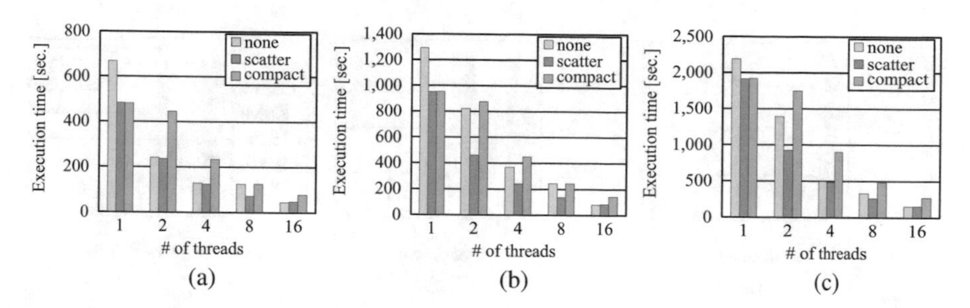


**Fig. 10** Execution times for computing the eigenpairs corresponding to the largest $\ell$ eigenvalues of $B_1$ by using the proposed eigensolver "Murata+BIR" in different `KMP_AFFINITY`. (**a**) Cases of $\ell = 250$. (**b**) Cases of $\ell = 500$. (**c**) Cases of $\ell = 1000$

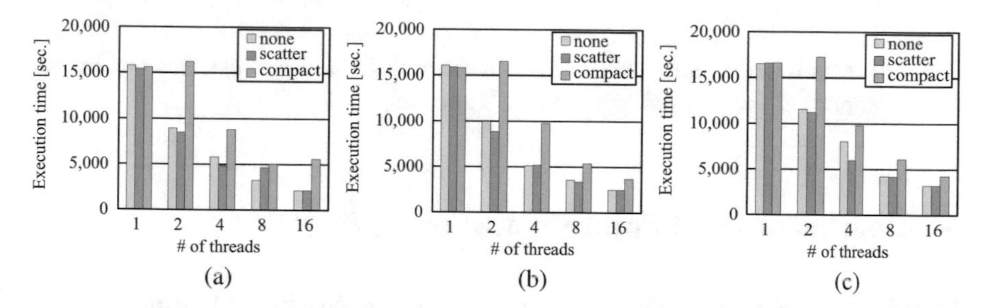


**Fig. 11** Execution times for computing the eigenpairs corresponding to the largest $\ell$ eigenvalues of $B_2$ by using "dsbevx" in different `KMP_AFFINITY`. (**a**) Cases of $\ell = 250$. (**b**) Cases of $\ell = 500$. (**c**) Cases of $\ell = 1000$

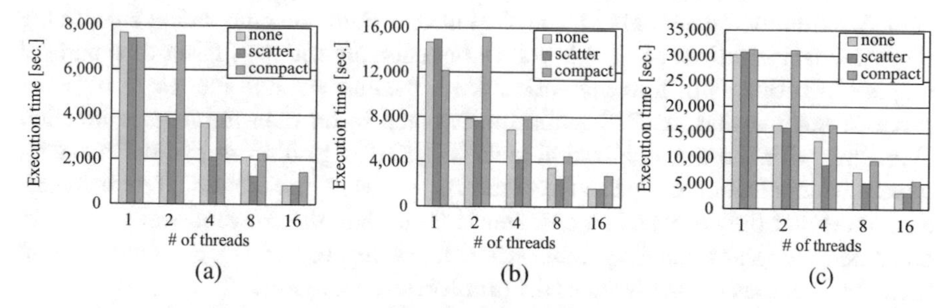


**Fig. 12** Execution times for computing the eigenpairs corresponding to the largest $\ell$ eigenvalues of $B_2$ by using "Murata+Inv" in different `KMP_AFFINITY`. (**a**) Cases of $\ell = 250$. (**b**) Cases of $\ell = 500$. (**c**) Cases of $\ell = 1000$

the proposed eigensolver "Murata+BIR" achieve the higher parallel efficiency than the conventional eigensolvers "dsbevx" and "Murata+Inv". We also observe that, if the number of threads is 16, each of the eigensolvers run with "none" achieves a competitive performance as that run with "scatter" does and the eigensolvers run with "compact" achieves the worst performance. From Figs. 10 and 13, the parallel efficiency of the proposed eigensolver run with "scatter" for $B_2$ is higher than that for $B_1$, which is smaller than $B_2$ in terms of both the matrix size and the bandwidth

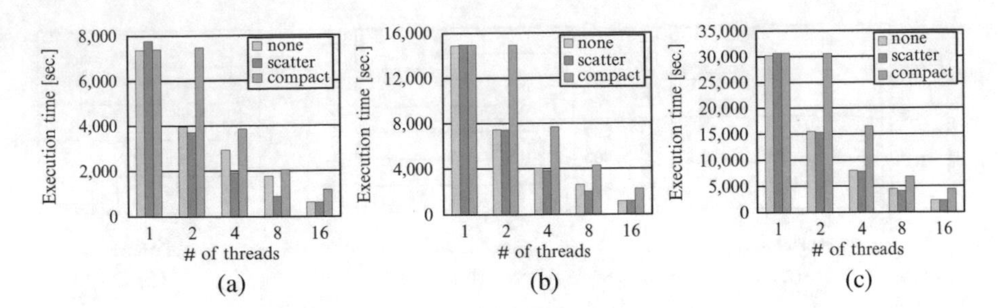


**Fig. 13** Execution times for computing the eigenpairs corresponding to the largest $\ell$ eigenvalues of $B_2$ by using the proposed eigensolver "Murata+BIR" in different `KMP_AFFINITY`. (**a**) Cases of $\ell = 250$. (**b**) Cases of $\ell = 500$. (**c**) Cases of $\ell = 1000$

size. Moreover, the parallel efficiency of it for both $B_1$ and $B_2$ becomes higher than as the number of the desired eigenpairs $\ell$ increases.

## 5 Conclusions and Future Work

In order to accelerate a subset computation of eigenpairs for real symmetric band matrices, the parallel symmetric band eigensolver is proposed, which computes directly the desired eigenvalues by using parallel Murata's bisection algorithm in Sect. 2 and the corresponding eigenvectors by using the BIR algorithm in Sect. 3.2. Employing not only Martin-Wilkinson's Gaussian elimination but also the block *LU* factorization, parallel Murata's bisection algorithm is faster than parallel Gupta's bisection algorithm. Numerical experiments on shared-memory multi-core processors show that the BIR algorithm is much faster than the inverse iteration algorithm with reorthogonalization since the BIR algorithm is parallelized with lower communication cost than the other. As the result, the numerical experiments also show that the proposed eigensolver is faster than the conventional solvers. In conclusion, we show that the parallel efficiency of the proposed eigensolver run with "scatter" becomes much higher as the problem size increases.

One of future work is to apply the proposed symmetric band eigensolver for computing a subset of eigenpairs of real symmetric dense matrices appearing in actual applications, such as the kernel principal component analysis. The number of the desired eigenpairs in such problems may be fewer than the number of the processing elements. In this case, the multi-section methods proposed in [18, 23] or the multi-section with multiple eigenvalues method [16] is expected to achieve the higher performance than the parallel bisection algorithm in Sect. 2.3. Thus, the development of the multi-section algorithm based on Murata's bisection algorithm is considered as the other future work.

**Acknowledgements** The authors would like to express their gratitude to reviewers for their helpful comments. In this work, we used the supercomputer of ACCMS, Kyoto University.

## References

1. Anderson, E., Bai, Z., Bischof, C., Blackford, L., Demmel, J., Dongarra, J., Du Croz, J., Greenbaum, A., Hammarling, S., McKenney, A., Sorensen, D.: LAPACK Users' Guide, 3rd edn. SIAM, Philadelphia, PA (1999)
2. Auckenthaler, T., Blum, V., Bungartz, H.J., Huckle, T., Johanni, R., Krämer, L., Lang, B., Lederer, H., Willems, P.: Parallel solution of partial symmetric eigenvalue problems from electronic structure calculations. Parallel Comput. **37**(12), 783–794 (2011)
3. Auckenthaler, T., Bungartz, H.J., Huckle, T., Krämer, L., Lang, B., Willems, P.: Developing algorithms and software for the parallel solution of the symmetric eigenvalue problem. J. Comput. Sci. **2**(3), 272–278 (2011)
4. Ballard, G., Demmel, J., Knight, N.: Avoiding communication in successive band reduction. ACM Trans. Parallel Comput. **1**(2), 11:1–11:37 (2015)
5. Barlow, J.L., Smoktunowicz, A.: Reorthogonalized block classical Gram-Schmidt. Numer. Math. **123**(3), 1–29 (2012)
6. Barth, W., Martin, R., Wilkinson, J.: Calculation of the eigenvalues of a symmetric tridiagonal matrix by the method of bisection. Numer. Math. **9**(5), 386–393 (1967)
7. Bischof, C., Sun, X., Lang, B.: Parallel tridiagonalization through two-step band reduction. In: Proceedings of the Scalable High-Performance Computing Conference, pp. 23–27. IEEE, New York (1994)
8. Bischof, C.H., Lang, B., Sun, X.: A framework for symmetric band reduction. ACM Trans. Math. Softw. **26**(4), 581–601 (2000)
9. Blackford, L.S., Demmel, J., Dongarra, J., Duff, I., Hammarling, S., Henry, G., Heroux, M., Kaufman, L., Lumsdaine, A., Petitet, A., Pozo, R., Remington, K., Whaley, R.C.: An updated set of basic linear algebra subprograms (BLAS). ACM Trans. Math. Softw. **28**(2), 135–151 (2002)
10. Chatelin, F.: Eigenvalues of Matrices. SIAM, Philadelphia, PA (2012)
11. Golub, G.H., van Loan, C.F.: Matrix Computations, 3rd edn. Johns Hopkins University Press, Baltimore, MD (1996)
12. Gupta, K.K.: Eigenproblem solution by a combined Sturm sequence and inverse iteration technique. Int. J. num. Meth. Eng. **7**(1), 17–42 (1973)
13. Hasegawa, H.: Symmetric band eigenvalue solvers for vector computers and conventional computers (in Japanese). J. Inf. Process. **30**(3), 261–268 (1989)
14. Ishigami, H., Kimura, K., Nakamura, Y.: A new parallel symmetric tridiagonal eigensolver based on bisection and inverse iteration algorithms for shared-memory multi-core processors. In: 2015 Tenth International Conference on P2P, Parallel, Grid, Cloud and Internet Computing (3PGCIC), pp. 216–223. IEEE, New York (2015)
15. Kahan, W.: Accurate eigenvalues of a symmetric tridiagonal matrix. Technical Report, Computer Science Dept. Stanford University (CS41) (1966)
16. Katagiri, T., Vömel, C., Demmel, J.W.: Automatic performance tuning for the multi-section with multiple eigenvalues method for symmetric tridiagonal eigenproblems. In: Applied Parallel Computing. State of the Art in Scientific Computing. Lecture Notes in Computer Science, vol. 4699, pp. 938–948. Springer, Berlin, Heidelberg (2007)
17. Kaufman, L.: Band reduction algorithms revisited. ACM Trans. Math. Softw. **26**(4), 551–567 (2000)
18. Lo, S., Philippe, B., Sameh, A.: A multiprocessor algorithm for the symmetric tridiagonal eigenvalue problem. SIAM J. Sci. Stat. Comput. **8**(2), s155–s165 (1987)

19. Martin, R., Wilkinson, J.: Solution of symmetric and unsymmetric band equations and the calculation of eigenvectors of band matrices. Numer. Math. **9**(4), 279–301 (1967)
20. Murata, K.: Reexamination of the standard eigenvalue problem of the symmetric matrix. II The direct sturm inverse-iteration for the banded matrix (in Japanese). Research Report of University of Library and Information Science **5**(1), 25–45 (1986)
21. Parlett, B.N.: The Symmetric Eigenvalue Problem. SIAM, Philadelphia, PA (1998)
22. Peters, G., Wilkinson, J.: The calculation of specified eigenvectors by inverse iteration. In: Bauer, F. (ed.) Linear Algebra. Handbook for Automatic Computation, vol. 2, pp. 418–439. Springer, Berlin, Heidelberg (1971)
23. Simon, H.: Bisection is not optimal on vector processors. SIAM J. Sci. Stat. Comput. **10**(1), 205–209 (1989)
24. Yokozawa, T., Takahashi, D., Boku, T., Sato, M.: Parallel implementation of a recursive blocked algorithm for classical Gram-Schmidt orthogonalization. In: Proceedings of the 9th International Workshop on State-of-the-Art in Scientific and Parallel Computing (PARA 2008) (2008)

# The Flexible ILU Preconditioning for Solving Large Nonsymmetric Linear Systems of Equations

**Takatoshi Nakamura and Takashi Nodera**

**Abstract** The ILU factorization is one of the most popular preconditioners for the Krylov subspace method, alongside the GMRES. Properties of the preconditioner derived from the ILU factorization are relayed onto the dropping rules. Recently, Zhang et al. (Numer Linear Algebra Appl 19:555–569, 2011) proposed a Flexible incomplete Cholesky (IC) factorization for symmetric linear systems. This paper is a study of the extension of the IC factorization to the nonsymmetric case. The new algorithm is called the Crout version of the flexible ILU factorization, and attempts to reduce the number of nonzero elements in the preconditioner and computation time during the GMRES iterations. Numerical results show that our approach is effective and useful.

## 1 Introduction

The preconditioned iterative methods for nonsymmetric linear systems [2, 4, 6, 10] are effective procedures for solving large and sparse linear systems of equations:

$$Ax = b, \tag{1}$$

arises from the discretization of elliptic partial differential equations. Two good preconditioners are known, such as the incomplete LU factorization(ILU) [1, 5, 8, 14] and the modified incomplete LU factorization [1], each of which makes use of an approximate factorization of the coefficient matrix into the product of a sparse lower triangular matrix $L$, and a sparse upper triangular matrix $U$. It has been observed

T. Nakamura
School of Fundamental Science and Technology, Graduate School of Science and Technology, Keio University, Yokohama, Japan
e-mail: takatoshi0161@gmail.com

T. Nodera (✉)
Faculty of Science and Technology, Department of Mathematics, Keio University, Yokohama, Japan
e-mail: nodera@math.keio.ac.jp



T. Sakurai et al. (eds.), *Eigenvalue Problems: Algorithms, Software and Applications in Petascale Computing*, Lecture Notes in Computational Science and Engineering 117, https://doi.org/10.1007/978-3-319-62426-6_4

on an empirical basis that it generates a linear system with eigenvalues that are mostly clustered near 1. The effectiveness of both techniques for nonsymmetric linear systems of equations derived from non-self-adjoint elliptic boundary value problems, has been demonstrated in many numerical experiments [3–6, 14].

There are now numerous Krylov subspace methods for solving nonsymmetric linear systems of equations, e.g. the GMRES, the Bi-CGSTAB, the QMR and the IDR(s) [9, 12, 13]. In order to be effective, these methods must be combined with a good preconditioner, and it is generally agreed that the choice of the preconditioner is even more critical than the choice of the Krylov subspace iterative methods. The GMRES [11] is useful for a thorough treatment of preconditioned iterative procedures. The preconditioning of a coefficient matrix is known as one of the methods for improving the convergence of the GMRES. The preconditioner of the ILU factorization applied to the GMRES is popular and is considered to be one of the fundamental preconditioners in the solution of large nonsymmetric linear systems of equations. The search for effective preconditioners is an active research topic in scientific computing. Several potentially successful methods of the ILU factorizations have been recently proposed [14]. The performance of the ILU factorization often is dependent on the dropping method to reduce fill-ins. There are some dropping strategies, for example, the dual dropping strategy which makes it possible to determine the sparsity of incomplete factorization preconditioners by two fill-in control parameters: (1) $\tau$: dropping tolerance and (2) $p$: the number of $p$ largest nonzero elements in the magnitude are kept. Recently, Zhang et al. [14] proposed using parameter $q$ to control the number of nonzero elements in preconditioner $L$, in the IC factorization. Their proposed scheme was called IC factorization with a multi-parameter strategy. The parallel implementation of the ILU factorization is investigated in [6, 7].

In this paper, the general framework of the dropping strategy in an ILU factorization will be proposed. Further to this, a method for overcoming the shortcomings that a calculating norm is needed to use diagonal elements, will be explored. In Sect. 2, the most promising approach for preconditioning will be discussed. In Sect. 3, two flexible ILU factorizations will be proposed and explored. In Sect. 4, the results of extensive numerical experimentations will be tabulated. The conclusion follows.

## 2 Preconditioning

The preconditioner $M$ can reduce the number of iterations, because the properties of this coefficient matrix can be improved through preconditioning [1]. One possibility is to solve the left preconditioned system of the equation:

$$M^{-1}A\boldsymbol{x} = M^{-1}\boldsymbol{b}. \tag{2}$$

In general, the preconditioning matrix $M$ is often chosen so that $\mathrm{cond}(M^{-1}A) \ll \mathrm{cond}(A)$, where $\mathrm{cond}(Z)$ is the condition number of matrix $Z$. A remedy exists when the preconditioner $M$ is available in factored form, e.g., as an incomplete LU factorization $M = LU$, where $L$ is a lower triangular matrix and $U$ is an upper triangular matrix.

## 2.1 *ILU Factorization*

The ILU factorization is an LU factorization with reducing fill-ins. The ILU factorization factorizes coefficient matrix $A$ as follows:

$$A = LU + R, \tag{3}$$

where $L$ is a lower triangular matrix, $U$ is an upper triangular matrix, and $R$ is an error matrix. In this paper, we use the Crout version ILU factorization which we usually call ILUC. The standard ILUC description of Algorithm 1 is due to Li et al. [4, pp. 717–719].

For lines 4 and 8 in the Algorithm 1, the unit $(k : n)$ of the $i$th row of $U$ is needed, and in the same way only the unit $(k + 1 : n)$ of the $i$th column of $L$ is required. Accessing entire rows of $U$ or columns of $L$ and then removing only the needed part is a costly option.

The dual dropping strategy was used in line 10 and 11. For a less complex problem, the effect of the dropping rule is not as important. For large scale problems, however, it is critically important. The number of iterations appears to be sensitive to the dropping tolerance. The basic idea of the dual dropping strategy is constituted by the following two steps:

1. Any elements of $L$ or $U$ whose magnitude is less than tolerance $\tau$ is dropped:

$$|u_{ij}| \le \tau \times \|z\| \;\Rightarrow\; u_{ij} = 0, \quad \text{or} \quad |l_{ij}| \le \tau \times \|w\| \;\Rightarrow\; l_{ij} = 0$$

where $\tau$ is a dropping tolerance.

**Algorithm 1** Crout version of the ILU factorization (ILUC)

1: **for** $k = 1 : n$ **do:**
2: Initialize row $z$: $z_{1:k-1} = 0,\ z_{k,k:n} = a_{k,k:n}$
3: **for** $\{\ i \mid 1 \le i \le k-1$ and $l_{ki} \ne 0\ \}$ **do:**
4: $z_{k:n} = z_{k:n} - l_{ki} * u_{i,k:n}$
5: **end for**
6: Initialize column $w$, as $w_{1:k} = 0,\ w_{k+1:n} = a_{k+1:n,k}$
7: **for** $\{i \mid 1 \le i \le k-1$ and $u_{ik} \ne 0\}$ **do:**
8: $w_{k+1:n} = z_{k+1:n} - u_{ik} * l_{k+1:n,i}$
9: **end for**
10: Apply a dropping rule to row $z$
11: Apply a dropping rule to column $w$
12: $u_{k,:} = z$
13: $l_{:,k} = w/u_{kk},\ l_{kk} = 1$
14: **end for**

2. In the $k$-th column of $L$, the number of the $p$ largest nonzero elements in the magnitude are kept. Similarly, the number of the $p$ largest nonzero elements in the $k$-th row of $U$, which includes the diagonal elements, are kept. This controls the total memory storage that can be used by the preconditioner.

To study parameter $p$, a new dropping strategy was proposed which changes $p$ by some parameters during the computation of the preconditioner. A dynamically changed parameter $q$ according to the magnitude of elements in the preconditioner $L$ , where $q$ is the number of nonzero elements kept in the corresponding column of $L$, was introduced for this exercise.

# 3 ILU Factorization with Multi-Parameter Strategies

In this section, we present some strategies to dynamically selected the number of nonzero elements in each column of preconditoner $L$ and each row of preconditioner $U$ in ILU factorization. Our consideration is focused on the performance of the Crout version of ILU factorization (ILUC) preconditioner (i.e. Algorithm 1) and choice of parameters.

## 3.1 Flexible ILU Factorization

Zhang [14] proposed a flexible IC factorization which changed parameter $p$ according to the norm of the already computed elements of preconditioner $L$. This idea was explored to propose a flexible ILU factorization with a new norm, and this will be referred to as the $n$-flexible ILU. In the $n$-flexible ILU factorization, $q$, the number of nonzero elements kept in each column of $L$ and each row of $U$, is determined as follows:

$$q = \begin{cases} \max\left(p_{\min},\ p + \left[c \log_{10} \mathrm{d}\frac{\|\boldsymbol{l}_j\|}{g_j}\right]\right), & (\|\boldsymbol{l}_j\| < g_j), \\ \min\left(p_{\max},\ p + \left[c \log_{10} \mathrm{d}\frac{\|\boldsymbol{l}_j\|}{g_j}\right]\right), & (\|\boldsymbol{l}_j\| \geq g_j), \end{cases} \tag{4}$$

where

$$g_j = \frac{\mathrm{d}\left(\sum_{k=1}^{j} \|\boldsymbol{l}_k\|\right)}{j}, \quad (j = 1, \ldots, n).$$

The parameter $p$ is selected as a basic parameter to control the number of nonzero elements in the preconditioner $p_{\min}$ and $p_{\max}$ that indicate the range of the number of nonzero elements kept in each column of $L$ and row of $U$. Moreover, the parameter $c$ is a proportion value to control the number of nonzero elements in each column of the preconditioner $L$ and row of the preconditioner $U$. We observe that parameter $c$ is quite important to obtain effective preconditioners by the flexible ILU factorization besides $p_{\min}$ and $p_{\max}$.

The nonzero elements of $L$ were compared with the nonzero elements of $\widetilde{L}$, where $L$ was generated by a fixed ILU factorization and $\widetilde{L}$ is generated by $n$-flexible ILU factorization, respectively:

$$\mathrm{nnz}(L) \approx np,$$

$$\mathrm{nnz}(\widetilde{L}) \approx np + \sum_{j=1}^{n} \left[ c \log_{10} \frac{\|\boldsymbol{l}_j\|}{g_j} \right].$$

This results in the following relation:

$$\sum_{j=1}^{n} \left[ c \log_{10} \frac{\|\boldsymbol{l}_j\|}{g_j} \right] < 0 \Rightarrow \mathrm{nnz}(L) > \mathrm{nnz}(\widetilde{L}). \tag{5}$$

The next step was to consider the logarithmic function $f(x) = \log_{10} x$. This function satisfied the following relation: (1) $f'(x)$ is a monotonic decreasing function, and (2) $f(1) = 0$. From these properties, it was not difficult to prove the following inequality:

$$\log_{10}(1+s) + \log_{10}(1-s) < 0 \quad (0 < s < 1). \tag{6}$$

Assuming that $\|\boldsymbol{l}_j\|/g_j$ is a symmetric distribution to 1, $\sum [c \log_{10} \|\boldsymbol{l}_j\|/g_j] < 0$ and $\mathrm{nnz}(L) > \mathrm{nnz}(\widetilde{L})$, were obtained. The upper matrix $U$ also satisfied the same relation. It was concluded that the $n$-flexible ILU factorization reduced the nonzero elements of the preconditioner.

## 3.2 Diagonal Flexible ILU Factorization

The $n$-flexible ILU factorization is characterized by the shortcoming that it needed to calculate the norm during each iteration and as a result, increased the computation time. To overcome this issue, diagonal elements were used instead of $\|\boldsymbol{l}_j\|/g_j$ and a diagonal flexible ILU factorization was proposed called the $d$-flexible ILU. The $d$-flexible ILU factorization determined the number of nonzero elements as follows:

$$q = \begin{cases} \max\left(p_{\min},\ p + \left[ c \log_{10} \mathrm{d} \dfrac{|d_j|}{\widetilde{g}_j} \right]\right), & (|d_j| < \widetilde{g}_j), \\ \min\left(p_{\max},\ p + \left[ c \log_{10} \mathrm{d} \dfrac{|d_j|}{\widetilde{g}_j} \right]\right), & (|d_j| \geq \widetilde{g}_j), \end{cases} \tag{7}$$

where

$$\widetilde{g}_j = \frac{\mathrm{d}\left(\sum_{k=1}^{j} |d_k|\right)}{j}, \quad (j = 1, \ldots, n).$$

In the next section, it will be verified that the $d$-flexible ILU factorization preconditioner based on ILUC is suitable for practical use.

# 4 Numerical Experiments

Numerical experiments were implemented, based on ITSOL packages [9]. In this section, the numerical results were used to compare the following methods for solving two examples: the $d$-flexible ILU, the $d$-fixed ILU (ILUC) and the $n$-flexible ILU. All numerical experiments were done on the DELL Precision T1700 with 3.50 GHz and a 16 GB main memory, using C language. In these experiments, $\boldsymbol{x}_0 = 0$ for an initial approximate solution, and solution $\boldsymbol{x}_i$ is considered to have converged if the norm of the residual, $\|\boldsymbol{r}_i\| = \|\boldsymbol{b} - A\boldsymbol{x}_i\|$, satisfied the following convergence criterion:

$$\|\boldsymbol{r}_i\| / \|\boldsymbol{r}_0\| < 1.0 \times 10^{-12} \tag{8}$$

where $\boldsymbol{r}_i$ is the residual vector of the $i$-th iteration. All the matrices were tested with the following parameter setting. $p_{\min} = p - 0.2p$, $p_{\max} = p + 0.2p$, $c = 8$. For thc choicc of parameter $p$, we firstly set up $p$ by 3 times $\mathrm{nnz}(A)/n$ which is usually appropriate for the most problems. Some sensitivity analysis of the flexible ILU factorization on parameter $c$ is performed in our examples, using the practical technique which has been given in Zhang et al. [14, Section 3.2]. When parameter $c$ varies from 4 to 20, we discovered that the convergence rate is not sensitive to $c$. Therefore, we may hold the parameter $c$ to some good fixed value such as 6 in our examples.

We denoted the computation time of the factorization as CPT, the computation time of GMRES as CGT, the total computation time as the Total, the rational of nonzero elements of $L$ and $U$ to nonzero elements of original coefficient matrix $A$ as nnz($LU$)/nnz($A$), and the iterations of GMRES as Its. These result were used to illustrate the efficiency of the flexible ILU preconditioning.

## 4.1 Example 1

The first matrix problem arising from the finite difference discretization of the boundary value problem of the two-dimensional partial differential equation with

Dirichlet boundary conditions in Joubert [3] is calculated as follows:

$$-\Delta u + D\left\{(y - \frac{1}{2})u_x + (x - \frac{1}{3})(x - \frac{2}{3})u_y\right\} - 43\pi^2 u$$
$$= G(x, y) \text{ on } \Omega = [0, 1]^2, \tag{9}$$

where $u(x, y) = 1 + xy$ on $\partial\Omega$. The operator is discretized using a five point centered finite difference scheme to discretize on a uniform grid with a mesh spacing $h = 1/128$ in either direction. For $D$ and $h$ is small, the generated matrix is a symmetric indefinite matrix with 16 distinct negative eigenvalues and the rest of the spectrum positive. The classical GMRES method with ILU(0) and MILU(0) preconditioning applied to this problem is difficult to converge. Sometimes these preconditioners fail to converge with GMRES method. In any case, this is a challenging test problem to solve.

In this problem, the parameters were set as follows: $p = 15$, $p_{\min} = p - 0.2p$, $p_{\max} = p + 0.2p$, $c = 6$. The dropping tolerance $\tau$ was also set up as $\tau = 10^{-4}$. Table 1 shows that in each example, the $d$-flexible ILU factorization reduces the nonzero elements of the preconditioning matrix without increasing total computation time. The results show that the $d$-flexible ILU factorization reduces memory usage efficiency.

**Table 1** Example 1—numerical results of the boundary value problem

| Preconditioner | $Dh$ | CPT (s) | CGT (s) | Total (s) | Its | nnz($LU$) / nnz($A$) |
|---|---|---|---|---|---|---|
| ILUC | $2^4$ | 0.760 | 1.990 | 2.750 | 35 | 7.549 |
| $n$-Flexible | | 0.970 | 1.700 | 2.670 | 25 | 8.690 |
| $d$-Flexible | | 0.640 | 2.020 | 2.660 | 38 | 6.624 |
| ILUC | $2^3$ | 0.720 | 2.740 | 3.460 | 45 | 7.883 |
| $n$-Flexible | | 0.960 | 2.240 | 3.200 | 34 | 9.266 |
| $d$-Flexible | | 0.620 | 2.680 | 3.300 | 47 | 6.988 |
| ILUC | $2^2$ | 0.710 | 3.910 | 4.620 | 60 | 8.034 |
| $n$-Flexible | | 0.970 | 3.100 | 4.070 | 44 | 9.707 |
| $d$-Flexible | | 0.620 | 3.980 | 4.600 | 66 | 7.399 |
| ILUC | $2^1$ | 0.660 | 10.610 | 11.270 | 172 | 7.776 |
| $n$-Flexible | | 0.980 | 4.310 | 5.290 | 58 | 9.826 |
| $d$-Flexible | | 0.620 | 10.530 | 11.150 | 175 | 7.441 |
| ILUC | $2^0$ | 0.590 | 27.750 | 28.340 | 473 | 7.282 |
| $n$-Flexible | | 0.980 | 7.780 | 8.760 | 107 | 9.865 |
| $d$-Flexible | | 0.590 | 27.730 | 28.320 | 473 | 7.261 |
| ILUC | $2^{-1}$ | 0.560 | – | – | – | 7.009 |
| $n$-Flexible | | 0.970 | 30.280 | 31.250 | 413 | 9.860 |
| $d$-Flexible | | 0.560 | – | – | – | 7.009 |

## 4.2 Example 2

The next test problem studied was the Poisson3Db, which is a computational fluid dynamics (CFD) problem from the University of Florida Matrix Collection [2]. This problem had a $85{,}623 \times 85{,}623$ real nonsymmetric matrix and the nonzero elements of this coefficient matrix were 2,374,949. The nonzero pattern of this matrix is shown in Fig. 1. For the choice of parameter $p$ in this problem, we used the above mentioned formula as $p = 3 \times \text{nnz}(A)/n \approx 83$. As a result, we now initially set $p = 80$, and reduce parameter $p$ to get a valuable evaluation. In fact, the modification of $p$ is expected to produce a more reasonable comparison. The other parameters were set as follows: $p_{\text{min}} = p - 0.2p$, $p_{\text{max}} = p + 0.2p$, $c = 6$, $\tau = 10^{-4}$.

Table 2 shows that the $d$-flexible ILU factorization was faster than other preconditioned methods and its preconditioner had the least nonzero elements.

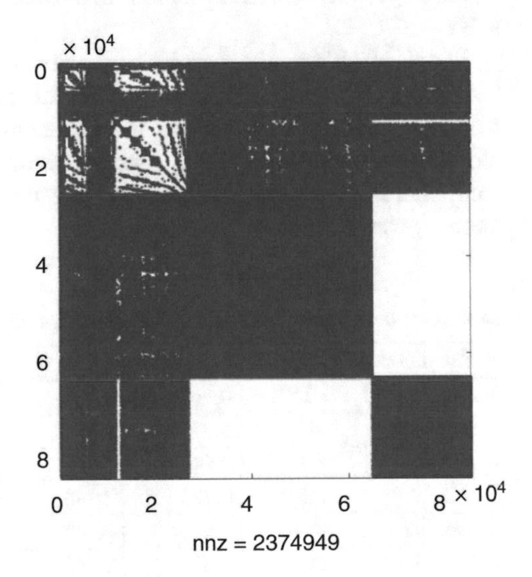


**Fig. 1** Example 2—number of nonzero elements of Poisson3Db's matrix

**Table 2** Example 2—numerical results of the Poisson3Db problem

| Preconditioner | $p$ | CPT (s) | CGT (s) | Total (s) | Its | nnz($LU$) / nnz($A$) |
|---|---|---|---|---|---|---|
| ILUC | 70 | 3.470 | 2.500 | 5.970 | 38 | 4.449 |
| $n$-Flexible | | 4.800 | 2.620 | 7.420 | 37 | 5.031 |
| $d$-Flexible | | 2.950 | 2.870 | 5.820 | 46 | 3.965 |
| ILUC | 75 | 3.660 | 2.580 | 6.240 | 38 | 4.639 |
| $n$-Flexible | | 5.260 | 2.670 | 7.930 | 36 | 5.385 |
| $d$-Flexible | | 3.190 | 2.530 | 5.720 | 40 | 4.161 |
| ILUC | 80 | 3.870 | 2.660 | 6.530 | 38 | 4.815 |
| $n$-Flexible | | 5.710 | 2.870 | 8.580 | 37 | 5.739 |
| $d$-Flexible | | 3.390 | 2.480 | 5.870 | 38 | 4.346 |

Judging from the computation time vs behavior of residual norm in Fig. 2, these were similar to other methods.

Figures 3 and 4 show the distribution of $\|\boldsymbol{l}_j\|/g_j$ for the $n$-Flexible ILU factorization and the distribution of $|d_j|/\widetilde{g}_j$ for the $d$-Flexible ILU factorization, respectively. These figures suggest that the $d$-Flexible ILU factorization is the best method for reducing the number of nonzero elements of the fill-in. We expect that the

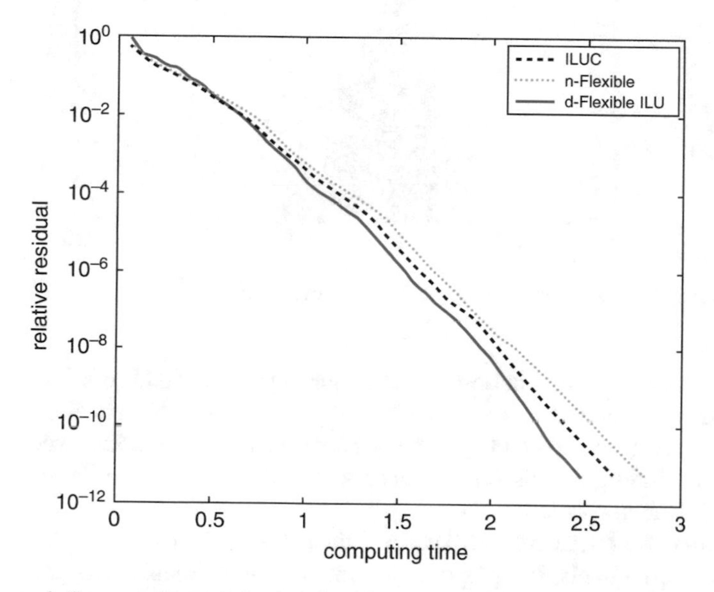


**Fig. 2** Example 2—convergence behavior of residual norm vs computation time, $p = 80$

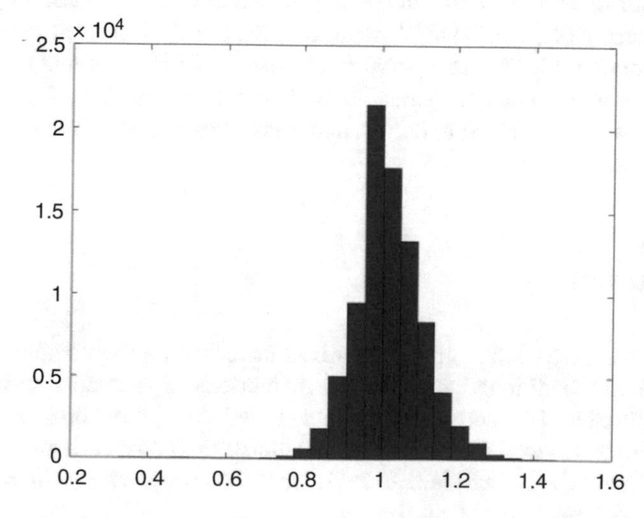


**Fig. 3** Example 2—distribution of $\|\boldsymbol{l}_j\|/g_j$ for $n$-Flexible ILU

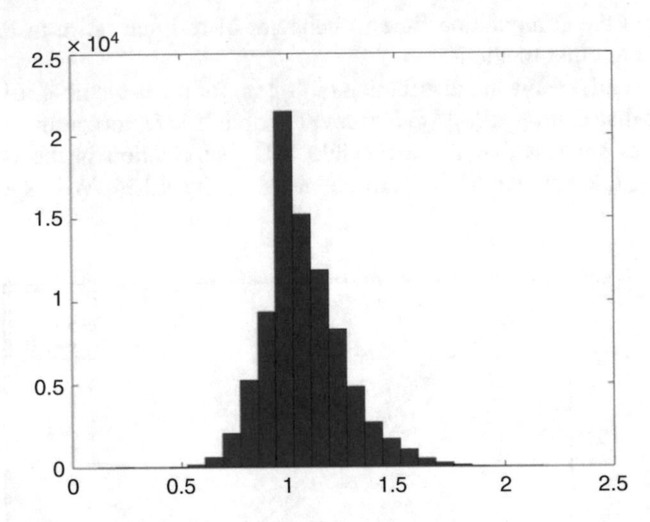


**Fig. 4** Example 2—distribution of $|d_j|/\widetilde{g_j}$ for $d$-Flexible ILU

almost same as the distribution of these figures can be found to solve the system arising from the boundary value problem of partial differential equation, especially convection diffusion and CFD problems. However, more numerical experiments are needed for finding the discretized matrices which can optimize the use of the $d$-flexible ILU factorization.

In summary, based on the data in Tables 1 and 2, it can be concluded that for these experiments, the proposed scheme of the $d$-flexible ILU preconditioner appears to be superior to other schemes in memory requirements especially in terms of the number of nonzero elements of the preconditioner. Furthermore, for many of these experiments, the GMRES with a $d$-flexible ILU preconditioner needed less total computation time compared to the fixed ILU (ILUC) and the $n$-flexible preconditioner with some exceptions. In Table 2, the results of the experiments are tabulated showing that the proposed scheme executes better or analogous in total time to the solution.

## 5 Conclusion

The dropping strategy is integral for the ILU factorization to generate an efficient and reliable preconditioned matrix. The numerical experiments show that the proposed $d$-flexible ILU factorization, is able to reduce certain nonzero elements of the preconditioner as well as the total computation time. The results also suggest that the GMRES with the $d$-flexible ILU factorization converges faster than a GMRES with the other classical ILU factorization.

It can be concluded that $d$-flexible ILU factorization is a practical and effective method for solving large sparse sets of nonsymmetric linear systems of equations.

Future studies are necessary for investigating and determining specific parameters, and finding matrices which can optimize the use of the $d$-flexible ILU factorization.

## References


1. Benzi, M.: Preconditioning techniques for large linear systems. J. Comp. Phys. 182, 418–477 (2002)
2. Davis, T., University of Florida Sparse Matrix Collection (online) (2005). Available from http://www.cise.ufl.edu/research/sparse/matrices/
3. Joubert, W.: Lanczos methods for the solution of nonsymmetric systems of linear equations. SIAM J. Matrix Anal. Appl. **13**, 926–943 (1992)
4. Li, N, Saad, Y., Chow, E.: Crout version of ILUT for sparse matrix. SIAM J. Sci. Comp. **25**, 716–728 (2003)
5. Mayer, J.: Alternating weighted dropping strategies for ILUTP. SIAM J. Sci. Comp. **4**, 1424–1437 (2006)
6. Moriya, K., Nodera, T.: Parallelization of IUL decomposition for elliptic boundary value problem of PDE on AP3000. In: ISHPC'99 Proceedings of the Second International Symposium on High Performance Computing. Lecture Notes in Computer Science, vol. 1615, pp. 344–353. Springer, London (1999)
7. Nodera, T., Tsuno, N.: The parallelization of incomplete LU factorization on AP1000. In: European Conference on Parallel Processing. Lecture Notes in Computer Science, vol. 1470, pp. 788–792. Springer, London (1998)
8. Saad, Y.: ILUT: a dual threshold incomplete LU factorization. Numer. Linear Algebra Appl. **1**, 387–402 (1994)
9. Saad, Y.: Iterative Methods for Sparse Linear Systems, 2nd edn. SIAM, Philadelphia (2003)
10. Saad, Y.: ITSOL collections (online). Available from http://www-users.cs.umn.edu/~saad/software/ITSOL/
11. Saad, Y., Schultz, M.H.: GMRES: a generalized minimal residual algorithm for solving nonsymmetric linear systems. SIAM J. Sci. Stat. Comput. **7**, 856–869 (1986)
12. Sonneveld, P., Van Gijzen, M.B.: IDR($s$): a family of simple and fast algorithms for solving large nonsymmetric systems of linear equations. SIAM J. Sci. Comput. **31**, 1035–1062 (2008)
13. Van Gijzen, M.B., Sonneveld, P.: Algorithm 913: an elegant IDR(s) variant that efficiently exploits biorthogonality properties. ACM Trans. Math. Softw. **38**, 5:1–5:19 (2011)
14. Zhang, Y., Huang, T.Z. , Jing, Y.F., Li, L.: Flexible incomplete Cholesky factorization with multi-parameters to control the number of nonzero elements in preconditioners. Numer. Linear Algebra Appl. **19**, 555–569 (2011)

# Improved Coefficients for Polynomial Filtering in ESSEX

**Martin Galgon, Lukas Krämer, Bruno Lang, Andreas Alvermann, Holger Fehske, Andreas Pieper, Georg Hager, Moritz Kreutzer, Faisal Shahzad, Gerhard Wellein, Achim Basermann, Melven Röhrig-Zöllner, and Jonas Thies**

**Abstract** The ESSEX project is an ongoing effort to provide exascale-enabled sparse eigensolvers, especially for quantum physics and related application areas. In this paper we first briefly summarize some key achievements that have been made within this project.

Then we focus on a projection-based eigensolver with polynomial approximation of the projector. This eigensolver can be used for computing hundreds of interior eigenvalues of large sparse matrices. We describe techniques that allow using lower-degree polynomials than possible with standard Chebyshev expansion of the window function and kernel smoothing. With these polynomials, the degree, and thus the number of matrix–vector multiplications, typically can be reduced by roughly one half, resulting in comparable savings in runtime.

## 1 The ESSEX Project

ESSEX—Equipping Sparse Solvers for Exascale—is one of the projects within the German Research Foundation (DFG) Priority Programme "Software for Exascale Computing" (SPPEXA). It is a joint effort of physicists, computer scientists and mathematicians to develop numerical methods and programming concepts for the solution of large sparse eigenvalue problems on extreme-scale parallel machines. ESSEX' goal is not to provide a general-purpose eigenvalue library directly usable

M. Galgon • L. Krämer • B. Lang (✉)
School of Mathematics and Natural Sciences, University of Wuppertal, Wuppertal, Germany
e-mail: lang@math.uni-wuppertal.de

A. Alvermann • H. Fehske • A. Pieper
Institute of Physics, University of Greifswald, Greifswald, Germany

G. Hager • M. Kreutzer • F. Shahzad • G. Wellein
Erlangen Regional Computing Center, Erlangen, Germany

A. Basermann • M. Röhrig-Zöllner • J. Thies
Simulation and Software Technology, German Aerospace Center (DLR), Cologne, Germany



T. Sakurai et al. (eds.), *Eigenvalue Problems: Algorithms, Software and Applications in Petascale Computing*, Lecture Notes in Computational Science and Engineering 117, https://doi.org/10.1007/978-3-319-62426-6_5

in the full range of eigenvalue computations. Rather, in the first funding period the focus was on appropriate methods for selected physical applications featuring rather different characteristics, such as the need for a few extremal eigenpairs, a bunch of interior eigenpairs, information about the whole eigenvalue distribution, and dynamics, of symmetric or Hermitian matrices. These methods include Jacobi–Davidson iteration, projection-based algorithms, the kernel polynomial method, and Chebyshev time propagation, respectively.

To obtain scalable high performance, ESSEX takes a holistic performance engineering approach encompassing the applications, algorithmic development, and underlying building blocks. One of the outcomes is an "Exascale Sparse Solver Repository" (ESSR) that provides high-performance implementations of several methods. In addition, experiences gained through ESSEX work can give guidelines for addressing structurally similar problems in numerical linear algebra. In the following we briefly summarize some of ESSEX' results so far.

Up to now, block variants of the Jacobi–Davidson algorithm have been considered worthwhile mainly for robustness reasons. Our investigations [20] have revealed that they also can be faster than non-blocked variants, provided that all basic operations (in particular sparse matrix times multiple vector multiplication and operations on tall and skinny matrices) achieve optimal performance. This is typically not the case if block vectors are stored in column-major ordering, and we showed that some care has to be taken when implementing algorithms using row-major block vectors instead. A blocked GMRES solver has been developed for solving the multiple Jacobi–Davidson correction equations occurring in the block algorithm.

An adaptive framework for projection-based eigensolvers has been developed [8], which allows the projection to be carried out with either polynomials or a contour integration as in the FEAST method [19]. The latter approach requires the solution of highly indefinite, ill-conditioned linear systems. A robust solver for these has been identified and implemented [9].

The eigensolvers in ESSEX are complemented by domain-specific algorithms for quantum physics computations such as the kernel polynomial method (KPM) [26] for the computation of eigenvalue densities and spectral functions. These algorithms, which are based on simple schemes for the iterative evaluation of (Chebyshev) polynomials of sparse matrices, are very attractive candidates for our holistic performance engineering approach. For example, in a large-scale KPM computation [14, 15] we achieved 11% of LINPACK efficiency on 4096 heterogeneous CPU–GPU nodes of Piz Daint at Swiss National Supercomputing Centre (CSCS). This is an unusually high value for sparse matrix computations whose performance is normally much more restricted by the main memory bandwidth.

Such progress is possible because in ESSEX the algorithmic work on the above methods goes hand-in-hand with the model-guided development of high-performance MPI+X hybrid parallel kernels for the computational hot spots. They are included in the "General, Hybrid, and Optimized Sparse Toolkit" (GHOST) [16]. Besides implementations of sparse matrix–(multiple) vector multiplication, optionally chained with vector updates and inner products to reduce data transfers, operations with "tall skinny" dense matrices, and other basic building blocks, GHOST also provides mechanisms for thread-level task management to utilize the aggregate computational power of heterogeneous nodes comprising standard multicore CPUs, Intel Xeon Phi manycore CPUs, and NVIDIA GPUs, and to enable asynchronous checkpointing with very low overhead.

A "Pipelined Hybrid-parallel Iterative Solver Toolkit" (PHIST) [25] provides an interface to the GHOST kernels and adapters to the Trilinos libraries Anasazi [1] and Belos [2]. It also contains higher-level kernels and a comprehensive test framework.

A unified sparse matrix storage format for heterogeneous compute environments, SELL-$C$-$\sigma$, has been proposed [13]. It is a generalization of the existing formats Compressed Row Storage and (Sliced) ELLPACK with row sorting and allows near-to-optimum performance for a wide range of matrices on CPU, Phi, and GPU, thus often obviating the former need to use different formats on different platforms.

While the methods and software developed in ESSEX are applicable to general eigenvalue problems, our project specifically addresses quantum physics research applications. Among these, the recent fabrication of graphene [4] and topological insulator [11] materials has renewed the interest in the solution of large scale interior eigenvalue problems. Realistic modeling of structured or composite devices made out of these materials directly leads to large sparse matrix representations of the Hamilton operator in the Schrödinger equation, and the eigenvalues and eigenvectors deep inside the spectrum, near the Fermi energy, determine the electronic structure and topological character and thus the functioning of such devices.

A more comprehensive description of the results obtained within the ESSEX project can be found in [25] and the references therein, and on the ESSEX homepage https://blogs.fau.de/essex/.

In the following sections we focus on the BEAST-P eigensolver that is available in the ESSR and in particular propose approaches for increasing its efficiency.

## 2 Accelerated Subspace Iteration with Rayleigh–Ritz Extraction

Eigenvalue problems (EVPs) $\mathsf{A}\mathbf{v} = \lambda\mathbf{v}$ with real symmetric or complex Hermitian matrices $\mathsf{A}$ arise in many applications, e.g., electronic structure computations in physics and chemistry [3].

Often the matrix is very large and sparse, and only a few extreme or interior eigenpairs $(\lambda, \mathbf{v})$ of the spectrum are required. In this case iterative solvers based on Krylov subspaces, such as Lanczos or Jacobi–Davidson-type methods [17, 24], or block variants of these [20, 21], tend to be most efficient. A common feature of such algorithms is a subspace that is expanded by a single vector or a block of vectors in each iteration, thus increasing its dimension.

**Algorithm 1** Key algorithmic steps for subspace iteration with Rayleigh–Ritz extraction

**Given:** $\mathsf{A} \in \mathbb{C}^{n\times n}$, $I_\lambda = [\alpha, \beta] \subset \mathbb{R}$
**Sought:** Those eigenpairs $(\lambda, \mathbf{v})$ of $\mathsf{A}$ such that $\lambda \in I_\lambda$
1: start with a subspace $\mathsf{Y} \in \mathbb{C}^{n\times m}$ of suitable dimension $m$
2: **while** not yet converged **do:**
3: compute $\mathsf{U} = f(\mathsf{A}) \cdot \mathsf{Y}$ for a suitable function $f$
4: compute $\mathsf{A_U} = \mathsf{U^*AU}$ and $\mathsf{B_U} = \mathsf{U^*U}$ and solve the size-$m$ generalized EVP $\mathsf{A_U W} = \lambda \mathsf{B_U W}$
5: replace $\mathsf{Y}$ with $\mathsf{U} \cdot \mathsf{W}$

In this work we focus on the situation when (1) the eigenpairs in a given interval are sought, $\lambda \in I_\lambda = [\alpha, \beta]$, (2) these eigenvalues are in the interior of the spectrum, and (3) their number is moderately large (some hundreds, say). Then subspace iteration, possibly coupled with a Rayleigh–Ritz extraction, may be competitive or superior to the afore-mentioned methods. The basic procedure is summarized in Algorithm 1; cf. also [8, 22].

The function $f$ can be chosen in many different ways, ranging from $f(x) = x$ (i.e., $\mathsf{U} = \mathsf{A} \cdot \mathsf{Y}$, "power iteration") to more sophisticated "filter functions;" cf., e.g., [22]. In particular, consider the "window function"

$$f(x) \equiv \chi_{I_\lambda}(x) = \begin{cases} 1, & x \in I_\lambda, \\ 0, & \text{otherwise} \end{cases}$$

and a column $\mathbf{y}_j$ of $\mathsf{Y}$, expanded w.r.t. an orthonormal system $\mathbf{v}_1, \ldots, \mathbf{v}_n$ of $\mathsf{A}$'s eigenvectors, $\mathbf{y}_j = \sum_{k=1}^{n} \eta_k \mathbf{v}_k$. Then

$$f(\mathsf{A}) \cdot \mathbf{y}_j = \sum_{k=1}^{n} \eta_k f(\lambda_k) \mathbf{v}_k = \sum_{\lambda_k \in I_\lambda} \eta_k \mathbf{v}_k ,$$

i.e., $y_j$ is projected onto the invariant subspace spanned by the desired eigenvectors. With this choice of $f$, the procedure in Algorithm 1 would terminate after just one iteration. Especially for large matrices, however, $\chi_{I_\lambda}(\mathsf{A}) \cdot \mathbf{y}_k$ can only be approximated, either by using specialized algorithms for matrix functions $f(\mathsf{A}) \cdot \mathbf{b}$ [12, 23] or by approximating the function: $f \approx \chi_{I_\lambda}$. We will focus on the latter approach, in particular on using polynomials for approximating $f$. Very good approximation can be achieved with rational functions, and methods for choosing these optimally have been investigated [10] (cf. also [19] for the FEAST algorithm, where the rational approximation is done via a numerical contour integration). However, rational functions require the solution of shifted linear systems $(\mathsf{A} - \sigma \mathsf{I})\mathbf{x}_j = \mathbf{y}_j$, and these can be very challenging if direct solvers are not feasible.

In this work we will instead consider polynomial functions $f(x) = p(x)$, where $p(x)$ is a polynomial of degree $d$. Then $f(\mathsf{A}) \cdot \mathbf{y}$ is easy to evaluate even if $\mathsf{A}$ is not available explicitly but its action $\mathsf{A} \cdot \mathbf{v}$ on any vector $\mathbf{v}$ can be determined (e.g., for very large sparse matrices whose nonzero entries are not stored but re-computed whenever they are needed).

An important observation is that, in order for the overall algorithm in Algorithm 1 to work, we must be able to determine the number of eigenvalues that are contained in $I_\lambda$, e.g., to decide whether all sought eigenpairs have been found and to adjust the dimension of the search space $\mathsf{Y}$. This can be done by counting the singular values of the current matrix $\mathsf{U}$ that are larger than some given bound $\tau_{\text{inside}}$. Thus we must have

$$p(x) \geq \tau_{\text{inside}} \quad \text{for } x \in [\alpha, \beta] \tag{1}$$

in order not to miss one of the sought eigenpairs. For compatibility with the FEAST realization of the projector we choose $\tau_{\text{inside}} = 0.5$ throughout. By contrast, *linearity* of the filter is not an issue: it is not necessary that $p(x) \approx 1$ in the interior of $I_\lambda$. (The values should, however, not be too large to avoid numerical problems in the SVD computation.)

For the approximate projector $p(\mathsf{A})$ to be effective, it must dampen the components $\mathbf{v}_j$ corresponding to unwanted eigenvalues $\lambda_j \notin I_\lambda$. Thus we require

$$|p(x)| \leq \tau_{\text{outside}} \quad \text{for } x \notin [\alpha - \delta, \beta + \delta] \tag{2}$$

for some threshold $\tau_{\text{outside}}$, e.g., $\tau_{\text{outside}} = 0.01$. The margin $\delta > 0$ is necessary because a continuous function $p$ cannot achieve 2 for all $x \notin I_\lambda$ while fulfilling 1. Note that if there are unwanted eigenvalues $\lambda_j \in (\alpha - \delta, \alpha) \cup (\beta, \beta + \delta)$ then the corresponding components $\mathbf{v}_j$ may not be sufficiently damped to yield satisfactory convergence of the overall method. In this case it may be necessary to increase the degree of the polynomial; cf. [8]. Anyway $p$ should be chosen such that a small margin can be achieved. This is the main focus of this work.

The remainder of the article is organized as follows. In Sect. 3 we will discuss how to reduce the margin while still trying to approximate the window function, $p(x) \approx \chi_{I_\lambda}$. In Sect. 4 we will see that the margin for $[\alpha, \beta]$ can be reduced further by approximating the window function for a *smaller* interval. Giving up the linearity constraint within the interval yields another type of filter that is discussed in Sect. 5. Numerical results presented in Sect. 6 show that the reduction of the margin also leads to a lower number of operations (measured by the number of matrix–vector multiplications) and faster execution for the overall eigensolver.

## 3 Polynomial Approximation of the Window Function

The Chebyshev approximation discussed in the following requires $x \in [-1, 1]$. To achieve this, the matrix $\mathsf{A}$ is shifted and scaled such that all its eigenvalues lie between $-1$ and $1$. The search interval $[\alpha, \beta]$ is transformed accordingly.

Throughout the following discussion we assume that this preprocessing has already been done.

It is well known [7] that the window function $\chi_{[\alpha,\beta]}$ can be approximated by a polynomial expansion

$$\chi_{[\alpha,\beta]}(x) \approx \sum_{k=0}^{d} c_k T_k(x) , \tag{3}$$

where the $T_k$ are the Chebyshev polynomials of the first kind,

$$T_0(x) \equiv 1 , \quad T_1(x) = x , \quad T_k(x) = 2x \cdot T_{k-1}(x) - T_{k-2}(x) \quad \text{for } k \geq 2,$$

and the coefficients are given by

$$\left.\begin{aligned} c_0 &= \tfrac{1}{\pi} \cdot (\arccos\alpha - \arccos\beta) , \\ c_k &= \tfrac{2}{k\pi} \cdot (\sin(k \cdot \arccos\alpha) - \sin(k \cdot \arccos\beta)) , \quad k \geq 1. \end{aligned}\right\} \tag{4}$$

Using a finite expansion, however, leads to so-called Gibbs oscillations [26] close to the jumps of the function, clearly visible in the left picture in Fig. 1. These oscillations can be reduced by using appropriate kernels [26], e.g., of Jackson, Fejér, Lorentz, or Lanczos type. Incorporating a kernel amounts to replacing the $c_k$ in 3 with modified coefficients $c'_k = g_k \cdot c_k$. For the Lanczos kernel, which has proved successful in the context of polynomial approximation [18], the corrections are given by

$$g_k = \left(\operatorname{sinc} \frac{k}{d+1}\right)^{\mu}, \quad k \geq 0 , \quad \text{where} \quad \operatorname{sinc}\xi = \frac{\sin(\pi\xi)}{\pi\xi} .$$

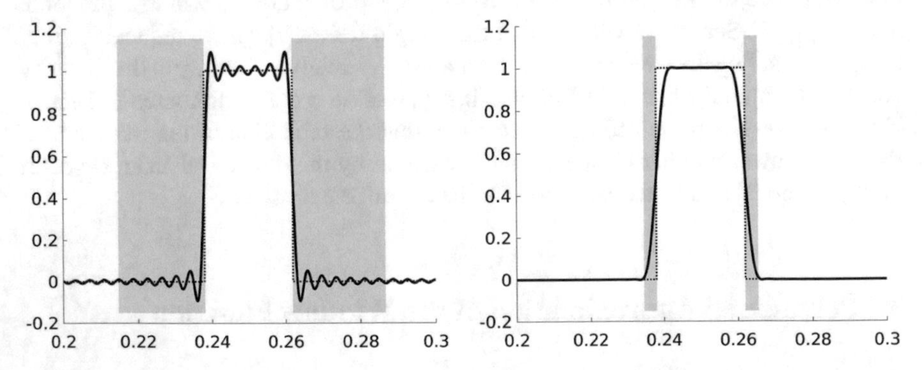


**Fig. 1** Dotted curves: window function $\chi_{[\alpha,\beta]}(x)$ for the interval $[\alpha, \beta] = [0.238, 0.262]$; solid curves: degree-1600 Chebyshev approximation $p(x)$ without Lanczos kernel (left picture) and with Lanczos kernel ($\mu = 2$; right picture). The "damping condition" $|p(x)| \leq \tau_{\text{outside}} = 0.01$ may be violated in the light gray areas

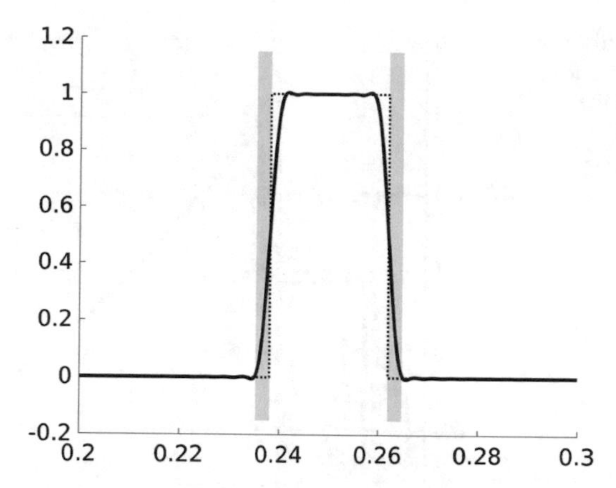


**Fig. 2** Window function $\chi_{[\alpha,\beta]}(x)$ for the interval $[\alpha, \beta] = [0.238, 0.262]$ (dotted line) and degree-1600 Chebyshev approximation $p(x)$ with $\mu = \sqrt{2}$ Lanczos kernel (solid line)

The parameter $\mu$ is assumed to be positive and integer.

The right picture in Fig. 1 demonstrates the smoothing effect of the Lanczos kernel ($\mu = 2$). As the oscillations are removed almost completely, the margin $\delta$ (width of the gray areas in Fig. 1) is reduced from approximately 0.02316 to 0.00334, even though the resulting $p$ has a much lower steepness in the points $\alpha$ and $\beta$. Roughly speaking, the right picture in Fig. 1 suggests bad damping throughout the whole (smaller) margin, whereas in the left picture good damping may be achieved even at some points within the (larger) margin.

To obtain a small margin $\delta$, the filter polynomial should be as steep as possible at the points $\alpha$ and $\beta$ and have oscillations $|p(x)| \geq \tau_{\text{outside}}$ only very close to the interval. In the following we will consider three different approaches for increasing the steepness while keeping the oscillations limited. The first approach—discussed below—still aims at approximating the window function $\chi_{[\alpha,\beta]}$. The other approaches are based on different target functions; they will be presented in Sects. 4 and 5.

For the further use as a filter in the eigensolver the integrality restriction for $\mu$ is not necessary. Indeed, non-integer $\mu$ values may lead to filters with smaller margin; see Fig. 2 for $\mu = \sqrt{2}$, with $\delta \approx 0.00276$. Thus the margin was reduced by another factor of 1.21 with respect to $\mu = 2$. In the remainder of the paper, this factor will be called the gain of a filter:

$$\text{gain} = \frac{\delta(\text{Chebyshev approximation with Lanczos kernel, } \mu = 2)}{\delta(\text{filter under consideration})} . \tag{5}$$

(A close look reveals that the amplitude of the oscillations outside the margin has increased w.r.t. $\mu = 2$, but not enough to violate the condition $|p(x)| \leq \tau_{\text{outside}}$.)

According to Fig. 3, a gain of roughly 1.4 may be achieved with an approximation to the window function if the kernel parameter $\mu$ is chosen appropriately, and

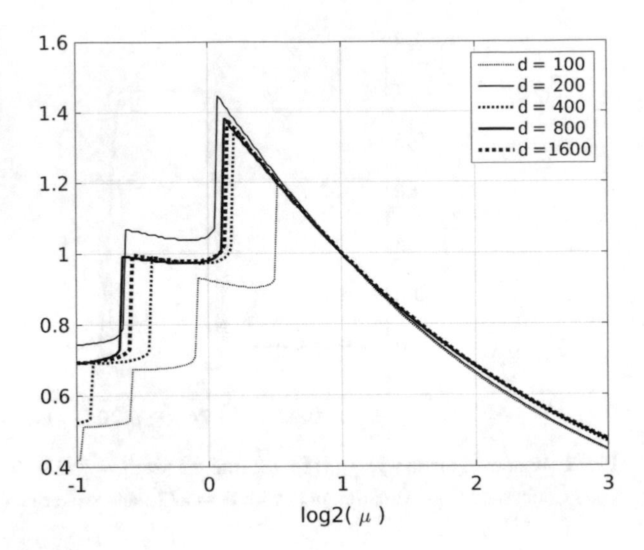


**Fig. 3** gain, as defined in 5, by using different values for the parameter $\mu$ in the Lanczos kernel for $[\alpha, \beta] = [0.238, 0.262]$

"appropriately" depends on the degree $d$. (The optimal value for $\mu$ also depends on $\alpha$ and $\beta$; this dependence is not shown in the picture.) Note that $\mu$ can be optimized *before* applying the filter $p(\mathsf{A})$ to some vectors $\mathbf{v}$ because $\delta$ depends only on the interval $[\alpha, \beta]$ and the degree $d$, but not on $\mathsf{A}$ and $\mathbf{v}$.

(For a given interval $[\alpha, \beta]$ and degree $d$, the value of the margin $\delta$, and thus the gain, depends on the threshold $\tau_{\text{outside}}$. Throughout this paper we use $\tau_{\text{outside}} = 0.01$, which is motivated by attempting to achieve a residual drop by a factor of 100 in one iteration of our BEAST-P eigensolver; cf. the discussion in [8].)

# 4 Shrinking the Interval

The second approach to improving the filter is based on shrinking the interval to a smaller one, $[\alpha, \beta] \mapsto [\widetilde{\alpha}, \widetilde{\beta}] = [\alpha + \Delta_1, \beta - \Delta_2] \subseteq [\alpha, \beta]$. If the Lanczos parameter $\mu$ is kept fixed, the shoulders of $p$ will follow the interval boundaries and move inwards. Thus, the resulting function values $\widehat{p}(\alpha)$ and $\widehat{p}(\beta)$ will drop below 0.5. We then try to restore the property 1 by scaling the polynomial (via its coefficients $\widehat{c}_k$):

$$\widetilde{p} = \varphi \cdot \widehat{p}\,, \quad \text{where} \quad \varphi = \frac{0.5}{\min\{\widehat{p}(\alpha), \widehat{p}(\beta)\}}\,. \tag{6}$$

Figure 4 shows the resulting polynomial for $\widetilde{\alpha} = 0.24032$ and $\widetilde{\beta} = 0.25969$.

It remains to determine by which amount the interval should be shrunk. The shift $\Delta_1$ is determined by the slope $p'(\alpha)$: we choose it proportional to $p(\alpha)/p'(\alpha)$, with a proportionality factor $\sigma \geq 0$, and analogously for $\beta$. Note that for low degrees $d$ and narrow intervals $[\alpha, \beta]$ the shoulder of $p$ may not be very steep, and thus the

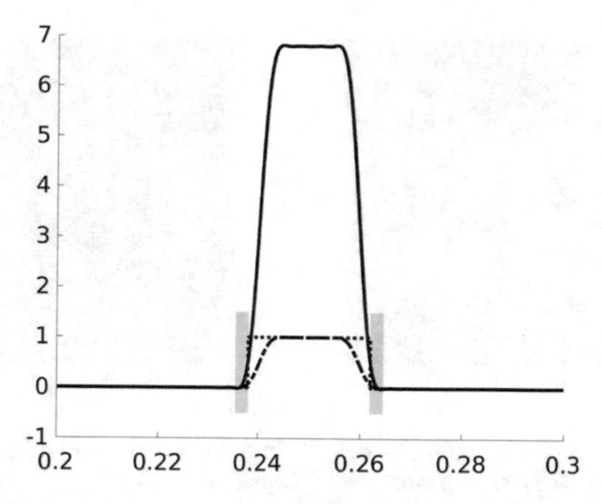


**Fig. 4** Window function $\chi_{[\alpha,\beta]}(x)$ for $[\alpha,\beta] = [0.238, 0.262]$ (dotted line) and degree-1600 Chebyshev approximation $\widetilde{p}(x)$ with $\mu = 2$ Lanczos kernel for the shrunken interval $[\widetilde{\alpha}, \widetilde{\beta}] = [0.24032, 0.25969]$ before (dash-dotted) and after (solid) scaling

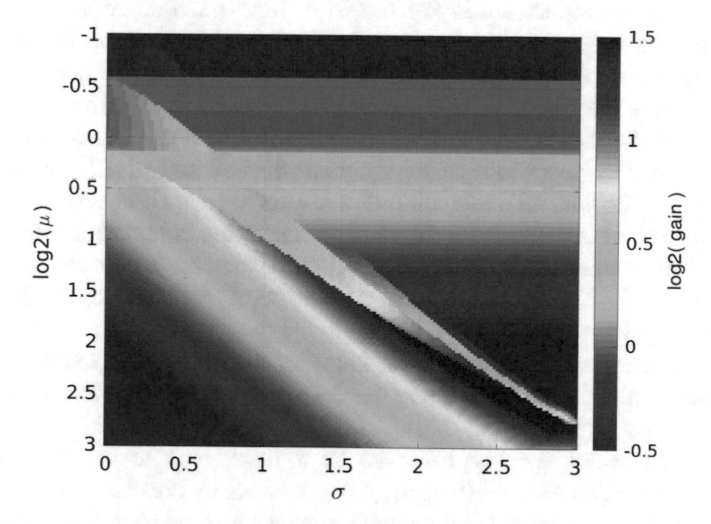


**Fig. 5** $\log_2$(gain), as defined in 5, by using different values for the parameter $\mu$ in the Lanczos kernel and different shrink factors $\sigma$ for $[\alpha,\beta] = [0.238, 0.262]$ and $d = 1600$

resulting interval can become empty: $\widetilde{\alpha} = \alpha + \sigma \frac{p(\alpha)}{p'(\alpha)} \geq \beta + \sigma \frac{p(\beta)}{p'(\beta)} = \widetilde{\beta}$. To avoid this situation and to preserve a certain width of the interval, the shifts are limited to $\Delta_i \leq 0.9 \cdot r$, where $r = (\beta - \alpha)/2$ is the radius of the original interval. In particular, none of the endpoints is moved over the original midpoint.

Figure 5 reveals that a gain of almost 3 can be obtained with suitable combinations $(\mu, \sigma)$ (note the logarithmic color coding). As $\sigma = 0$ disables shrinking, the left border of the plot corresponds to the "$d = 1600$" curve in Fig. 3.

The search for a suitable $(\mu, \sigma)$ combination is simplified by the observation that there are just three "essentially different" patterns for the dependence gain$(\mu, \delta)$:

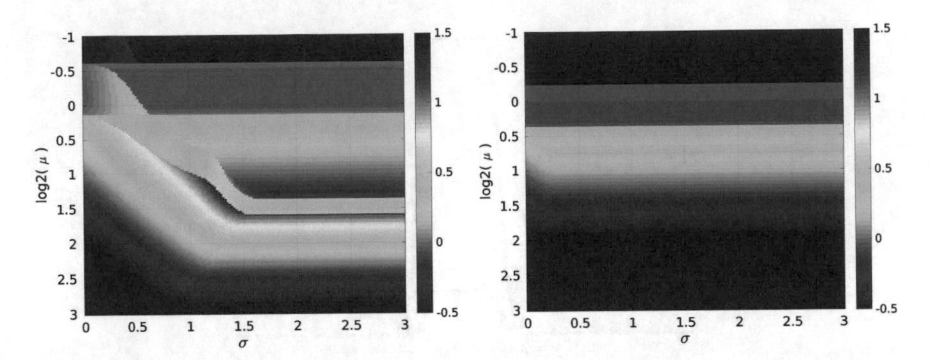


**Fig. 6** Typical pattern of $\log_2(\mathsf{gain})$, as defined in 5, for $[\alpha, \beta] = [0.238, 0.262]$. Left picture: "critical" degree ($d = 565$); right picture: "low" degree ($d = 141$)

while Fig. 5 gives a typical pattern for "high" degrees, the patterns for "critical" and "low" degrees are shown in Fig. 6. (With "low" fixed degrees, our BEAST-P eigensolver converges slowly, whereas "high" degrees cannot improve convergence sufficiently to compensate for the increased work in the computation of $p(\mathsf{A}) \cdot \mathsf{Y}$. Therefore our adaptive scheme [8] tries to work with the "critical" degrees in between.) Whether a given degree $d$ is to be considered high, critical, or low, depends mainly on the width of the interval, $\beta - \alpha$, and to a lesser degree on the location of the interval's midpoint with respect to $[-1, 1]$: intervals close to the origin require higher values of $d$ than intervals near the boundaries $\pm 1$; see Fig. 7 for very similar patterns corresponding to different interval widths and locations.

Regardless of whether the degree is low, critical or high, the $(\mu, \sigma)$ combination yielding the optimal $\mathsf{gain}$ is located close to the diagonal $\log_2(\mu) = \sigma$. Therefore our search considers only those combinations on a $(\Delta \log_2 \mu, \Delta\sigma)$-equispaced grid that lie within a specified band along the diagonal (see Fig. 8) and selects the one giving the highest $\mathsf{gain}$.

This BAND search may be followed by a closer look at the vicinity of the selected combination $P_{\text{best}} = (\log_2 \mu_{\text{best}}, \sigma_{\text{best}})$, either by considering the points on an equispaced GRID centered at $P_{\text{best}}$ (with smaller step sizes $\Delta' \log_2 \mu \ll \Delta \log_2 \mu$, $\Delta'\sigma \ll \Delta\sigma$), or by following a PATH originating at $P_{\text{best}}$: consider the eight neighbors of $P_{\text{best}}$ at distances $\pm\frac{1}{2}\Delta \log_2 \mu$, $\pm\frac{1}{2}\Delta\sigma$, go to the one giving the best $\mathsf{gain}$, and repeat until none of the neighbors is better (then halve the step sizes and repeat until a prescribed minimum step size is reached).

The GRID and PATH search might also be used without a preceding BAND search, but the two-phase approach tends to be more efficient.

In a parallel setting, the evaluation of the $\mathsf{gain}$ at the different points in a BAND or GRID search can be done concurrently, thus requiring only one global reduction operation to determine the optimum combination $(\log_2 \mu, \sigma)$. Then each process recomputes the coefficients $\widetilde{c}_k$ corresponding to this combination to avoid global communication involving a length-$(d + 1)$ vector. The PATH search has lower potential for parallelization, but tends to be more efficient serially, in particular if

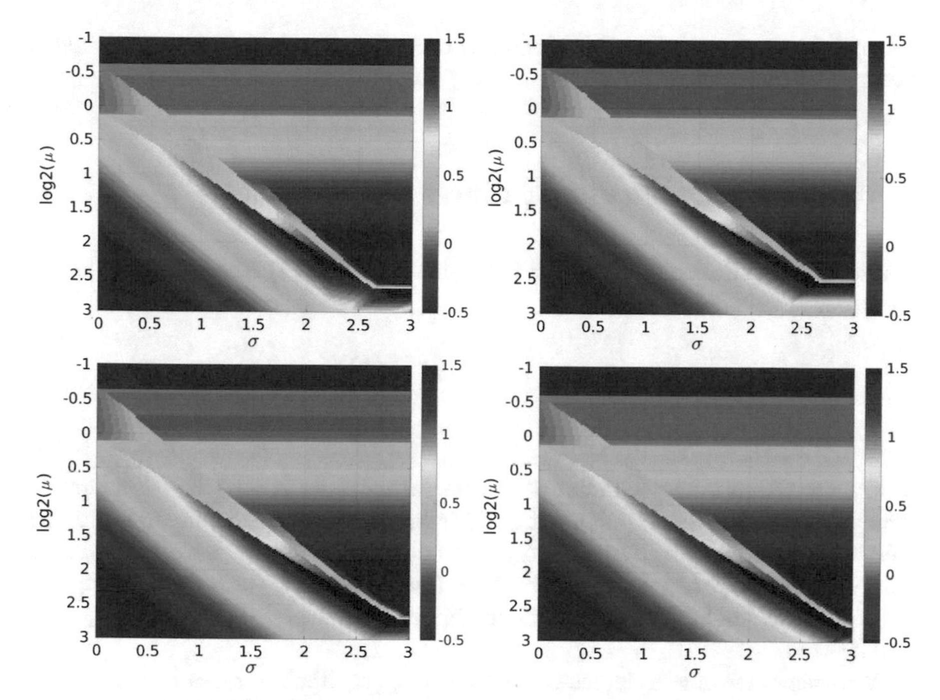


**Fig. 7** Intervals $[\alpha, \beta]$ with the same width, but different location, or different width may lead to almost the same pattern as in Fig. 5 if a suitable degree $d$ is considered. Top left: $[\alpha, \beta] = [-0.984, -0.960]$, $d = 400$; top right: $[\alpha, \beta] = [0.560, 0.584]$, $d = 1131$; bottom left: $[\alpha, \beta] = [-0.012, 0.012]$, $d = 1600$; bottom right: $[\alpha, \beta] = [0.150, 0.350]$, $d = 200$

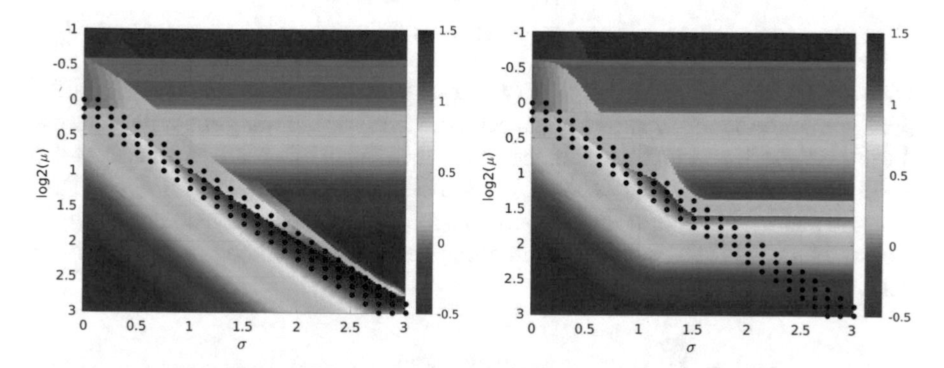


**Fig. 8** The points mark the $(\log_2 \mu, \sigma)$ combinations that are considered in the BAND search; cf. Figs. 5 and 6 for the underlying $\log_2(\text{gain})$ patterns

we avoid to re-evaluate gain for combinations that already had been considered before along the path.

## 5 Iteratively Compensating Filters

As already mentioned in Sect. 2, it is not necessary that $f(x) \approx 1$ within $[\alpha, \beta]$. Consider instead the function

$$f_1(x) = \begin{cases} f_{\max} - (f_{\max} - 0.5) \cdot \left(\frac{x-m}{r}\right)^{d_f}, & x \in [\alpha, \beta] \\ 0, & \text{otherwise} \end{cases},$$

where $r = (\beta - \alpha)/2$ is the radius of the interval and $m = (\alpha + \beta)/2$ is its midpoint. Thus, $f_1(x)$ is a degree-$d_f$ polynomial within $[\alpha, \beta]$, taking its maximum $f_{\max}$ at the midpoint and $f_1(\alpha) = f_1(\beta) = 0.5$ at the boundaries, and $f_1(x) \equiv 0$ outside the interval. The two parameters may be chosen freely such that $f_{\max} > \tau_{\text{inside}}$ and $d_f \geq 0$ is an even integer. See the top left picture in Fig. 9 for the function $f_1$ with $f_{\max} = 5$ and $d_f = 8$.

We then determine a degree-$d$ Chebyshev approximation $p_1$ to $f_1$ and scale it to achieve $\min\{p_1(\alpha), p_1(\beta)\} = 0.5$ (top right picture in Fig. 9). To reduce the oscillations outside $[\alpha, \beta]$, we "compensate" for them by taking the negative error $-p_1(x)$ as a target in the second step, i.e., we now approximate the function

$$f_2 = \begin{cases} f_1(x), & x \in [\alpha, \beta] \\ -\rho \cdot p_1(x), & \text{otherwise} \end{cases}$$

with a relaxation parameter $\rho > 0$ (we used $\rho = 0.75$). This is repeated until a prescribed number of iterations (e.g., 50) is reached or the margin did not improve during the last 3, say, iterations. In the example in Fig. 9 this procedure takes $34 + 3$ iterations to reduce the margin from $\delta \approx 0.01305$ for $y_1$ to 0.00150 for $y_{\text{best}} = y_{34}$.

Note that in this approach the coefficients $c_k$ cannot be computed cheaply with a closed formula such as 4. Using the orthogonality of the Chebyshev polynomials w.r.t. the inner product

$$\langle p, q \rangle = \int_{-1}^{+1} w(\xi) p(\xi) q(\xi)\, d\xi \quad \text{with} \quad w(\xi) = \frac{1}{\pi\sqrt{1-\xi^2}},$$

the coefficients are given by

$$c_k = \frac{\langle f, T_k \rangle}{\langle T_k, T_k \rangle}$$

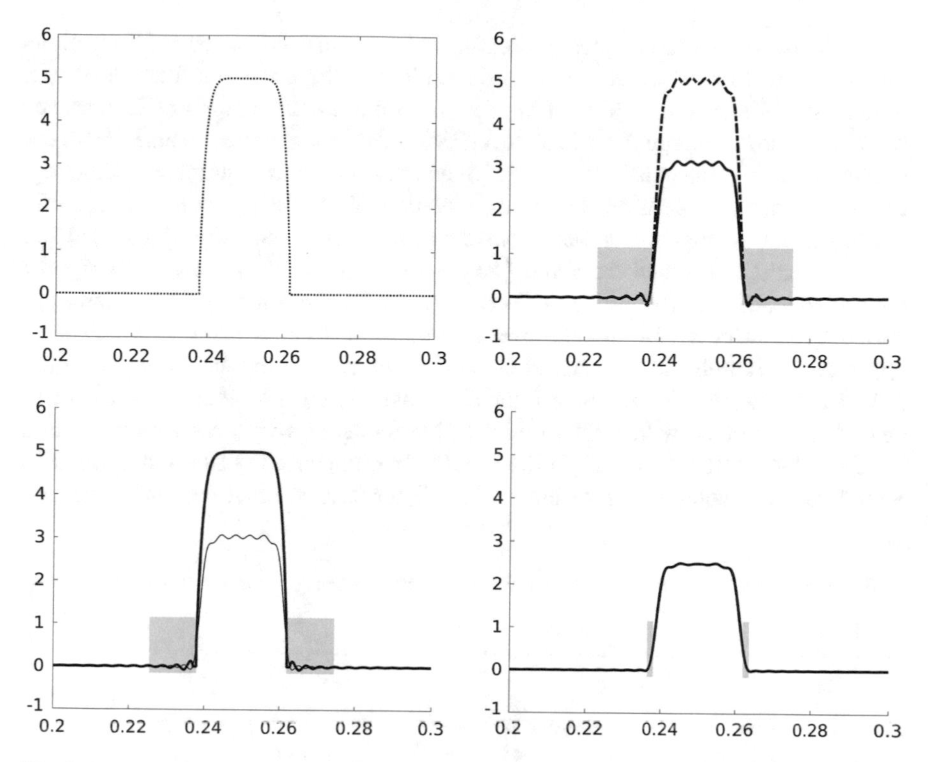


**Fig. 9** Top left: The function $f_1$ for $[\alpha, \beta] = [0.238, 0.262]$, $f_{\max} = 5$ and $d_f = 8$. Top right: The resulting degree-1600 approximation $p_1(x)$ before (dash-dotted) and after (solid) scaling to achieve $\min\{p(\alpha), p(\beta)\} = 0.5$. Bottom left: "Compensating" filter function $f_2$ (thick line) and resulting approximation $p_2$ (thin line). Bottom right: Final filter polynomial $p = p_{34}$

and can be obtained by numerical integration. Here, $f$ is the target function for the current iteration. This makes determining iteratively compensating filters more expensive than optimized shrunken Lanczos filters. (Cf. also [5] for an alternative approach for approximating functions working with a modified inner product.)

## 6 Numerical Experiments

So far we have focused on the gain of the improved filters, i.e. on the reduction of the margin $\delta$ where sufficient damping of unwanted eigenvalues may be violated. When applying these filters in, e.g., iterative eigensolvers then the ultimate goal is to speed up the computations. In our experiments we use the filters in a polynomial-accelerated subspace iteration with Rayleigh–Ritz extraction and adaptive control of the polynomial degrees; see [8] for a detailed description of this algorithm (BEAST-P in the ESSR).

As the overall work is typically dominated by the matrix–vector multiplications (MVMs), our first set of experiments determines if the improved filters lead to a reduction of the overall number of MVMs. We use a test set comprising 21 matrices with dimensions ranging from 1152 to 119,908. Twelve matrices are taken from the University of Florida matrix collection [6] and nine come from graphene modeling. For each matrix we consider two search intervals $I_\lambda = [\alpha, \beta]$ containing roughly 300 eigenvalues. Eigenpairs were considered converged (and were locked) when they reached the residual threshold $\|\mathsf{A}\mathbf{x}_j - \lambda_j\mathbf{x}_j\| \leq 10^{-12} \cdot n \cdot \max\{|\alpha|, |\beta|\}$ for the scaled matrix (cf. beginning of Sect. 3). In Fig. 10 the resulting 42 problems are sorted by "hardness," i.e., by the overall number of MVMs taken by Chebyshev approximation with Lanczos kernel ($\mu = 2$). The plots show the reduction of the MVM count w.r.t. this reference filter if we use (1) the shrunken Lanczos filters described in Sect. 4 with BAND optimization alone or with BAND optimization, followed by PATH search, or (2) the iteratively compensating filters described in Sect. 5, or (3) a combination of both. In the latter case, we first determine the best

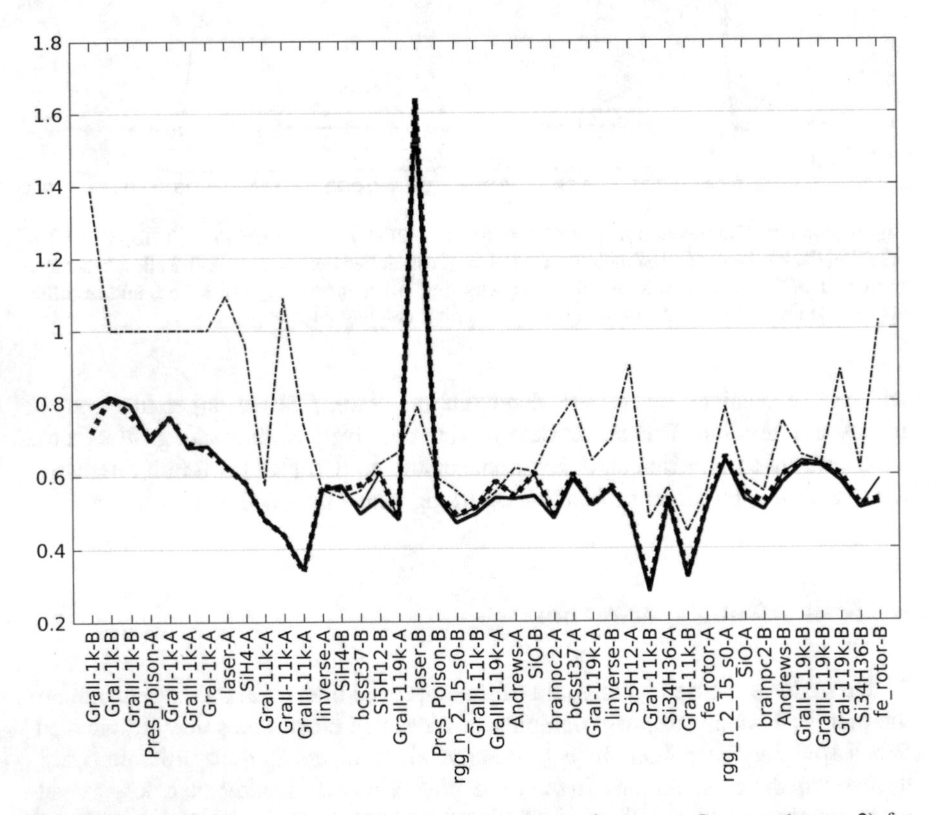


**Fig. 10** Ratios of the overall number of matrix-vector products w.r.t. Lanczos ($\mu = 2$) for iteratively compensating filters (dash-dotted thin line), shrunken Lanczos filters with BAND optimization (dotted line), BAND and PATH optimization (solid thin line), and combined filters (see main text; solid thick line)

shrunken Lanczos filter by a BAND+PATH search, and only if this did not yield a gain $\geq 2.0$ then an iteratively compensating filter is determined as well, and the gain-maximal of the two filters is taken.

The data indicate that roughly 40–50% of the MVMs can be saved in most cases with the combined approach and that most of the improvement can already be achieved with just GRID optimization of shrunken Lanczos filters turned on. The effectiveness of the iteratively compensating filters as a stand-alone method tends to be inferior to filters involving shrunken Lanczos. (In general, the latter are better for critical and high degrees, whereas iteratively compensating filters may be superior for low degrees.) In a single case the improved filters involving shrunken Lanczos led to an increase of the MVMs (by 60%). This seems to be an artefact of our adaptive scheme: All variants take seven iterations to find all 304 eigenpairs contained in the search interval. With improved coefficients, 40% fewer MVMs are needed so far, but then the adaptive control fails to detect completeness and triggers additional iterations with increasing degree; we will investigate this issue further.

The reduced MVM count through the optimized filters did not come with a significant degradation of the quality of the computed eigensystems. On average, the maximum residual for each problem, $\max_j \|\mathbf{A}\mathbf{x}_j - \lambda_j \mathbf{x}_j\|$, and the maximum deviation from orthonormality, $\max_{i,j} |\mathbf{x}_i^* \mathbf{x}_j - \delta_{i,j}|$, incurred a slight increase (by 25% and 37%, resp.), whereas the maximum deviation of the eigenvalues, $\max_j |\lambda_{j,\text{computed}} - \lambda_j|$, (w.r.t. values $\lambda_j$ obtained with a direct eigensolver) decreased by 13%.

The above experiments were done with Matlab on machines with varying load and therefore do not provide reliable timing information. Timings were obtained with substantially larger matrices on the Emmy cluster (two 2.2GHz 10-core Xeon 2260v2 per node) at Erlangen Regional Computing Center with a parallel implementation featuring the performance-optimized kernels described in [14]. The data in Table 1 indicate that using the improved coefficients allowed the adaptive scheme to settle at a much lower degree for the polynomials, yielding a reduction of the MVM count and overall runtime by roughly one half. Even with the complete optimization of the coefficients done redundantly in each node, their computation took only a small amount of the overall time. Parallelizing this step as described at the end of Sect. 4 will reduce its time consumption even further.

**Table 1** Final degree in the adaptive scheme (starting with $d = 100$ and increasing by factors of 2 or $\lfloor\sqrt{2}\rfloor$), overall number of MVMs, overall time, and time required for computing the coefficients of the polynomials, for two problems from modeling topological insulators, using standard Chebyshev approximation with Lanczos kernel ($\mu = 2$) or the improved coefficients for the filters

| Filter | Final degree | Overall MVMs | Overall time | Time for coeffs |
|---|---|---|---|---|
| *Topological insulator, n* = 268,435,456, 148 *evals*, 128 *nodes à* 20 *cores* | | | | |
| Lanczos ($\mu = 2$) | 4525 | 5,598,502 | 7.11 h | 0.00 h |
| Improved (combined) | 2255 | 2,602,360 | 3.44 h | 0.02 h |
| *Topological insulator, n* = 67,108,864, 148 *evals*, 64 *nodes à* 20 *cores* | | | | |
| Lanczos ($\mu = 2$) | 2262 | 2,726,112 | 1.97 h | 0.00 h |
| Improved (combined) | 1127 | 1,482,035 | 1.10 h | 0.01 h |

## 7 Conclusions

After a very brief overview of the ESSEX project and some of its results in the first funding period, we have focused on one method used for computing a moderate number (some hundreds, say) of interior eigenpairs for very large symmetric or Hermitian matrices: subspace iteration with polynomial acceleration and Rayleigh-Ritz extraction. We have presented two techniques for reducing the degree of the polynomials. One of them was based on determining standard Chebyshev approximations with suitable Lanczos kernels to a window function, but for a *shrunken* interval. The other technique was iterative, starting with a polynomial-shaped target function and trying to compensate the error made in the previous approximation. In both cases the optimization of the coefficients of the final polynomial was done with respect to the margin, i.e., the width of the area where sufficient damping of unwanted eigenpairs cannot be guaranteed. Numerical experiments showed that the polynomials thus obtained indeed also reduce the overall number of matrix–vector multiplications in the eigensolver, and thus its runtime.

**Acknowledgements** This work was supported by the German Research Foundation (DFG) through the Priority Programme 1648 "Software for Exascale Computing" (SPPEXA) under project ESSEX. The authors would like to thank the unknown referees for their helpful comments.

## References

1. Baker, C.G., Hetmaniuk, U.L., Lehoucq, R.B., Thornquist, H.K.: Anasazi software for the numerical solution of large-scale eigenvalue problems. ACM Trans. Math. Softw **36**(3), 13:1–13:23 (2009)
2. Bavier, E., Hoemmen, M., Rajamanickam, S., Thornquist, H.: Amesos2 and Belos: direct and iterative solvers for large sparse linear systems. Sci. Prog. **20**(3), 241–255 (2012)
3. Blum, V., Gehrke, R., Hanke, F., Havu, P., Havu, V., Ren, X., Reuter, K., Scheffler, M.: Ab initio molecular simulations with numeric atom-centered orbitals. Comput. Phys. Commun. **180**(11), 2175–2196 (2009)
4. Castro Neto, A.H., Guinea, F., Peres, N.M.R., Novoselov, K.S., Geim, A.K.: The electronic properties of graphene. Rev. Mod. Phys. **81**, 109–162 (2009)
5. Chen, J., Anitescu, M., Saad, Y.: Computing $f(A)b$ via least squares polynomial approximations. SIAM J. Sci. Comput. **33**(1), 195–222 (2011)
6. Davis, T.A., Hu, Y.: The University of Florida sparse matrix collection. ACM Trans. Math. Softw. **38**(1), 1:1–1:25 (2011)
7. Druskin, V., Knizhnerman, L.: Two polynomial methods to compute functions of symmetric matrices. U.S.S.R. Comput. Math. Math. Phys. **29**(6), 112–121 (1989)
8. Galgon, M., Krämer, L., Lang, B.: Adaptive choice of projectors in projection based eigensolvers. Preprint BUW-IMACM 15/07, University of Wuppertal (2015)
9. Galgon, M., Krämer, L., Thies, J., Basermann, A., Lang, B.: On the parallel iterative solution of linear systems arising in the FEAST algorithm for computing inner eigenvalues. Parallel Comput. **49**, 153–163 (2015)

10. Güttel, S., Polizzi, E., Tang, P.T.P., Viaud, G.: Zolotarev quadrature rules and load balancing for the FEAST eigensolver. SIAM J. Sci. Comput. **37**(4), A2100–A2122 (2015)
11. Hasan, M.Z., Kane, C.L.: Topological insulators. Rev. Mod. Phys. **82**, 3045–3067 (2010)
12. Higham, N.J.: Functions of Matrices: Theory and Computation. SIAM, Philadelphia, PA (2008)
13. Kreutzer, M., Hager, G., Wellein, G., Fehske, H., Bishop, A.R.: A unified sparse matrix data format for efficient general sparse matrix-vector multiplication on modern processors with wide SIMD units. SIAM J. Sci. Comput. **36**(5), C401–C423 (2014)
14. Kreutzer, M., Hager, G., Wellein, G., Pieper, A., Alvermann, A., Fehske, H.: Performance engineering of the kernel polynomial method on large-scale CPU-GPU systems. In: Proceedings of the IDPDS 2015, 29th International Parallel and Distributed Processing Symposium, pp. 417–426. IEEE Computer Society, Washington (2015)
15. Kreutzer, M., Pieper, A., Alvermann, A., Fehske, H., Hager, G., Wellein, G., Bishop, A.R.: Efficient large-scale sparse eigenvalue computations on heterogeneous hardware (2015). http://sc15.supercomputing.org/sites/all/themes/SC15images/tech_poster/tech_poster_pages/post205.html. Poster at the 2015 ACM/IEEE Int. Conf. for High Performance Computing, Networking, Storage and Analysis
16. Kreutzer, M., Thies, J., Röhrig-Zöllner, M., Pieper, A., Shahzad, F., Galgon, M., Basermann, A., Fehske, H., Hager, G., Wellein, G.: GHOST: building blocks for high performance sparse linear algebra on heterogeneous systems. Preprint arXiv:1507.08101v2 (2015). http://arxiv.org/pdf/1507.08101v2.pdf
17. Lanczos, C.: An iteration method for the solution of the eigenvalue problem of linear differential and integral operators. J. Res. Nat. Bur. Stand. **45**(4), 255–282 (1950)
18. Pieper, A., Kreutzer, M., Galgon, M., Alvermann, A., Fehske, H., Hager, G., Lang, B., Wellein, G.: High-performance implementation of Chebyshev filter diagonalization for interior eigenvalue computations. Preprint arXiv:1510.04895 (2015)
19. Polizzi, E.: Density-matrix-based algorithm for solving eigenvalue problems. Phys. Rev. B **79**(11), 115112 (2009)
20. Röhrig-Zöllner, M., Thies, J., Kreutzer, M., Alvermann, A., Pieper, A., Basermann, A., Hager, G., Wellein, G., Fehske, H.: Increasing the performance of the Jacobi–Davidson method by blocking. SIAM J. Sci. Comput. **37**(6), C697–C722 (2015)
21. Ruhe, A.: Implementation aspects of band Lanczos algorithms for computation of eigenvalues of large sparse symmetric matrices. Math. Comp. **33**(146), 680–687 (1979)
22. Saad, Y.: Numerical Methods for Large Eigenvalue Problems, 2nd edn. SIAM, Philadelphia, PA (2011)
23. Schweitzer, M.: Restarting and error estimation in polynomial and extended Krylov subspace methods for the approximation of matrix functions. Ph.D. thesis, University of Wuppertal (2015)
24. Sleijpen, G.L.G., Van der Vorst, H.A.: A Jacobi–Davidson iteration method for linear eigenvalue problems. SIAM J. Matrix Anal. Appl. **17**(2), 401–425 (1996)
25. Thies, J., Galgon, M., Shahzad, F., Alvermann, A., Kreutzer, M., Pieper, A., Röhrig-Zöllner, M., Basermann, A., Fehske, H., Hager, G., Lang, B., Wellein, G.: Towards an exascale enabled sparse solver repository. Preprint (2015). http://elib.dlr.de/100211/
26. Weiße, A., Wellein, G., Alvermann, A., Fehske, H.: The kernel polynomial method. Rev. Mod. Phys. **78**(1), 275–306 (2006)

# Eigenspectrum Calculation of the $O(a)$-Improved Wilson-Dirac Operator in Lattice QCD Using the Sakurai-Sugiura Method

**Hiroya Suno, Yoshifumi Nakamura, Ken-Ichi Ishikawa, Yoshinobu Kuramashi, Yasunori Futamura, Akira Imakura, and Tetsuya Sakurai**

**Abstract** We have developed a computer code to find eigenvalues and eigenvectors of non-Hermitian sparse matrices arising in lattice quantum chromodynamics (lattice QCD). The Sakurai-Sugiura (SS) method (Sakurai and Sugiura, J Comput Appl Math 159:119, 2003) is employed here, which is based on a contour integral, allowing us to obtain desired eigenvalues located inside a given contour of the complex plane. We apply the method here to calculating several low-lying eigenvalues of the non-Hermitian $O(a)$-improved Wilson-Dirac operator $D$ (Sakurai et al., Comput Phys Commun 181:113, 2010). Evaluation of the low-lying eigenvalues is crucial since they determine the sign of its determinant $\det D$, important quantity in lattice QCD. We are particularly interested in such cases as finding the lowest eigenvalues to be equal or close to zero in the complex plane. Our implementation is tested for the Wilson-Dirac operator in free case, for which the eigenvalues are analytically

---

H. Suno (✉)
RIKEN Advanced Institute for Computational Science, Kobe, Hyogo 650-0047, Japan

RIKEN Nishina Center for Accelerator-Based Science, Wako, Saitama 351-0198, Japan
e-mail: suno@riken.jp

Y. Nakamura
RIKEN Advanced Institute for Computational Science, Kobe, Hyogo 650-0047, Japan

K.-I. Ishikawa
Department of Physical Science, Hiroshima University, Higashi-Hiroshima, Hiroshima 739-8526, Japan

Y. Kuramashi
RIKEN Advanced Institute for Computational Science, Kobe, Hyogo 650-0047, Japan

Center for Computational Science, University of Tsukuba, Tsukuba, Ibaraki 305-8577, Japan

Faculty of Pure and Applied Science, University of Tsukuba, Tsukuba, Ibaraki 305-8571, Japan

Y. Futamura • A. Imakura • T. Sakurai
Department of Computer Science, University of Tsukuba, Tsukuba, Ibaraki 305-8573, Japan


T. Sakurai et al. (eds.), *Eigenvalue Problems: Algorithms, Software and Applications in Petascale Computing*, Lecture Notes in Computational Science and Engineering 117, https://doi.org/10.1007/978-3-319-62426-6_6

known. We also carry out several numerical experiments using different sets of gauge field configurations obtained in quenched approximation as well as in full QCD simulation almost at the physical point. Various lattice sizes $L_xL_yL_zL_t$ are considered from $8^3 \times 16$ to $96^4$, amounting to the matrix order $12L_xL_yL_zL_t$ from 98,304 to 1,019,215,872.

## 1 Introduction

The determinant of the Wilson-Dirac operator, $\det D$, plays an important role in lattice QCD. In general, the determinant of an $\mathcal{L} \times \mathcal{L}$ matrix $D$ can be written, in terms of its eigenvalues $\lambda_l$, as

$$\det D = \prod_{l=1}^{\mathcal{L}} \lambda_l, \tag{1}$$

so that all the information about the Wilson fermion determinant, or the fermion measure, is concentrated in the eigenspectrum. This makes the eigenspectrum calculation of the Wilson-Dirac operator an interesting subject. Because the eigenspectrum possesses the vertical and horizontal symmetries and that the complex eigenvalues are therefore always paired, the determinant can be further expressed in the form:

$$\det D = \prod_{\lambda_l \in \mathbb{R}} \lambda_l \prod_{\lambda_{l'} \in \mathbb{C}} |\lambda_{l'}|^2. \tag{2}$$

We are thus particularly interested in calculating several low-lying real eigenvalues of the Wilson-Dirac operator, since the sign problem may occur due to the real, negative eigenvalues.

In this work, we develop exploratorily a computer code for calculating the low-lying eigenspectrum of the Wilson-Dirac operator. For such sparse eigenproblems as the Wilson-Dirac equation, the Implicitly Restarted Arnoldi Method (IRAM) [1] is one of the conventional choices. It is noteworthy mentioning earlier attempts to improve eigenvalue computations of the non-Hermiltian Dirac operator, such as in [2]. We adopt here the Sakurai-Sugiura (SS) method since this makes it possible to set more flexibly the region for searching eigenvalues. The SS method is based on contour integrals, allowing us to calculate eigenvalues located in a given domain of the complex plane as well as the associated eigenvectors. Our computer code will be applied to calculating low-lying eigenvalues of the non-Hermitian $O(a)$-improved Wilson-Dirac operator. We consider the spatiotemporal lattice sizes $8^3 \times 16$ and $96^4$, amounting to the matrix order of 98,304 and 1,019,215,872, respectively. Eigenvalue calculations will be performed using gauge field configurations for the free case, those generated in quenched approximation as well as those generated by a full QCD simulation, focusing on such cases as finding the low-lying eigenvalues to be localized very close to zero in the complex plane.

## 2 Sakurai-Sugiura Method for the Wilson-Dirac Operator

The lattice QCD is defined on a hypercubic four-dimensional lattice of finite extent expressed as $L_x \times L_y \times L_z \times L_t$ with $L_{x,y,z}$ being the three-dimensional spatial extent and $L_t$ the temporal one. The lattice spacing is set to unity for notational convenience. The fields are defined on the sites $n$ with periodic boundary conditions. We define two types of fields on the lattice. One is the gauge field represented by $U_\mu(n)^{a,b}$ with $\mu = 1, 2, 3, 4$ (corresponding respectively to $x, y, z, t$) and $a, b = 1, 2, 3$, which is a $3 \times 3$ SU(3) matrix assigned on each link. The other is the quark field $q(n)^a_\alpha$ which resides on each site carrying the Dirac index $\alpha = 1, 2, 3, 4$ and the color index $a = 1, 2, 3$. The $O(a)$-improved Wilson-Dirac operator is written as

$$
\begin{aligned}
D^{a,b}_{\alpha,\beta}(n,m) = \delta_{\alpha,\beta}\delta^{a,b}\delta(n,m) - \kappa \sum_{\mu=1}^{4} &[(1-\gamma_\mu)_{\alpha,\beta}(U_\mu(n))^{a,b}\delta(n+\hat{\mu},m) \\
&+ (1+\gamma_\mu)_{\alpha,\beta}((U_\mu(m))^{b,a})^*\delta(n-\hat{\mu},m)] \\
&+ \kappa c_{\mathrm{SW}} \sum_{\mu,\nu=1}^{4} \frac{i}{2}(\sigma_{\nu\mu})_{\alpha,\beta}(F_{\nu\mu}(n))^{a,b}\delta(n,m),
\end{aligned} \tag{3}
$$

where $\hat{\mu}$ denotes the unit vector in the $\mu$ direction. $\kappa$ is the hopping parameter, and the coefficient $c_{\mathrm{SW}}$ is a parameter to be adjusted for the $O(a)$-improvement. The Euclidean gamma matrices are defined in terms of the Minkowski ones in the Bjorken-Drell convention: $\gamma_j = -i\gamma^j_{BD}$ $(j = 1, 2, 3)$, $\gamma_4 = \gamma^0_{BD}$, $\gamma_5 = \gamma^5_{BD}$, and $\sigma_{\mu\nu} = \frac{1}{2}[\gamma_\mu, \gamma_\nu]$. The explicit representation for $\gamma_{1,2,3,4,5}$ are given in [3], together with the expression for the field strength $F_{\mu\nu}(n)$ in terms of the gauge field $U_\mu(n)$. The $O(a)$-improved Wilson-Dirac operator defined in Eq. (3) is a sparse, complex non-Hermitian square matrix $D \in \mathbb{C}^{\mathscr{L}\times\mathscr{L}}$, where only 51 out of $\mathscr{L} = L_x \times L_y \times L_z \times L_t \times 3 \times 4$ entries in each row have nonzero values.

In this work, we consider an eigenproblem

$$
Dx_l = \lambda_l x_l, \ (l = 1, 2, \ldots, \mathscr{L}), \tag{4}
$$

where $\lambda_l$ and $x_l$ are eigenvalues and eigenvectors, respectively. In order to extract eigenpairs $(\lambda_l, x_l)$ from the matrix $D$ in Eq. (4), we adopt the Sakurai-Sugiura method, proposed in [4–6]. This method is based on contour integrals, allowing us to calculate eigenvalues located in a given domain of the complex plane, together with the associated eigenvectors. In this method, we first define matrices $S_k \in \mathbb{C}^{\mathscr{L}\times L}$ as

$$
S_k \equiv \frac{1}{2\pi i}\oint_\Gamma z^k(zI-D)^{-1}Vdz, \ k = 0, 1, \ldots, M-1. \tag{5}
$$

Here, $\Gamma$ is a positively oriented closed curve in the complex plane inside which we seek for eigenvalues, and $M$ denotes the maximum moment degree. The matrix

$V \in \mathbb{C}^{\mathscr{L}\times L}$, called source matrix, contains $L$ column vectors $V = [v_1, v_2, \ldots, v_L]$, and we take a random vector for each of these source vectors. We assume that $\Gamma$ is given by a circle and the integration is evaluated via a trapezoidal rule on $\Gamma$. By designating the center and the radius of the circle as $\gamma$ and $\rho$, respectively, and by defining the quadrature points as $z_j = \gamma + \rho e^{2\pi i(j-1/2)/N}$, $j = 1, 2, \ldots, N$, we approximate the integral in Eq. (5) via an $N$-point trapezoidal rule:

$$S_k \approx \hat{S}_k \equiv \frac{1}{N}\sum_{j=1}^{N} z_j^k (z_j I - D)^{-1} V. \tag{6}$$

We then carry out the singular value decomposition for the matrix $\hat{S} = [\hat{S}_0, \hat{S}_1, \ldots, \hat{S}_{M-1}] \in \mathbb{C}^{\mathscr{L}\times LM}$ as follows

$$\hat{S} = \tilde{Q}\Sigma W, \tag{7}$$

$$\tilde{Q} = [q_1, q_2, \ldots, q_{LM}] \in \mathbb{C}^{\mathscr{L}\times LM}, \tag{8}$$

$$\Sigma = \mathrm{diag}(\sigma_1, \sigma_2, \ldots, \sigma_{LM}). \tag{9}$$

We next determine the numerical rank $m$ of the matrix $\hat{S}$. The value of $m$ is fixed as the number of singular values satisfying $\sigma_i > \delta$, with $\delta$ being the threshold for determining the numerical rank and usually set to $\delta = 10^{-12}$. The original eigenproblem is transformed to a smaller eigenproblem via the transformation matrix $Q = [q_1, q_2, \ldots, q_m]$, with only the first $m$ column vectors from $\tilde{Q}$ are incorporated. We finally solve the smaller eigenequation $Q^{\mathrm{H}} D Q u_l = \mu_l u_l$ and obtain an approximation to the eigenpairs $\lambda_l \approx \mu_l$ and $x_l \approx Q u_l$. Although our purpose is to know the eigenvalue distribution and that the eigenvectors are not necessary, we choose to calculate them since we need to check the accuracy via the relative residual norms.

As can be seen from Eq. (6), the Sakurai-Sugiura method produces a subspace with the matrix basis involving the inverses of the matrices $(z_j I - D)$. The matrix inversion can be performed solving the shifted linear equations

$$(z_j I - D) y_{jl} = v_l. \tag{10}$$

There exist several implementations based on direct methods such as LAPACK and MUMPS libraries. These implementations, however, are hardly applicable to such linear problems as arising in lattice QCD due to their large matrix sizes, and some iterative methods are desirable to solve such large sparse linear equations. In this work, we implement exploratorily the BiCGStab algorithm as is presented in Algorithm 1. The BiCGStab algorithm will be found to converge very slowly, suffering from the ill-condition problem of the shifted linear equations. We choose, however, to employ the solution vectors as are obtained from a sufficiently large number of BiCGStab iterations. The shift-invariance property of the Krylov subspace of $D$ under any translation indicates that substantial time saving can

**Algorithm 1** BiCGStab algorithm for solving $Ay \equiv (zI - D)y = v$

1: **initial guess** $y \in \mathbb{C}^{\mathcal{L}}$,
2: **compute** $r = v - Ay$,
3: **set** $p = r' = r$,
4: **choose** $\tilde{r}$ such that $(\tilde{r}, r) \neq 0$,
5: **while** $||r||_2/||v||_2 > \epsilon$ **do:**
6: $\quad \alpha = (\tilde{r}, r)/(\tilde{r}, Ap)$,
7: $\quad y \leftarrow y + \alpha p$,
8: $\quad r \leftarrow r - \alpha Ap$,
9: $\quad \zeta = (Ar, r)/(Ar, Ar)$,
10: $\quad y \leftarrow y + \zeta r$,
11: $\quad r \leftarrow r - \zeta Ar$,
12: $\quad \beta = (\alpha/\zeta) \cdot (\tilde{r}, r)/(\tilde{r}, r')$,
13: $\quad p \leftarrow r + \beta(p - \zeta Ap)$,
14: $\quad r' = r$,
15: **end while**.

be achieved by solving all shifted equations with only one sequence of Krylov subspaces, which can be a subject of the future development. In practice, the center $\gamma$ and the radius $\rho$ of the contour can be determined by running the code beforehand with small values of $N$, $L$ and $M$, which also allows us to estimate approximatively the multiplicity of eigenvalues inside $\Gamma$. Then, the default values of $N = 32$ and $M = 16$ can be mostly used, but the value of $L$ is crucial, and need to be large in case of large multiplicity or in order to obtain higher accuracy.

Code development is carried out based on the lattice QCD simulation program LDDHMC/ ver1.3K0.52ovlpcomm1.2 developed for the K computer [7, 8]. The K computer, at the RIKEN Advanced Institute for Computational Science, consists of 82,944 computational nodes and 5184 I/O nodes connected by the so-called "Tofu" network, providing 11.28 Pflops of computing capability. The Tofu network topology is six-dimensional one with 3D-mesh times 3D-torus shape. Each node has a single 2.0 GHz SPARC64 VIIIfx processor equipping 8 cores with SIMD enabled 256 registers, 6 MB shared L2 cache and 16 GB of memory. The L1 cache sizes per each core are 32 KB/2WAY (instruction) and 32 KB/2WAY (data).

## 3 Simulation Results

Our implementation is tested here for the free-case Wilson-Dirac operator, of which the eigenspectrum can be analytically obtained. For the free case, $(U_\mu(n))^{a,b} = \delta_{ab}$, the Wilson-Dirac action, in the momentum space ($\psi(x) = \int dk \exp(ikx)\tilde{\psi}(k)$), turns out to be

$$\tilde{D}(k) = 1 - \kappa \sum_{\mu=1}^{4} [2\cos(k_\mu) - 2i\gamma_\mu \sin(k_\mu)]. \tag{11}$$

Since the allowed momentum components must be the discrete elements of the Brillouin zone ($k_\mu = \frac{2\pi}{L_\mu} l_\mu$), we then obtain the expression for the free Wilson-Dirac eigenvalues as

$$\lambda^{\text{free}}_{\{l_\mu\}} = 1 - 2\kappa \left[ \sum_{\mu=1}^{4} \cos\left(\frac{2\pi}{L_\mu} l_\mu\right) \pm i \sqrt{\sum_{\mu=1}^{4} \sin^2\left(\frac{2\pi}{L_\mu} l_\mu\right)} \right], \; l_\mu = 0, 1, \ldots, L_\mu - 1. \tag{12}$$

Each set of the numbers $\{l_\mu\}$ correspond to 12 eigenvalues, 2 sets of 6 multiple eigenvalues with the plus and minus signs in Eq. (12). Note also that for the hopping parameter $\kappa = 1/8$, the minimum eigenvalue coincide with the origin, $z = 0$.

Figure 1 shows the eigenvalues of the free-case Wilson-Dirac operator calculated by the SS method for the lattice size $8^3 \times 16$ and the hopping parameter $\kappa = 1/8$. We have used the number of quadrature $N = 32$, the number of source vector $L = 64$ and the maximum moment degree $M = 16$. We notice three sets of eigenvalues inside the integration contour of which the quadrature points $z_j$ are indicated by asterisks. Each of these three sets contains 12 multiple eigenvalues, so that in total 36 eigenvalues are found inside the integration contour. The BiCGStab algorithm used for solving the shifted linear equation (10) is found to converge very slowly for some quadrature points $z_j$, due to the ill-condition problem: for the quadrature point the most on the right-hand side $z_1$, the residual $||r||_2/||v||_2$ decreases only to about $10^{-8} \sim 10^{-10}$ with 1000 BiCGStab iterations, while it decreases less than $10^{-14}$ with about 200 BiCGStab iterations for the quadrature point the most on the left-hand side $z_{17}$. However, using the solution vectors obtained

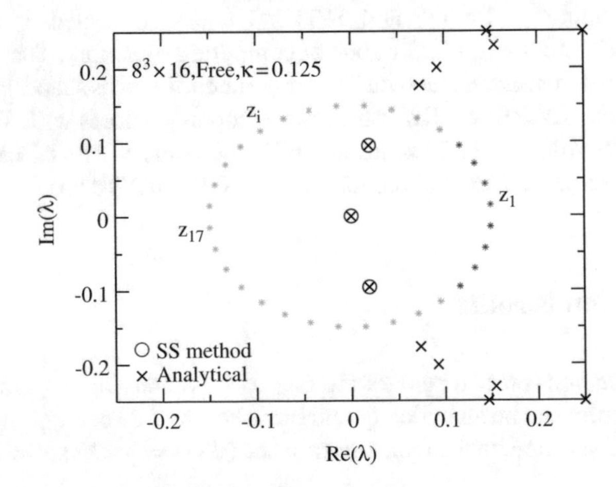


**Fig. 1** Eigenvalues of the free-case Wilson-Dirac operator in a $8^3 \times 16$ lattice for the hopping parameter $\kappa = 1/8$. The eigenvalues obtained from the SS method are indicated as *red circles*, those obtained from the analytical expression are indicated as *black crosses*. *Green* (*blue*) *asterisks* indicate the quadrature points $z_j$ for which the BiCGStab algorithm converges (does not converge) to less than $10^{-14}$ of the relative residuals norms with 1000 iterations

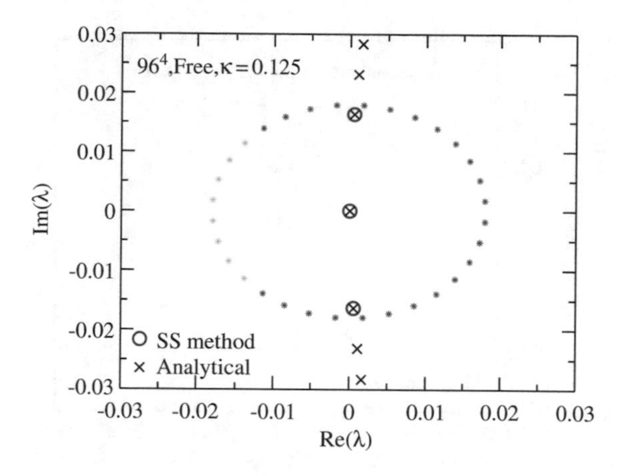


**Fig. 2** The same as in Fig. 1, but for the lattice size $96^4$

with about 1000 iterations, we have been able to obtain the above eigenvalues accurately with the relative residual norms about $10^{-7}$. This may indicate that even if the shifted equations (10) are not solved with high precision, we can obtain a certain precision about the eigenvalues. If we decrease the number of BiCGStab iterations, we could still obtain acceptable values for the eigenvalues but with more or less lower accuracy. Our eigenvalues obtained by the SS method are visually indistinguishable from those obtained from the analytical expression in Eq. (12). Figure 2 shows the same results but for the larger lattice size, $96^4$. We have used the parameters for the SS method $(N, L, M) = (32, 128, 16)$. Here, inside the integration contour, we have found one set of 12 multiple eigenvalues at the origin and 2 sets of 36 multiple eigenvalues above and below the origin, 84 eigenvalues in total. The relative residual norms have been found to be around $5 \times 10^{-4}$. These eigenvalues calculated by the SS method are also shown to be indistinguishable from those from the analytical expression in Eq. (12). Note that, we set $L = 128$ since it must be greater than or equal to the maximum multiplicity of the eigenvalues in $\Gamma$. If there is no multiplicity, we usually set a small value to $L$ e.g. $L = 8$.

We have also carried out eigenvalue calculations with a sample of gauge field configurations generated in quenched approximation. We generate these configurations on a $8^3 \times 16$ lattice with the Iwasaki gauge action $\beta = 1.9$ using the lattice QCD program LDDHMC/ver1.3K0.52ovlpcomm 1.2 [7, 8]. Figure 3 shows the eigenvalues of the $O(a)$-improved Wilson-Dirac operator with these gauge configurations in quenched approximation for the parameters $\kappa = 0.1493$ and $c_{SW} = 1.6$ . These values of $\kappa$ and $c_{SW}$ have been chosen so that the minimum real eigenvalue is close to the origin in the complex plane. We have used the parameters for the SS method $(N, L, M) = (32, 96, 16)$. The relative residual norms of the eigenvalues have been found to be around $10^{-5}$.

Finally, we have performed eigenvalue calculations using a set gauge field configurations generated in a full QCD calculation. These gauge field configurations

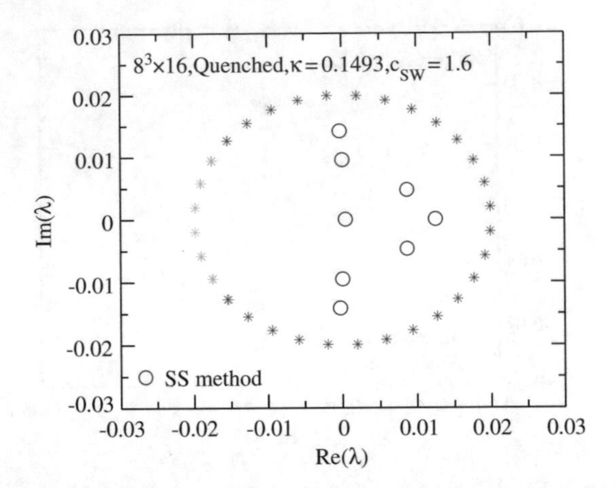


**Fig. 3** Eigenvalues of the $O(a)$-improved Wilson-Dirac operator in a $8^3 \times 16$ lattice for the hopping parameter $\kappa = 0.1493$ and the improvement parameter $c_{\mathrm{SW}} = 1.6$, with gauge field configurations generated in quenched approximation. The eigenvalues obtained from the SS method are indicated as *red circles*. *Green* (*blue*) *asterisks* indicate the quadrature points $z_j$ for which the BiCGStab algorithm converges (does not converge) to less than $10^{-14}$ of the relative residuals norms with 1000 iterations

are generated by a 2 + 1 flavor QCD simulation near the physical point on a $96^4$ lattice [9] with the hopping parameters $(\kappa_{\mathrm{ud}}, \kappa_{\mathrm{s}}) = (0.126117, 0.124700)$ and $c_{\mathrm{SW}} = 1.110$. The other details are given in [9]. Figure 4 shows the eigenvalues calculated by the SS method for the hopping and improvement parameters $\kappa = 0.126117$ and $c_{\mathrm{SW}} = 1.110$. We have used the parameters for the SS method $(N, L, M) = (32, 16, 16)$, and the maximum number of BiCGStab iterations 30,000 for solving the shifted linear equations. The relative residual norms of the eigenvalues have been found to be around $5 \times 10^{-4}$.

The above eigenvalue calculations have been performed on the K computer, using 16 nodes for the lattice size $8^3 \times 16$ and 16,384 nodes for the lattice size $96^4$. Almost the whole computer time is consumed in the multiplication of the quark field by the Wilson-Dirac operator. The time spent by each matrix-vector multiplication is found to be $1.29 \times 10^{-3}$ s for the lattice size $8^3 \times 16$, and $5.5 \times 10^{-3}$ s for the lattice size $96^4$. For the $8^3 \times 16$ free case, the elapsed time amounts to 2160 s with about $1.67 \times 10^6$ matrix-vector multiplications being performed. For the $96^4$ free case, the elapsed time is 21,500 s with about $3.93 \times 10^6$ matrix-vector multiplications. For the $8^3 \times 16$ quenched-approximation case, the elapsed time is found to be 3480 s with about $3.01 \times 10^6$ matrix-vector multiplications, while for the $96^4$ Full QCD case, we have found the elapsed time to be 65,300 s with $1.27 \times 10^7$ matrix-vector multiplications. In Fig. 5, we show the performance per node of the matrix-vector multiplication as a function of the lattice size.

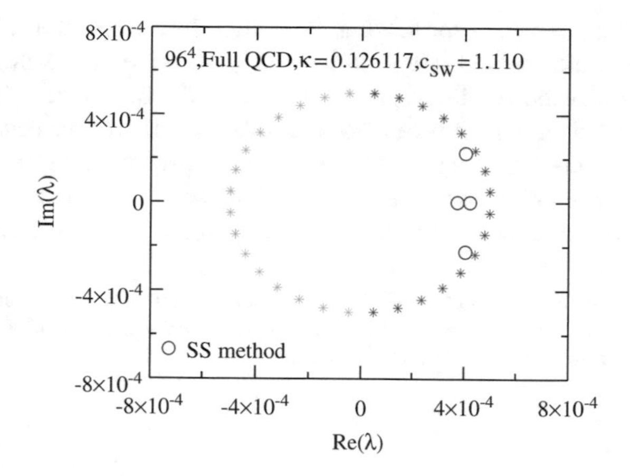


**Fig. 4** Eigenvalues of the $O(a)$-improved Wilson-Dirac operator in a $96^4$ lattice for the hopping parameter $\kappa = 0.126117$ and the improvement parameter $c_{SW} = 1.110$, with gauge field configurations generated in a full QCD simulation. The eigenvalues obtained from the SS method are indicated as *red circles*. *Green* (*blue*) *asterisks* indicate the quadrature points $z_j$ for which the BiCGStab algorithm converges (does not converge) to less than $10^{-14}$ of the relative residuals norms with 30,000 iterations

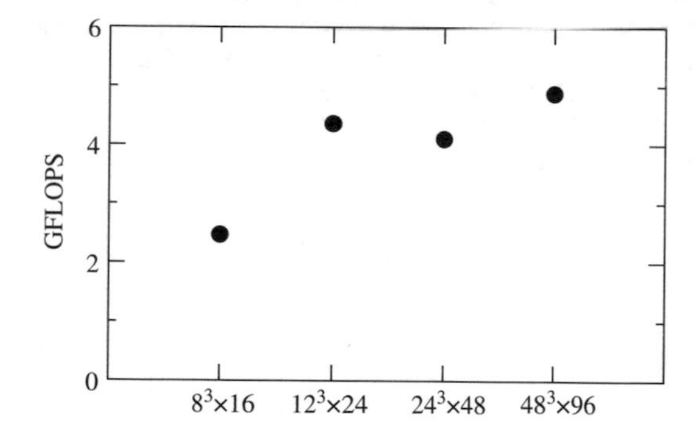


**Fig. 5** Performance (in GFLOPS) per node of the matrix-vector multiplication as a function of the lattice size. Note that for the lattice sizes $8^3 \times 16$, $12^3 \times 24$, $24^3 \times 48$, and $48^3 \times 96$, we use respectively 16, 16, 256, and 2048 nodes of the K computer

## 4 Summary

In this work, we have exploratorily developed a computer code for calculating eigenvalues and eigenvectors of non-Hermitian matrices arising in lattice QCD, using the Sakurai-Sugiura method. We have applied our implementation to calculating low-lying eigenvalues of the $O(a)$-improved Wilson-Dirac operator with gauge field configurations for the free case and those generated in quenched approximation and in full QCD. Eigenvalues have been obtained with relative residual norms from about $10^{-7}$ to $5 \times 10^{-4}$, with the accuracy being limited by the slow convergence of the BiCGStab algorithm for solving the shifted linear equations. We think that the Sakurai-Sugiura method is a promising way to solve eigenvalue problems in lattice QCD under the condition that it is combined with a more efficient shifted linear equation solver, which is desirable in order to improve the accuracy of these eigenvalues. For the future work, some preconditioner might be necessary for the iterative linear solver. We are actually carrying out implementation of a Krylov subspace iterative method specifically for solving shifted linear systems [10].

**Acknowledgements** This research used computational resources of the K computer provided by the RIKEN Advanced Institute for Computational Science through the HPCI System Research project (Project ID:hp120170, hp140069, hp150248).

## References

1. Sorensen, D.: SIAM J. Matrix Anal. Appl. **13**, 357 (1992)
2. Neff, H.: Nucl. Phys. B **106**, 1055 (2002)
3. Sakurai, T., Tadano, H., Kuramashi, Y.: Comput. Phys. Commun. **181**, 113 (2010)
4. Sakurai, T., Sugiura, H.: J. Comput. Appl. Math. **159**, 119 (2003)
5. Sakurai, T., Futamura, Y., Tadano, H.: J. Algorithms Comput. Technol. **7**, 249 (2013)
6. A software named "z-Pares" is available at the web site http://zpares.cs.tsukuba.ac.jp (2014)
7. PACS-CS Collaboration, Aoki, S., et al.: Phys. Rev. D **79**, 034503 (2009)
8. Boku, T., et al.: PoS **LATTICE2012**, 188 (2012)
9. Ukita, N., et al.: PoS **LATTICE2015**, 194 (2015)
10. Ogasawara, M., Tadano, H., Sakurai, T., Itoh, S.: J. Soc. Ind. Appl. Math. **14**, 193 (2004) (in Japanese)

# Properties of Definite Bethe–Salpeter Eigenvalue Problems

**Meiyue Shao and Chao Yang**

**Abstract** The Bethe–Salpeter eigenvalue problem is solved in condense matter physics to estimate the absorption spectrum of solids. It is a structured eigenvalue problem. Its special structure appears in other approaches for studying electron excitation in molecules or solids also. When the Bethe–Salpeter Hamiltonian matrix is definite, the corresponding eigenvalue problem can be reduced to a symmetric eigenvalue problem. However, its special structure leads to a number of interesting spectral properties. We describe these properties that are crucial for developing efficient and reliable numerical algorithms for solving this class of problems.

## 1 Introduction

Discretization of the *Bethe–Salpeter equation* (BSE) [15] leads to an eigenvalue problem $Hz = \lambda z$, where the coefficient matrix $H$ has the form

$$H = \begin{bmatrix} A & B \\ -\overline{B} & -\overline{A} \end{bmatrix}. \tag{1}$$

The matrix $A$ and $B$ in (1) satisfy

$$A^* = A, \qquad \overline{B}^* = B. \tag{2}$$

Here $A^*$ and $\overline{A}$ are the conjugate transpose and complex conjugate of $A$, respectively. In this paper, we call $H$ a *Bethe–Salpeter Hamiltonian matrix*, or, in short, a BSE Hamiltonian. In condense matter physics, the Bethe–Salpeter eigenvalue problem is derived from a Dyson's equation for a 2-particle Green's function used to describe excitation events that involve two particles simultaneously. It is a special case of the $J$-symmetric eigenvalue problem [3]. This type of eigenvalue problem also appears in linear response (LR) time-dependent density functional theory, and the random

M. Shao • C. Yang (✉)
Lawrence Berkeley National Laboratory, Berkeley, CA 94720, USA
e-mail: myshao@lbl.gov; cyang@lbl.gov



T. Sakurai et al. (eds.), *Eigenvalue Problems: Algorithms, Software and Applications in Petascale Computing*, Lecture Notes in Computational Science and Engineering 117, https://doi.org/10.1007/978-3-319-62426-6_7

phase approximation theory. In these approaches, $H$ is sometimes called a Casida Hamiltonian, a linear response Hamiltonian, or a random phase approximation (RPA) Hamiltonian.

The dimension of $A$ and $B$ can be quite large, because it scales as $\mathcal{O}(N^2)$, where $N$ is number of degrees of freedom required to represent a three-dimensional single particle wavefunction. As a result, efficient numerical algorithms must be developed to solve the Bethe–Salpeter eigenvalue problem. To gain computational efficiency, these methods should take advantage of the special structure of the Hamiltonian in (1).

Let

$$C_n = \begin{bmatrix} I_n & 0 \\ 0 & -I_n \end{bmatrix}, \qquad \Omega = \begin{bmatrix} A & B \\ \overline{B} & \overline{A} \end{bmatrix}. \tag{3}$$

Then $H = C_n\Omega$, with both $C_n$ and $\Omega$ Hermitian. In most physics problems, the condition

$$\Omega \succ 0 \tag{4}$$

holds, that is, the matrix $\Omega$ is positive definite. We call $H$ a *definite* Bethe–Salpeter Hamiltonian matrix when (4) is satisfied. It has been shown in [16] that, in general, solving a Bethe–Salpeter eigenvalue problem is equivalent to solving a real Hamiltonian eigenvalue problem. However, a definite Bethe–Salpeter eigenvalue problem, which is of most interest in practice, has many additional properties. In this paper we restrict ourselves to this special case, i.e., we assume that the condition (4) holds.

There are several ways to reformulate the definite Bethe–Salpeter eigenvalue problem. One equivalent formulation of $Hz = \lambda z$ yields a generalized eigenvalue problem (GEP) $C_n z = \lambda^{-1}\Omega z$. As $\Omega$ is positive definite, $C_n z = \lambda^{-1}\Omega z$ is a Hermitian–definite GEP and hence has real eigenvalues. Another equivalent formulation is $(\Omega - \lambda C_n)z = 0$, where $\Omega - \lambda C_n$ is a *definite pencil* [7, 20] with a *definitizing shift* $\lambda_0 = 0$. In addition, the eigenvalue problem $H^2 z = \lambda^2 z$ can be written as a product eigenvalue problem $(C_n\Omega C_n)\Omega z = \lambda^2 z$ in which both $C_n\Omega C_n$ and $\Omega$ are positive definite. These formulations suggest that a definite Bethe–Salpeter eigenvalue problem can be transformed to symmetric eigenvalue problems. As a result, we can analyze various properties of the Bethe–Salpeter eigenvalue problem by combining existing theories of symmetric eigenvalue problems (see, e.g., [14, 20]) with the special structure of $H$.

In this paper, we describe several spectral properties of a definite BSE Hamiltonian. These properties include the orthogonality of eigenvectors, the Courant–Fischer type of min–max characterization of the eigenvalues, the Cauchy type interlacing properties, and the Weyl type inequalities for establishing bounds on a structurely perturbed definite BSE Hamiltonian. Most properties take into account the special structure of the BSE Hamiltonian. Although the derivations are relatively straightforward, these properties are important for developing efficient and reliable algorithms for solving the definite Bethe–Salpeter eigenvalue problem.

The rest of this paper is organized as follows. In Sect. 2, we analyze the spectral decomposition of $H$ and derive two types of orthogonality conditions on the eigenvectors. Variational properties based on these two types of orthogonality conditions are established in Sect. 3. Finally, we provide several eigenvalue perturbation bounds in Sect. 4.

# 2 Preliminaries

## 2.1 Spectral Decomposition

As a highly structured matrix, a definite BSE Hamiltonian admits a structured spectral decomposition as stated in the following theorem.

**Theorem 1 ([16, Theorem 3])** *A definite Bethe–Salpeter Hamiltonian matrix is diagonalizable and has real spectrum. Furthermore, it admits a spectral decomposition of the form*

$$H = \begin{bmatrix} X & \overline{Y} \\ Y & \overline{X} \end{bmatrix} \begin{bmatrix} \Lambda & 0 \\ 0 & -\Lambda \end{bmatrix} \begin{bmatrix} X & -\overline{Y} \\ -Y & \overline{X} \end{bmatrix}^*, \tag{5}$$

*where* $\Lambda = \operatorname{diag}\{\lambda_1, \ldots, \lambda_n\} \succ 0$, *and*

$$\begin{bmatrix} X & -\overline{Y} \\ -Y & \overline{X} \end{bmatrix}^* \begin{bmatrix} X & \overline{Y} \\ Y & \overline{X} \end{bmatrix} = I_{2n}. \tag{6}$$

As the eigenvalues of a definite BSE Hamiltonian appear in pairs $\pm\lambda$, we denote by $\lambda_i^+(U)$ ($\lambda_i^-(U)$) the $i$th smallest *positive* (largest negative) eigenvalue of a matrix $U$ with real spectrum. When the matrix is omitted, $\lambda_i^+$ (or $\lambda_i^-$) represents $\lambda_i^+(H)$ (or $\lambda_i^-(H)$), where $H$ is a definite BSE Hamiltonian. Thus the eigenvalues of $H$ are labeled as

$$\lambda_n^- \leq \cdots \leq \lambda_1^- < \lambda_1^+ \leq \cdots \leq \lambda_n^+.$$

To represent the structure of the eigenvectors of $H$, we introduce the notation

$$\phi(U, V) := \begin{bmatrix} U & \overline{V} \\ V & \overline{U} \end{bmatrix},$$

where $U$ and $V$ are matrices of the same size. The structure is preserved under summation, real scaling, complex conjugation, transposition, as well as matrix multiplication

$$\phi(U_1, V_1)\phi(U_2, V_2) = \phi(U_1U_2 + \overline{V}_1V_2, V_1U_2 + \overline{U}_1V_2).$$

The Eqs. (5) and (6) can be rewritten as

$$C_n\Omega = \phi(X, Y)C_n\phi(\Lambda, 0)\phi(X, -Y)^*, \qquad \phi(X, -Y)^*\phi(X, Y) = I_{2n}.$$

The converse of Theorem 1 is also true in the following sense: If (5) holds, then $H$ is a definite BSE Hamiltonian because

$$H = C_n\big[\phi(X, -Y)\phi(\Lambda, 0)\phi(X, -Y)^*\big].$$

As a result, $H^{-1} = C_n\big[\phi(X, -Y)\phi(\Lambda^{-1}, 0)\phi(X, -Y)^*\big]$ is also a definite BSE Hamiltonian.

## 2.2 *Orthogonality on the Eigenvectors*

From the spectral decomposition of a definite BSE Hamiltonian $H$, we immediately obtain two types of orthogonality conditions on the eigenvectors of $H$.

First, the fact $\Omega = \phi(X, -Y)\phi(\Lambda, 0)\phi(X, -Y)^*$ implies that

$$\phi(X, Y)^*\Omega\phi(X, Y) = \phi(\Lambda, 0).$$

Therefore, the eigenvectors of $H$ are orthogonal with respect to the $\Omega$-inner product defined by $\langle u, v\rangle_\Omega := v^*\Omega u$. The eigenvectors can be normalized as

$$\phi(\tilde{X}, \tilde{Y})^*\Omega\phi(\tilde{X}, \tilde{Y}) = I_{2n}.$$

through a diagonal scaling $\phi(\tilde{X}, \tilde{Y}) = \phi(X, Y)\phi(\Lambda^{-1/2}, 0)$.

Second, it follows directly from (6) that

$$\phi(X, Y)^*C_n\phi(X, Y) = C_n.$$

This indicates that the eigenvectors of $H$ are also orthogonal with respect to the $C$-inner product defined by $\langle u, v\rangle_C := v^*C_nu$, which is an *indefinite scalar* *product* [20]. Furthermore, the positive (negative) eigenvalues of $H$ are also the $C$-positive ($C$-negative) eigenvalues of the definite pencil $\Omega - \lambda C_n$.[1]

These two types of orthogonal properties can be used to construct structure preserving projections that play a key role in Krylov subspace based eigensolvers. Suppose that $\phi(X_k, Y_k) \in \mathbb{C}^{2n\times 2k}$ is orthonormal with respect to the $\Omega$-inner product. Then projection using $\phi(X_k, Y_k)$ yields a $2k \times 2k$ Hermitian matrix of the form

$$H_k := \phi(X_k, Y_k)^* \Omega H \phi(X_k, Y_k) = \phi(X_k, Y_k)^* \Omega C_n \Omega \phi(X_k, Y_k) =: C_n \phi(A_k, \overline{B}_k). \tag{7}$$

It can be easily shown that the eigenvalues of the projected Hermitian matrix $H_k$ also occur in pairs $\pm\theta$, as $H_k$ admits a structured spectral decomposition $H_k = \phi(U_k, V_k) C_k \phi(\Theta_k, 0) \phi(U_k, V_k)^*$, where $\phi(U_k, V_k)^* \phi(U_k, V_k) = I_{2k}$. Furthermore, the matrix $\phi(X_k, Y_k)\phi(U_k, V_k)$ is again orthonormal with respect to the $\Omega$-inner product. Thus we regard (7) as a structure preserving projection. But we remark that $\Theta_k$ is not always positive definite here as $H_k$ can sometimes be singular.

Similarly, if $\phi(X_k, Y_k) \in \mathbb{C}^{2n\times 2k}$ is orthonormal with respect to the $C$-inner product, that is, $\phi(X_k, Y_k)^* C_n \phi(X_k, Y_k) = C_k$. Then

$$H_k := C_k \phi(X_k, Y_k)^* C_n H \phi(X_k, Y_k) = C_k\big[\phi(X_k, Y_k)^* \Omega \phi(X_k, Y_k)\big] \tag{8}$$

is a $2k \times 2k$ definite BSE Hamiltonian. Therefore the projection (8) in $C$-inner product can also be regarded as structure preserving.

## 3 Variational Properties

### 3.1 Min–Max Principles

The $i$th smallest eigenvalues of the Hermitian–definite pencil $C_n - \mu\Omega$, denoted by $\mu_i$, can be characterized by the Courant–Fischer min–max principle

$$\mu_i = \min_{\dim(\mathcal{V})=i} \max_{\substack{z\in\mathcal{V}\\ z\neq 0}} \frac{z^* C_n z}{z^* \Omega z} \tag{9}$$

$$= \max_{\dim(\mathcal{V})=2n-i+1} \min_{\substack{z\in\mathcal{V}\\ z\neq 0}} \frac{z^* C_n z}{z^* \Omega z}, \tag{10}$$

[1] A vector $v$ is called $C$-positive, $C$-negative, $C$-neutral, respectively, if $v^* C_n v > 0$, $v^* C_n v < 0$, $v^* C_n v = 0$. An eigenvalue of $\Omega - \lambda C_n$ is called $C$-positive ($C$-negative) if its associated eigenvector is $C$-positive ($C$-negative).

where $\mathscr{V}$ is a linear subspace of $\mathbb{C}^{2n}$. Notice that

$$\lambda_i^+ = \frac{1}{\mu_{2n-i+1}} > 0, \qquad (1 \le i \le n).$$

Taking the reciprocal of (9) and (10) yields Theorem 2 below. The theorem is also a direct consequence of the Wielandt min–max principle discussed in [10, Theorem 2.2] for the definite pencil $\Omega - \lambda C_n$.

**Theorem 2** *Let* $H = C_n \Omega$ *be a definite Bethe–Salpeter Hamiltonian matrix as defined in* (1). *Then*

$$\lambda_i^+ = \max_{\dim(\mathscr{V})=2n-i+1} \min_{\substack{z\in\mathscr{V} \\ z^*C_nz>0}} \frac{z^*\Omega z}{z^*C_nz} \tag{11}$$

$$= \min_{\dim(\mathscr{V})=i} \max_{\substack{z\in\mathscr{V} \\ z^*C_nz>0}} \frac{z^*\Omega z}{z^*C_nz} \tag{12}$$

*for* $1 \le i \le n$.

An important special case is $i = 1$, for which we have the following Corollary 1.

**Corollary 1 ([19])** *The smallest positive eigenvalue of a definite Bethe–Salpeter Hamiltonian matrix* $H = C_n \Omega$ *satisfies*

$$\lambda_1^+ = \min_{x^*x-y^*y\neq 0} \varrho(x,y), \tag{13}$$

*where*

$$\varrho(x,y) = \frac{\begin{bmatrix} x \\ y \end{bmatrix}^* \begin{bmatrix} A & B \\ \overline{B} & \overline{A} \end{bmatrix} \begin{bmatrix} x \\ y \end{bmatrix}}{|x^*x - y^*y|}. \tag{14}$$

*is the* Thouless functional.

Thanks to this result, the computation of $\lambda_1^+$ can be converted to minimizing the Thouless functional (14). Thus optimization based eigensolvers, such as the Davidson algorithm [6] and the LOBPCG algorithm [8], can be adopted to compute $\lambda_1^+$.

Finally, we remark that, from a computational point of view, the use of (12) requires additional care, because for an arbitrarily chosen subspace $\mathscr{V} \subset \mathbb{C}^{2n}$ the quantity

$$\sup_{\substack{z\in\mathscr{V} \\ z^*C_nz>0}} \frac{z^*\Omega z}{z^*C_nz} = \sup_{\substack{z\in\mathscr{V} \\ z^*C_nz=1}} z^*\Omega z$$

can easily become $+\infty$ when $\mathscr{V}$ contains $C$-neutral vectors.

### 3.2 Trace Minimization Principles

In many applications, only a few smallest positive eigenvalues of $H$ are of practical interest. The computation of these interior eigenvalues requires additional care since interior eigenvalues are in general much more difficult to compute compared to external ones. Recently, (13) has been extended to a trace minimization principle for real BSE Hamiltonians [1], so that several eigenvalues can be computed simultaneously using a blocked algorithm [2, 9]. In the following, we present two trace minimization principles, corresponding to the two types of structured preserving projections discussed in Sect. 2.2.

**Theorem 3** *Let $H = C_n\Omega$ be a definite Bethe–Salpeter Hamiltonian matrix as defined in* (1). *Then*

$$-\left(\frac{1}{\lambda_1^+} + \cdots + \frac{1}{\lambda_k^+}\right) = \min_{\phi(X,Y)^*\Omega\phi(X,Y)=I_{2k}} \operatorname{trace}(X^*X - Y^*Y) \tag{15}$$

*holds for* $1 \le k \le n$.

*Proof* We rewrite the eigenvalue problem $Hz = \lambda z$ as $C_n z = \lambda^{-1}\Omega z$. Then by the trace minimization principle for Hermitian–definite GEP, we obtain

$$-\left(\frac{1}{\lambda_1^+} + \cdots + \frac{1}{\lambda_k^+}\right) = \min_{Z^*\Omega Z=I_k} \operatorname{trace}(Z^*C_nZ).$$

Notice that

$$\mathscr{S}_1 := \left\{ \begin{bmatrix} X \\ Y \end{bmatrix} \in \mathbb{C}^{2n\times k} : \phi(X,Y)^*\Omega\phi(X,Y) = I_{2k} \right\}$$

is a subset of

$$\mathscr{S}_2 := \left\{ Z \in \mathbb{C}^{2n\times k} : Z^*\Omega Z = I_k \right\}.$$

We have

$$\min_{Z\in\mathscr{S}_2} \operatorname{trace}(Z^*C_nZ) \le \min_{Z\in\mathscr{S}_1} \operatorname{trace}(Z^*C_nZ).$$

The equality is attainable, since the minimizer in $\mathscr{S}_2$ can be chosen as the eigenvectors of $H$, which is also in $\mathscr{S}_1$. As a result, (15) follows directly from the fact that $[X^*, Y^*]C_n[X^*, Y^*]^* = X^*X - Y^*Y$. □

**Theorem 4** *Let $H = C_n\Omega$ be a definite Bethe–Salpeter Hamiltonian matrix as defined in* (1). *Then*

$$\lambda_1^+ + \cdots + \lambda_k^+ = \min_{\phi(X,Y)^* C_n \phi(X,Y) = C_k} \operatorname{trace}(X^*AX + X^*BY + Y^*\overline{B}X + Y^*\overline{A}Y) \tag{16}$$

*holds for* $1 \le k \le n$.

*Proof* As the eigenvalues of $H$ are also the eigenvalues of the definite pencil $\Omega - \lambda C_n$, by the trace minimization property of definite pencils (see, for example, [9, Theorem 2.4]), we obtain

$$\begin{aligned}\lambda_1^+ + \cdots + \lambda_k^+ &= \min_{Z^* C_n Z = I_k} \operatorname{trace}(Z^*\Omega Z) \\ &= \frac{1}{2} \min_{Z^* C_n Z = C_k} \operatorname{trace}(Z^*\Omega Z).\end{aligned}$$

The rest of the proof is nearly identical to that of Theorem 3. Because

$$\mathscr{S}_1 := \left\{ \begin{bmatrix} X \\ Y \end{bmatrix} \in \mathbb{C}^{2n\times k} : \phi(X,Y)^* C_n \phi(X,Y) = C_k \right\}$$

is a subset of

$$\mathscr{S}_2 := \left\{ Z \in \mathbb{C}^{2n\times k} : Z^* C_n Z = C_k \right\},$$

we have

$$\min_{Z\in\mathscr{S}_2} \operatorname{trace}(Z^*\Omega Z) \le \min_{Z\in\mathscr{S}_1} \operatorname{trace}(Z^*\Omega Z).$$

The equality is attainable by choosing the corresponding eigenvectors of $H$, which belong to both $\mathscr{S}_1$ and $\mathscr{S}_2$. □

Theorems 3 and 4 can both be used to derive structure preserving optimization based eigensolvers. We shall discuss the computation of eigenvalues in separate publications. We also refer the readers to [10] for more general variational principles.

### *3.3 Interlacing Properties*

We have already seen that the two types of orthogonality conditions on the eigenvectors of $H$ can both be used to construct structure preserving projections that can be used for eigenvalue computations. In this subsection we point out some difference on the location of the Ritz values.

When the $\Omega$-inner product is used for projection, we have the following Cauchy type interlacing property.

**Theorem 5** *Let $H = C_n\Omega$ be a definite Bethe–Salpeter Hamiltonian matrix as defined in* (1). *Suppose that $\phi(X,Y)^*\Omega\phi(X,Y) = I_{2k}$, where $1 \le k \le n$. Then the eigenvalues of $\phi(X,Y)^*\Omega H\phi(X,Y)$ are real and appear in pairs $\pm\theta$. Moreover*[2]

$$\lambda_i^+\big(\phi(X,Y)^*\Omega H\phi(X,Y)\big) \le \lambda_{n+i-k}^+(H), \qquad (1 \le i \le k). \tag{17}$$

*Proof* The first half of the theorem follows from the discussions in Sect. 2.2. We only show the interlacing property. Notice that $U := \Omega^{1/2}\phi(X,Y)$ has orthonormal columns in the standard inner product, that is, $U^*U = I_{2k}$. By the Cauchy interlacing theorem, we have

$$\begin{aligned}\lambda_i^+\big(\phi(X,Y)^*\Omega H\phi(X,Y)\big) &= \lambda_i^+\big(U^*\Omega^{1/2}C_n\Omega^{1/2}U\big)\\ &\le \lambda_{n+i-k}^+\big(\Omega^{1/2}C_n\Omega^{1/2}\big)\\ &= \lambda_{n+i-k}^+(H).\end{aligned}$$

□

In contrast to the standard Cauchy interlacing theorem, there is no nontrivial lower bound on the Ritz value $\lambda_i^+\big(\phi(X,Y)^*\Omega H\phi(X,Y)\big)$ here. In fact, the projected matrix $\phi(X,Y)^*\Omega H\phi(X,Y)$ can even be zero. For instance,

$$A = I_n, \quad B = 0, \quad X = \frac{1}{\sqrt{2}}\begin{bmatrix} I_k \\ I_k \\ 0 \end{bmatrix}, \quad Y = \frac{1}{\sqrt{2}}\begin{bmatrix} I_k \\ -I_k \\ 0 \end{bmatrix}$$

is an example for such an extreme case (assuming $2k \le n$).

For projections based on the $C$-inner product, we establish Theorem 6 below. Similar to Theorem 5, Ritz values are only bounded in one direction. However, in this case, it is possible to provide a meaningful (though complicated) upper bound for the Ritz value. We refer the readers to [1, Theorem 4.1] for the case of real BSE. Further investigation in this direction is beyond the scope of this paper.

**Theorem 6** *Let $H = C_n\Omega$ be a definite Bethe–Salpeter Hamiltonian matrix as defined in* (1). *Suppose that $\phi(X,Y)^*C_n\phi(X,Y) = C_k$, where $1 \le k \le n$. Then the eigenvalues of $C_k\phi(X,Y)^*C_nH\phi(X,Y)$ appear in pairs $\pm\theta$. Moreover*

$$\lambda_i^+\big(C_k\phi(X,Y)^*C_nH\phi(X,Y)\big) \ge \lambda_i^+(H), \qquad (1 \le i \le k). \tag{18}$$

*Proof* Notice that the eigenvalues of $C_k\big(\phi(X,Y)^*\Omega\phi(X,Y)\big)$ can also be regarded as the eigenvalues of the definite pencil $\phi(X,Y)^*(\Omega - \lambda C_n)\phi(X,Y)$. Then the

[2] In the case when $\phi(X,Y)^*\Omega H\phi(X,Y)$ is singular, we assign half of the zero eigenvalues with the positive sign in the notation $\lambda_i^+\big(\phi(X,Y)^*\Omega H\phi(X,Y)\big)$.

conclusion follows from the Cauchy interlacing property of definite pencils [9, Theorem 2.3]. □

From a computational perspective, (18) provides more useful information than (17), because the Ritz value $\lambda_i^+\big(C_k\phi(X,Y)^*C_nH\phi(X,Y)\big)$ is bounded in terms of the corresponding eigenvalue $\lambda_i^+(H)$ to be approximated. The inequality (17) gives an upper bound of the Ritz value. But we have less control over the location of $\lambda_i^+\big(\phi(X,Y)^*\Omega H\phi(X,Y)\big)$.

Finally, we remark that the trace minimization principle (16) can also be derived by the interlacing property (18).

# 4 Eigenvalue Perturbation Bounds

## 4.1 Weyl Type Inequalities

In the perturbation theory of symmetric eigenvalue problems, Weyl's inequality implies that the eigenvalues of a Hermitian matrix are well conditioned when a Hermitian perturbation is introduced. In the following we establish similar results for definite Bethe–Salpeter eigenvalue problems.

**Theorem 7** *Let $H$ and $H+\Delta H$ be definite Bethe–Salpeter Hamiltonian matrices. Then*

$$\left|\frac{\lambda_i^+(H+\Delta H)-\lambda_i^+(H)}{\lambda_i^+(H)}\right| \le \kappa_2(H)\frac{\|\Delta H\|_2}{\|H\|_2}, \qquad (1\le i\le n),$$

*where $\kappa_2(H)=\|H\|_2\|H^{-1}\|_2$.*

*Proof* Let $\Delta\Omega = C_n\Delta H$. Then $\Omega+\Delta\Omega$ is positive definite. We rewrite $Hz=\lambda z$ as the GEP $C_nz=\lambda^{-1}\Omega z$. It follows from the Weyl inequality on Hermitian–definite GEP [12, Theorem 2.1] that

$$\left|\frac{1}{\lambda_i^+(H)}-\frac{1}{\lambda_i^+(H+\Delta H)}\right| \le \frac{\|\Omega^{-1}\|_2\|\Delta\Omega\|_2}{\lambda_i^+(H+\Delta H)}.$$

By simple arithmetic manipulations, we arrive at

$$\left|\frac{\lambda_i^+(H+\Delta H)-\lambda_i^+(H)}{\lambda_i^+(H)}\right| \le \kappa_2(\Omega)\frac{\|\Delta\Omega\|_2}{\|\Omega\|_2} = \kappa_2(H)\frac{\|\Delta H\|_2}{\|H\|_2}.$$

□

Theorem 7 characterizes the sensitivity of the eigenvalues of $H$ when a structured perturbation is introduced—the relative condition number of $\lambda_i^+(H)$ is bounded by $\kappa_2(H)$. When the perturbation is also a definite BSE Hamiltonian, the eigenvalues are perturbed monotonically. We have the following result.

**Theorem 8** *Let $H$, $\Delta H \in \mathbb{C}^{2n\times 2n}$ be definite Bethe–Salpeter Hamiltonian matrices. Then*

$$\lambda_i^+(H+\Delta H) \geq \lambda_i^+(H) + \lambda_1^+(\Delta H), \qquad (1 \leq i \leq n).$$

*Proof* Let $\Delta\Omega = C_n \Delta H$. Then by Theorem 2 we have

$$\begin{aligned}
\lambda_i^+(H+\Delta H) &= \max_{\dim(\mathscr{V})=2n-i+1} \min_{\substack{z\in\mathscr{V}\\ z^*C_n z>0}} \left(\frac{z^*\Omega z}{z^*C_n z} + \frac{z^*\Delta\Omega z}{z^*C_n z}\right)\\
&\geq \max_{\dim(\mathscr{V})=2n-i+1} \min_{\substack{z\in\mathscr{V}\\ z^*C_n z>0}} \left(\frac{z^*\Omega z}{z^*C_n z} + \lambda_1^+(\Delta H)\right)\\
&= \lambda_i^+(H) + \lambda_1^+(\Delta H).
\end{aligned}$$

□

A special perturbation in the context of Bethe–Salpeter eigenvalue problems is to drop the off-diagonal blocks in $H$. Such a perturbation is known as the *Tamm–Dancoff approximation* (TDA) [5, 18]. Similar to the monotonic perturbation behavior above, it has been shown in [16] that TDA overestimates all positive eigenvalues of $H$. In the following, we present a simpler proof of this property than the one given in [16].

**Theorem 9 ([16, Theorem 4])** *Let $H$ be a definite Bethe–Salpeter Hamiltonian matrix as defined in* (1). *Then*

$$\lambda_i^+(H) \leq \lambda_i^+(A), \qquad (1 \leq i \leq n).$$

*Proof* Notice that $H^2 = (C_n\Omega C_n)\Omega$ with both $C_n\Omega C_n$ and $\Omega$ positive definite. By the arithmetic–geometric inequality on positive definite matrices [4, Sect. 3.4], we obtain

$$\lambda_i^+(H) = \sqrt{\lambda_{2i}^+\big((C_n\Omega C_n)\Omega\big)} \leq \lambda_{2i}^+\left(\frac{C_n\Omega C_n + \Omega}{2}\right) = \lambda_{2i}^+\left(\begin{bmatrix} A & 0\\ 0 & \bar{A}\end{bmatrix}\right) = \lambda_i^+(A).$$

□

Combining Theorems 7 and 9, we obtain the following corollary. It characterizes to what extent existing results in the literature obtained from TDA are reliable.

**Corollary 2** *If $H$ is a Bethe–Salpeter Hamiltonian matrix as defined in* (1), *then*

$$0 \leq \frac{\lambda_i^+(A) - \lambda_i^+(H)}{\lambda_i^+(H)} \leq \kappa_2(H) \frac{\|B\|_2}{\|H\|_2}, \qquad (1 \leq i \leq n).$$

## 4.2 Residual Bounds

Another type of perturbation bounds on eigenvalues measures the accuracy of approximate eigenvalues in terms of the residual norm. These bounds are of interest in eigenvalue computations. In the following we discuss several residual bounds for the definite Bethe–Salpeter eigenvalue problem.

**Theorem 10** *Let $H = C_n \Omega$ be a definite Bethe–Salpeter Hamiltonian matrix. Suppose that $X, Y \in \mathbb{C}^{n\times k}$ satisfy*

$$\phi(X,Y)^* \Omega \phi(X,Y) = I_{2k}, \qquad \phi(X,Y)^* C_n \phi(X,Y) = \begin{bmatrix} \Theta & 0 \\ 0 & -\Theta \end{bmatrix}^{-1},$$

*for some $k$ between* 1 *and $n$, where $\Theta = \operatorname{diag}\{\theta_1, \ldots, \theta_k\} \succ 0$. Then there exists a BSE Hamiltonian $\Delta H = C_n \Delta\Omega = C_n \phi(\Delta A, \overline{\Delta B})$ such that*

$$(H + \Delta H)\phi(X,Y) = \phi(X,Y) \begin{bmatrix} \Theta & 0 \\ 0 & -\Theta \end{bmatrix}. \tag{19}$$

*and*

$$\|\Delta H\|_2 \leq 2\|H\|_2^{1/2} \|R\|_2, \tag{20}$$

*where*

$$R = H\phi(X,Y) - \phi(X,Y) \begin{bmatrix} \Theta & 0 \\ 0 & -\Theta \end{bmatrix}.$$

*Proof* It follows from the definition of $R$ that

$$\phi(X,Y)^* C_n R = I_{2k} - \phi(X,Y)^* C_n \phi(X,Y) \begin{bmatrix} \Theta & 0 \\ 0 & -\Theta \end{bmatrix} = 0.$$

Let

$$\Delta\Omega = C_n R \phi(X,Y)^* \Omega + \Omega \phi(X,Y) R^* C_n.$$

Then $\Delta\Omega$ is Hermitian. Since $\Theta$ is real, we have

$$R = C_n\phi(A,\overline{B})\phi(X,Y) - \phi(X,Y)C_n\phi(\Theta,0) = C_n\phi(AX+BY-X\Theta, A\overline{Y}+B\overline{X}+Y\Theta),$$

indicating that

$$\Omega\phi(X,Y)R^*C_n = \phi(A,\overline{B})\phi(X,Y)\phi(AX+BY-X\Theta, A\overline{Y}+B\overline{X}+Y\Theta)^*$$

has the block structure $\phi(\cdot,\cdot)$. Thus $\Delta H := C_n\Delta\Omega$ is a BSE Hamiltonian. It can be easily verified that (19) is satisfied. Finally,

$$\|\Delta H\|_2 = \|\Delta\Omega\|_2 \le 2\|\Omega^{1/2}\|_2\|\Omega^{1/2}\phi(X,Y)\|_2\|R^*\|_2 = 2\|H\|_2^{1/2}\|R\|_2. \qquad \square$$

Roughly speaking, (20) implies that for definite BSE, Rayleigh–Ritz based algorithms that produce small residual norms are backward stable. When $\kappa_2(H)$ is of modest size, backward stability implies forward stability according to Theorem 7. The following theorem provides a slightly better estimate compared to simply combining Theorems 7 and 10.

**Theorem 11** *Under the same assumption of Theorem 10, there exist k positive eigenvalues of H, $\lambda_{j_1} \le \cdots \le \lambda_{j_k}$, such that*

$$|\theta_i - \lambda_{j_i}| \le \|H\|_2^{1/2}\|R\|_2, \qquad (1 \le i \le k).$$

*Proof* Notice that $U := \Omega^{1/2}\phi(X,Y)$ has orthonormal columns (in the standard inner product), and

$$\Omega^{1/2}R = \Omega^{1/2}C_n\Omega^{1/2}U - U\begin{bmatrix}\Theta & 0\\ 0 & -\Theta\end{bmatrix}.$$

By the residual bound for standard Hermitian eigenvalue problems (see [14, Theorem 11.5.1] or [17, Sect. IV.4.4]), we obtain that there are $2k$ eigenvalues of $\Omega^{1/2}C_n\Omega^{1/2}$, $\tilde{\lambda}_{-j_k} \le \cdots \le \tilde{\lambda}_{-j_1} \le \tilde{\lambda}_{j_1} \le \cdots \le \tilde{\lambda}_{j_k}$, such that

$$\max\left\{|\theta_i + \tilde{\lambda}_{-j_i}|, |\theta_i - \tilde{\lambda}_{j_i}|\right\} \le \|\Omega\|_2^{1/2}\|R\|_2 = \|H\|_2^{1/2}\|R\|_2, \qquad (1 \le i \le k).$$

Note that at least one of the inequalities $\tilde{\lambda}_{j_1} > 0$ and $\tilde{\lambda}_{-j_1} < 0$ holds. As the eigenvalues of $\Omega^{1/2}C_n\Omega^{1/2}$ are identical to those of $H$, the conclusion follows immediately by choosing

$$\lambda_{j_i} = \begin{cases}\tilde{\lambda}_{j_i}, & \text{if } \tilde{\lambda}_{j_1} > 0,\\ -\tilde{\lambda}_{-j_i}, & \text{otherwise,}\end{cases} \qquad (1 \le i \le k). \qquad \square$$

Finally, we end this section by a Temple–Kato type quadratic residual bound as stated in Theorem 12. The quadratic residual bound explains the fact that the accuracy of a computed eigenvalues is in general much higher compared to that of the corresponding eigenpair. Such a behavior has been reported for real Bethe–Salpeter eigenvalue problem in [9]. It is certainly possible to extend Theorem 12 to a subspace manner using techniques in [11, 13].

**Theorem 12** *Let* $(\theta, \hat{z})$ *be an approximate eigenpair of a definite BSE Hamiltonian* $H = C_n\Omega$ *satisfying*

$$\frac{\hat{z}^*\Omega\hat{z}}{\hat{z}^*C_n\hat{z}} = \theta.$$

*Then the eigenvalue of* $H$ *closest to* $\theta$*, denoted by* $\lambda$*, satisfies*

$$\left|\theta^{-1} - \lambda^{-1}\right| \le \frac{\|H^{-1}\|_2\|H\hat{z} - \theta\hat{z}\|_2^2}{\operatorname{gap}(\theta)\hat{z}^*\Omega\hat{z}},$$

*where*

$$\operatorname{gap}(\theta) := \min_{\lambda_i(H)\neq\theta} \left|\theta^{-1} - \lambda_i(H)^{-1}\right|.$$

*Proof* The theorem is a direct consequence of [14, Theorem 11.7.1] on the equivalent Hermitian eigenvalue problem $\left(\Omega^{-1/2}C_n\Omega^{-1/2}\right)\left(\Omega^{1/2}z\right) = \lambda\left(\Omega^{1/2}z\right)$. □

## 5 Summary

The Bethe–Salpeter eigenvalue problem is an important class of structured eigenvalue problems arising from several physics and chemistry applications. The most important case, the definite Bethe–Salpeter eigenvalue problem, has a number of interesting properties. We identified two types of orthogonality conditions on the eigenvectors, and discussed several properties of the corresponding structure preserving projections. Although most of our theoretical results can be derived by extending similar results for general symmetric eigenvalue problems to this class of problems, they play an important role in developing and analyzing structure preserving algorithms for solving this type of problems. Numerical algorithms will be discussed in a separate publication.

**Acknowledgements** The authors thank Ren-Cang Li for helpful discussions. Support for this work was provided through Scientific Discovery through Advanced Computing (SciDAC) program funded by U.S. Department of Energy, Office of Science, Advanced Scientific Computing Research.

## References

1. Bai, Z., Li, R.C.: Minimization principles for the linear response eigenvalue problem I: theory. SIAM J. Matrix Anal. Appl. **33**(4), 1075–1100 (2012). doi:10.1137/110838960
2. Bai, Z., Li, R.C.: Minimization principles for the linear response eigenvalue problem II: computation. SIAM J. Matrix Anal. Appl. **34**(2), 392–416 (2013). doi:10.1137/110838972
3. Benner, P., Fassbender, H., Yang, C.: Some remarks on the complex $J$-symmetric eigenproblem. Preprint MPIMD/15–12, Max Planck Institute Magdeburg. Available from http://www.mpi-magdeburg.mpg.de/preprints/ (2015)
4. Bhatia, R., Kittaneh, F.: Notes on matrix arithmetic–geometric mean inequalities. Linear Algebra Appl. **308**, 203–211 (2000). doi:10.1016/S0024-3795(00)00048-3
5. Dancoff, S.M.: Non-adiabatic meson theory of nuclear forces. Phys. Rev. **78**(4), 382–385 (1950). doi:10.1103/PhysRev.78.382
6. Davidson, E.R.: The iterative calculation of a few of the lowest eigenvalues and corresponding eigenvectors of large real symmetric matrices. J. Comput. Phys. **17**(1), 87–94 (1975). doi:10.1016/0021-9991(75)90065-0
7. Gohberg, I., Lancaster, P., Rodman, L.: Matrices and Indefinite Scalar Products. Operator Theory: Advances and Applications, vol. 8. Birkhäuser, Basel (1983)
8. Knyazev, A.V.: Toward the optimal preconditioned eigensolver: locally optimal block preconditioned conjugate gradient method. SIAM J. Sci. Comput. **23**(2), 517–541 (2001). doi:10.1137/S1064827500366124
9. Kressner, D., Miloloža Pandur, M., Shao, M.: An indefinite variant of LOBPCG for definite matrix pencils. Numer. Algorithms **66**(4), 681–703 (2014). doi:10.1007/s11075-013-9754-3
10. Liang, X., Li, R.C.: Extensions for Wielandt's min–max principles for positive semi-definite pencils. Linear Multilinear Algebra **62**(8), 1032–1048 (2014). doi:10.1080/03081087.2013.803242
11. Mathias, R.: Quadratic residual bounds for the Hermitian eigenvalue problem. SIAM J. Matrix Anal. Appl. **19**(2), 541–550 (1998). doi:10.1137/S0895479896310536
12. Nakatsukasa, Y.: Perturbation behavior of a multiple eigenvalue in generalized Hermitian eigenvalue problems. BIT Numer. Math. **50**, 109–121 (2010). doi:10.1007/s10543-010-0254-8
13. Nakatsukasa, Y.: Eigenvalue perturbation bounds for Hermitian block tridiagonal matrices. Appl. Numer. Math. **62**, 67–78 (2012). doi:10.1016/j.apnum.2011.09.010
14. Parlett, B.N.: The Symmetric Eigenvalue Problem. Classics in Applied Mathematics, vol. 20. SIAM, Philadelphia, PA (1998). Corrected reprint of the 1980 original
15. Salpeter, E.E., Bethe, H.A.: A relativistic equation for bounded-state problems. Phys. Rev. **84**(6), 1232–1242 (1951). doi:10.1103/PhysRev.84.1232
16. Shao, M., H. da Jornada, F., Yang, C., Deslippe, J., Louie, S.G.: Structure preserving parallel algorithms for solving the Bethe–Salpeter eigenvalue problem. Linear Algebra Appl. **488**, 148–167 (2016). doi:10.1016/j.laa.2015.09.036
17. Stewart, G.W., Sun, J.: Matrix Perturbation Theory. Academic, Boston, MA (1990)
18. Tamm, I.Y.: Relativistic interaction of elementary particles. J. Phys. (USSR) **9**, 449–460 (1945)
19. Thouless, D.J.: Vibrational states of nuclei in the random phase approximation. Nucl. Phys. **22**, 78–95 (1961). doi:10.1016/0029-5582(61)90364-9
20. Veselić, K.: Damped Oscillations of Linear Systems—A Mathematical Introduction. Lecture Notes in Mathematics, vol. 2023. Springer, Heidelberg (2011)

# Preconditioned Iterative Methods for Eigenvalue Counts

**Eugene Vecharynski and Chao Yang**

**Abstract** We describe preconditioned iterative methods for estimating the number of eigenvalues of a Hermitian matrix within a given interval. Such estimation is useful in a number of applications. It can also be used to develop an efficient spectrum-slicing strategy to compute many eigenpairs of a Hermitian matrix. Our method is based on the Lanczos- and Arnoldi-type of iterations. We show that with a properly defined preconditioner, only a few iterations may be needed to obtain a good estimate of the number of eigenvalues within a prescribed interval. We also demonstrate that the number of iterations required by the proposed preconditioned schemes is independent of the size and condition number of the matrix. The efficiency of the methods is illustrated on several problems arising from density functional theory based electronic structure calculations.

## 1 Introduction

The problem of estimating the number of eigenvalues of a large and sparse Hermitian matrix $A$ within a given interval $[\xi, \ \eta]$ has recently drawn a lot of attention, e.g., [12–14]. One particular use of this estimation is in the implementation of a "spectrum slicing" technique for computing many eigenpairs of a Hermitian matrix [1, 11]. Approximate eigenvalue counts are used to determine how to divide the desired spectrum into several subintervals that can be examined in parallel. In large-scale data analytics, efficient means of obtaining approximate eigenvalue counts is required for estimating the generalized rank of a given matrix; see, e.g., [22].

A traditional approach for counting the number of eigenvalues of $A$ in $[\xi, \ \eta]$ is based on the Sylevester's law of inertia [15]. The inertia of the shifted matrices $A-\xi I$ and $A - \eta I$ are obtained by performing $LDL^T$ factorizations of these matrices [1]. This approach, however, is impractical if $A$ is extremely large or not given explicitly.

E. Vecharynski (✉) • C. Yang
Computational Research Division, Lawrence Berkeley National Laboratory, Berkeley, CA 94720, USA
e-mail: evecharynski@lbl.gov; cyang@lbl.gov



T. Sakurai et al. (eds.), *Eigenvalue Problems: Algorithms, Software and Applications in Petascale Computing*, Lecture Notes in Computational Science and Engineering 117, https://doi.org/10.1007/978-3-319-62426-6_8

Several techniques that avoid factoring $A$ have recently been described in [12, 14]. These methods only require multiplying $A$ with a number of vectors. In [12], a survey that describes several approaches to approximating the so-called density of states (DOS), which measures the probability of finding eigenvalues near a given point on the real line is presented. The DOS approximation can then be used to obtain an estimate of the number of eigenvalues in $[\xi, \eta]$. The potential drawback of a DOS estimation based approach is that, instead of directly targeting the specific interval $[\xi, \eta]$, it always tries to approximate the eigenvalue distribution on the entire spectrum first.

Conceptually, the approaches in [12, 14] are based on constructing a least-squares polynomial approximation of a spectral filter. Such approximations, however, often yield polynomials of a very high degree if $A$ is ill-conditioned or the eigenvalues to be filtered are tightly clustered. These are common issues in practical large-scale computations. In particular, matrices originating from the discretization of partial differential operators tend to become more ill-conditioned as the mesh is refined. As a result, the polynomial methods of [12, 14] can become prohibitively expensive. The overall cost of the computation becomes even higher if the cost of multiplying $A$ with a vector is relatively high.

In this work we explore the possibility of using preconditioned iterative methods to reduce the cost of estimating the number of eigenvalues within an interval. By applying the Lanczos or Arnoldi iteration to preconditioned matrices with properly constructed Hermitian positive definite (HPD) preconditioners, we can significantly reduce the number of matrix-vector multiplications required to obtain accurate eigenvalue counts. Furthermore, when a good preconditioner is available, we can keep the number of matrix-vector multiplications (roughly) constant even as the problem size and conditioning of $A$ increase. The methods we present in this paper do not require the lower and upper bounds of the spectrum of $A$ to be estimated a priori. This feature compares favorably with the methods of [12, 14] since obtaining such bounds can by itself be a challenging task.

This paper is organized as following. Section 2 outlines the main idea, followed by derivation of the preconditioned Lanczos-type estimator based on Gauss quadrature in Sect. 3. The preconditioned Arnoldi-type algorithm is presented in Sect. 4. In Sect. 5, we discuss the proposed methods from the polynomial perspective. The performance of the introduced schemes depends to a large extent on the quality of the HPD preconditioner associated with the matrix $A - \tau I$. While the development of such a preconditioner is outside the scope of this paper, we point to several available options in Sect. 6. Several numerical experiments are reported in Sect. 7.

## 2 Basic Idea

To simplify our presentation, let us assume that the endpoints $\xi$ and $\eta$ are different from any eigenvalue of $A$. Then the number of eigenvalues $c(\xi, \eta)$ of $A$ in $[\xi, \eta]$ is given by the difference $c(\xi, \eta) = n\_(A - \eta I) - n\_(A - \xi I)$, where $n\_(A - \tau I)$ denotes the negative inertia (i.e., the number of negative eigenvalues) of $A - \tau I$. Hence, in order to approximate $c(\xi, \eta)$, it is sufficient to estimate $n\_(A - \tau I)$ for a given real number $\tau$.

The problem of estimating $n\_(A - \tau I)$ can be reformulated as that of approximating the trace of a matrix step function. Namely, let

$$h(x) = \begin{cases} 1, x < 0\,; \\ 0, \text{otherwise}\,. \end{cases} \tag{1}$$

Then

$$n\_(A - \tau I) = \text{trace}\,\{h(A - \tau I)\}\,. \tag{2}$$

Now let us assume that $T$ is an HPD preconditioner for the shifted matrix $A - \tau I$ in the sense that the spectrum of $TA$ is clustered around a few distinct points on the real line. Specific options for constructing such preconditioners will be discussed in Sect. 6.

If $T$ is available in a factorized form $T = M^*M$, estimating $n\_(A - \tau I)$ is equivalent to estimating $n\_(M(A - \tau I)M^*)$, i.e., transforming $A - \tau I$ to $C = M(A - \tau I)M^*$ preserves the inertia. Hence, we have

$$n\_(A - \tau I) = \text{trace}\,\{h(C)\}\,. \tag{3}$$

If $T = MM^*$ is chosen in such a way that its spectrum has a favorable distribution, i.e., the eigenvalues of $C$ is clustered in a few locations, then estimating trace $\{h(C)\}$ can be considerably easier than estimating trace $\{h(A - \tau I)\}$.

If the multiplication of $C$ with a vector can be performed efficiently, then the trace of $C$ can be estimated as

$$\text{trace}\,\{C\} \approx \frac{1}{m}\sum_{j=1}^{m} v_j^* C v_j, \tag{4}$$

where the entries of each vector $v_j$ are i.i.d. random variables with zero mean and unit variance; see [2, 10]. It follows that

$$n\_(A - \tau I) = \text{trace}\,\{h(C)\} \approx \frac{1}{m}\sum_{j=1}^{m} v_j^* h(C) v_j, \tag{5}$$

for a sufficiently large sample size $m$.

The variance of the stochastic trace estimator is known to depend on the magnitude of off-diagonal entries of the considered matrix [2], which is $h(C)$ in (5).[1] Clearly, different choices of the preconditioned operator $C$ yield different matrices $h(C)$, and hence lead to different convergence rates of the estimator (5).

# 3 Preconditioned Lanczos

If $A$ is large, then the exact evaluation of $h(C)$ in (5) can be prohibitively expensive, because it requires a full eigendecomposition of the preconditioned matrix. A more practical approach in this situation would be to (approximately) compute $v^*h(C)v$ for a number of randomly sampled vectors $v$ without explicitly evaluating the matrix function.

## *3.1 The Gauss Quadrature Rule*

Let us assume that $T = M^*M$ is available in the factorized form and let $C = M(A - \tau I)M^*$ in (5). We also assume that the Hermitian matrix $C$ has $p \leq n$ distinct eigenvalues $\mu_1 < \mu_2 < \ldots < \mu_p$.

Consider the orthogonal expansion of $v$ in terms of the eigenvectors of $C$, i.e., $v = \sum_{i=1}^{p} \alpha_i u_i$, where $u_i$ is an normalized eigenvector associated with the eigenvalue $\mu_i$, and $\alpha_i = u_i^* v$. It is then easy to verify that

$$v^*h(C)v = \sum_{i=1}^{p} \alpha_i^2 h(\mu_i) \equiv \sum_{i=1}^{p_-} \alpha_i^2, \quad \alpha_i^2 = |u_i^* v|^2, \tag{6}$$

where $p_-$ denotes the number of negative eigenvalues. The right-hand side in (6) can be viewed as a Stieltjes integral of the step function $h$ with respect to the measure defined by the piecewise constant function

$$\alpha_{C,v}(x) = \begin{cases} 0, & \text{if } x < \mu_1, \\ \sum_{j=1}^{i} \alpha_j^2, & \text{if } \mu_i \leq x < \mu_{i+1}, \\ \sum_{j=1}^{p} \alpha_j^2, & \text{if } \mu_p \leq x. \end{cases} \tag{7}$$

[1] The variance of a Gaussian trace estimator applied to a Hermitian matrix also depends on the magnitude of the diagonal elements and can be expressed only in terms of the eigenvalues of $A$.

Therefore, using (7), we can write (6) as

$$v^* h(C) v = \int h(x) d\alpha_{C,v}(x) \equiv \int_{\mu_1}^{0} d\alpha_{C,v}(x). \tag{8}$$

Computing the above integral directly is generally infeasible because the measure (7) is defined in terms of the unknown eigenvalues of $C$. Nevertheless, the right-hand side of (8) can be approximated by using the Gauss quadrature rule [6], so that

$$v^* h(C) v \approx \sum_{i=1}^{k} w_i h(\theta_i) \equiv \sum_{i=1}^{k\_} w_i, \tag{9}$$

where the $k$ nodes $\theta_1 \leq \theta_2 \leq \ldots \leq \theta_k$ and weights $w_1, w_2, \ldots, w_k$ of the quadrature are determined from $k$ steps of the Lanczos procedure (see Algorithm 1) applied to the preconditioned matrix $C$ with the starting vector $v$. In (9), $k\_$ denotes the number of negative nodes $\theta_i$.

Specifically, given $q_1 = v/\|v\|$, running $k$ steps of the Lanczos procedure in Algorithm 1 yields the relation

$$CQ_k = Q_{k+1} J_{k+1,k}, \quad Q_{k+1}^* Q_{k+1} = I, \tag{10}$$

where $J_{k+1,k}$ is the tridiagonal matrix

$$J_{k+1,k} = \begin{bmatrix} \alpha_1 & \beta_2 & & \\ \beta_2 & \alpha_2 & \ddots & \\ & \ddots & \ddots & \beta_k \\ & & \beta_k & \alpha_k \\ & & & \beta_{k+1} \end{bmatrix} \in \mathbf{R}^{(k+1)\times k}. \tag{11}$$

**Algorithm 1** The Lanczos procedure for $M(A - \tau I)M^*$

**Input**: Matrix $A - \tau I$, $T = M^* M$, starting vector $v$, and number of steps $k$.
**Output**: Tridiagonal matrix $J_{k+1,k}$ and the Lanczos basis $Q_{k+1} = [q_1, q_2, \ldots, q_{k+1}]$.

1: $q_1 \leftarrow v/\|v\|$; $q_0 \leftarrow 0$; $\beta_1 \leftarrow 0$; $Q_1 \leftarrow q_1$;
2: **for** $i = 1 \rightarrow k$ **do:**
3: $\quad w \leftarrow M(A - \tau I)M^* q_i - \beta_i q_{i-1}$;
4: $\quad \alpha_i \leftarrow q_i^* w$; $w \leftarrow w - \alpha_i q_i$;
5: $\quad$ Reorthogonalize $w \leftarrow w - Q_i(Q_i^* w)$;
6: $\quad \beta_{i+1} \leftarrow \|w\|$; $q_{i+1} \leftarrow w/\beta_{i+1}$; $Q_{i+1} \leftarrow [Q_i,\ q_{i+1}]$;
7: **end for**

The eigenvalues of the leading $k \times k$ submatrix of $J_{k+1,k}$, denoted by $J_k$, are ordered so that $\theta_1 \leq \theta_2 \leq \ldots \leq \theta_{k\_} < 0 \leq \theta_{k\_+1} \leq \ldots \leq \theta_k$. Then the Gauss quadrature rule on the right-hand side of (9) is defined by eigenvalues and eigenvectors of $J_k$, i.e.,

$$v^*h(C)v \approx \|v\|^2 e_1^* h(J_k) e_1 = \sum_{i=1}^{k} w_i h(\theta_i) \equiv \sum_{i=1}^{k\_} w_i, \quad w_i = \|v\|^2 |z_i(1)|^2, \tag{12}$$

where $z_i$ is the eigenvector of $J_k$ associated with the eigenvalue $\theta_i$, $z_i(1)$ denotes its first component [6], and $k\_$ denotes the number of negative Ritz values.

If the preconditioner $T = MM^*$ is chosen in such a way that the spectrum of $C = M(A - \tau I)M^*$ is concentrated within small intervals $[a, b] \subset (-\infty, 0)$ and $[c, d] \subset (0, \infty)$, then, by (7), the measure $\alpha_{M(A-\tau I)M^*,v}$ will have jumps inside $[a, b]$ and $[c, d]$, and will be constant elsewhere. Hence, the integral in (8) will be determined only by integration over $[a, b]$ because $h$ vanishes in $[c, d]$. Therefore, in order for quadrature rule (9) to be a good approximation to (8), its nodes should be chosen inside $[a, b]$.

In the extreme case in which clustered eigenvalues of $C$ coalesce into a few eigenvalues of higher multiplicities, the number of Lanczos steps required to obtain an accurate approximation in (12) is expected to be very small.

**Proposition 1** *Let the preconditioned matrix $C = M(A - \tau I)M^*$ have $p$ distinct eigenvalues. Then the Gauss quadrature* (12) *will be exact with at most $k = p$ nodes.*

*Proof* Let $v = \sum_{i=1}^{p} \alpha_i u_i$, where $u_i$ is an eigenvector of $C$ associated with the eigenvalue $\mu_i$. Then $p$ steps of Lanczos process with $v$ as a starting vector produce a tridiagonal matrix $J_p$ and an orthonormal basis $Q_p$, such that the first column of $Q_p$ is $\hat{v} = v/\|v\|$. The eigenvalues $\theta_i$ of $J_p$ are exactly the $p$ distinct eigenvalues of $C$. The eigenvectors $z_i$ of $J_p$ are related to those of $C$ as $u_i = Q_p z_i$. Thus, we have $w_i = \|v\|^2 |z_i(1)|^2 = \|v\|^2 |\hat{v}^* u_i|^2 = |v^* u_i|^2$, and, by comparing with (6), we see that the quadrature (12) gives the exact value of $v^* h(M(A - \tau I)M^*)v$.

Proposition 1 implies that in the case of an ideal preconditioner, where $M(A - \tau I)M^*$ has only two distinct eigenvalues, the Gauss quadrature rule (12) is guaranteed to be exact within at most two Lanczos steps.

Finally, note that, in exact arithmetic, the Lanczos basis $Q_i$ should be orthonormal [15]. However, in practice, the orthogonality may be lost; therefore, we reorthogonalize $Q_i$ at every iteration of Algorithm 1 (see step 5).

### 3.2 The Algorithm

Let $J_k^{(j)}$ denote the $k$-by-$k$ tridiagonal matrix resulting from the $k$-step Lanczos procedure applied to $C = M(A - \tau I)M^*$ with a random starting vector $v_j$. Assume

**Algorithm 2** The preconditioned Lanczos-type estimator for $n\_(A - \tau I)$

**Input**: Matrix $A$, shift $\tau$, HPD preconditioner $T = M^*M$ for $A - \tau I$, number of steps $k$, and parameter $m$.

**Output**: approximate number $C_\tau$ of eigenvalues of $A$ that are less than $\tau$;

1: $L_\tau \leftarrow 0$.
2: **for** $j = 1 \to m$ **do:**
3: Generate $v \sim \mathcal{N}(0, I)$.
4: Run $k$ steps of Lanczos process in Algorithm 1 with the starting vector $v$ to obtain tridiagonal matrix $J_k$.
5: Find the eigendecomposition of $J_k$. Let $z_1, \ldots, z_{k\_}$ be unit eigenvectors associated with negative eigenvalues of $J_k$.
6: Set $L_\tau \leftarrow L_\tau + \|v\|^2 \sum_{i=1}^{k\_} w_i$, where $w_i = |z_i(1)|^2$.
7: **end for**
8: Return $L_\tau \leftarrow [L_\tau / m]$.

that $k_j$ is the number of its negative eigenvalues. Then, by (5) and (12), the quantity $n\_(A - \tau I)$ can be approximated from the estimator

$$L_\tau(k, m) = \frac{1}{m} \sum_{j=1}^{m} \sum_{i=1}^{k_j} w_i^{(j)}, \quad w_i^{(j)} = \|v_j\|^2 |z_i^{(j)}(1)|^2, \quad v_j \in \mathcal{N}(0, I), \tag{13}$$

where $z_i^{(j)}(1)$ denotes the first components of a normalized eigenvector $z_i^{(j)}$ of $J_k^{(j)}$ associated with the negative eigenvalues. It is expected that, for a sufficiently large $m$, $L_\tau(k, m) \approx n\_(A - \tau I)$. The expression (13) is what Algorithm 2 uses to estimate the number of eigenvalues of $A$ that are to the left of $\tau$. Here and throughout, the notation $v_j \in \mathcal{N}(0, I)$ means that the entries of each sampling vector $v_j$ are i.i.d. random variables chosen from normal distribution with zero mean and unit variance.

In order to estimate the number of eigenvalues in a given interval $[\xi, \eta]$, Algorithm 2 should be applied twice with $\tau = \xi$ and $\tau = \eta$. The difference between the estimated $n\_(A - \xi I)$ and $n\_(A - \eta I)$ yields the desired count. The two runs of Algorithm 2 generally require two different HPD preconditioners, one for $A - \xi I$ and the other for $A - \eta I$. In some cases, however, it can be possible to come up with a single preconditioner that works well for both runs.

The cost of Algorithm 2 is dominated by computational work required to perform the preconditioned matrix-vector multiplication of $M(A - \tau I)M^* v$ at each iteration of the Lanczos procedure. The eigenvalue decomposition of the tridiagonal matrix $J_k$, as well as reorthogonalization of the Lanczos basis in step 5 of Algorithm 1, is negligibly small for small values of $k$, which can be ensured by a sufficiently high quality preconditioner.

### 3.3 Bias of the Estimator

A relation between the Gauss quadrature (12) and matrix functional $v^*h(C)v$ can be expressed as

$$\sum_{i=1}^{k\_} w_i = v^*h(C)v + \epsilon_k,$$

where $\epsilon_k$ is the error of the quadrature rule. Thus, (13) can be written as

$$L_\tau(k,m) = \frac{1}{m}\sum_{j=1}^{m} v_j^* h(C) v_j + \frac{1}{m}\sum_{j=1}^{m} \epsilon_k^{(j)}, \tag{14}$$

where $\epsilon_k^{(j)}$ denotes the error of the quadrature rule for $v_j^* h(C) v_j$. As $m$ increases, the first term in the right-hand side of (14) converges to trace $\{h(C)\} = n\_(A - \tau I)$. Thus, $L_\tau(k,m)$ is a biased estimate of $n\_(A-\tau I)$, where the bias is determined by the (average) error of the quadrature rule, given by the second term in the right-hand side of (14). In other words, the accuracy of $L_\tau(k,m)$ generally depends on how well the Gauss quadrature captures the value of the matrix functional $v^*h(M(A - \tau I)M^*)v$.

Bounds on the quadrature error for a matrix functional $v^*f(C)v$, where $f$ is a sufficiently smooth function and $C$ is a Hermitian matrix, are well known. In particular, the result of [3] gives the bound

$$|\epsilon_k| \le \frac{N_k}{2k!}\beta_{k+1}^2\beta_k^2 \ldots \beta_2^2, \tag{15}$$

where the constant $N_k$ is such that $|f^{(2k)}(x)| \le N_k$ for $x$ in the interval containing spectrum of $C$, and $\beta_j$ are the off-diagonal entries of (11).

Function $h(x)$ in (1) is discontinuous. Therefore, bound (15) does not directly apply to measure the quadrature error for the functional $v^*h(M(A - \tau I)M^*)v$. However, since the rule (12) depends on the values of $h(x)$ only at the Ritz values $\theta_i$ generated by the Lanczos process for $M(A - \tau I)M^*$, it will yield exactly the same result for any function $\tilde{h}(x)$, such that $\tilde{h}(\theta_i) = h(\theta_i)$ for all $\theta_i$. If, additionally, $\tilde{h}(x)$ assumes the same values as $h(x)$ on the spectrum of $M(A - \tau I)M^*$, then, by (6), the functionals $v^*h(M(A - \tau I)M^*)v$ and $v^*\tilde{h}(M(A - \tau I)M^*)v$ will also be identical. Hence, the quadrature errors for $v^*\tilde{h}(M(A - \tau I)M^*)v$ and $v^*h(M(A - \tau I)M^*)v$ will coincide. But then we can choose $\tilde{h}(x)$ as a $2k$ times continuously differentiable function and apply (15) to bound the quadrature error for $v^*\tilde{h}(M(A-\tau I)M^*)v$. This error will be exactly the same as that of the quadrature (12) for $v^*h(M(A-\tau I)M^*)v$, which we are interested in.

In particular, let us assume that the eigenvalues of $M(A - \tau I)M^*$ and Ritz values $\theta_i$ are located in intervals $[a, b)$ and $(c, d]$ to the left and right of origin, respectively. Then we can choose $\tilde{h}(x)$ such that it is constant one on $[a, b)$ and constant zero on $(c, d]$. On the interval $[b, c]$, which contains zero, we let $\tilde{h}(x)$ to be a polynomial $p(x)$ of degree $4k + 1$, such that $p(b) = 1$, $p(c) = 0$, and $p^{(l)}(b) = p^{(l)}(c) = 0$ for $l = 1, \ldots, 2k$. This choice of polynomial will ensure that the piecewise function $\tilde{h}(x)$ is $2k$ times continuously differentiable. (Note that $p(x)$ can always be constructed by (Hermite) interpolation with the nodes $b$ and $c$; see, e.g., [16].) We then apply (15) to obtain the bound on the quadrature error for $v^*\tilde{h}(M(A - \tau I)M^*)v$. As discussed above, this yields the estimate of the error $\epsilon_k$ of quadrature rule (12) for functional $v^*h(M(A - \tau I)M^*)v$. Thus, we can conclude that the latter is bounded by (15), where $N_k$ is the maximum of $|p^{(2k)}(x)|$ on the interval $[b, c]$.

This finding shows that we can expect that (12) provides a better approximation of $v^*h(M(A - \tau I)M^*)v$ when the intervals $[a, b)$ and $(c, d]$, containing eigenvalues of $M(A - \tau I)M^*$ along with the Ritz values produced by the Lanczos procedure, are bounded away from zero. In this case, the rate of change of the polynomial $p(x)$ on $[b, c]$ will not be too high, resulting in a smaller value of $N_k$ in (15).

Fortunately, a good choice of the preconditioner $T = M^*M$ can ensure that eigenvalues of $M(A - \tau I)M^*$ are clustered and away from zero. In this case, the Ritz values typically converge rapidly to these eigenvalues after a few Lanczos steps. Thus, with a good preconditioner, the Gauss quadrature (12) can effectively approximate the matrix functional $v^*h(M(A - \tau I)M^*)v$, yielding small errors $\epsilon_k$ for a relatively small number of quadrature nodes. As a result, the bias of the estimator $L_\tau(k, m)$ in (14) will be small and, as confirmed by numerical experiments in Sect. 7.

## 3.4 The Generalized Averaged Gauss Quadrature Rule

The Gauss quadrature rule (12) is exact for all polynomials of degree at most $2k - 1$; e.g., [6].

In the recent work of [17] (and references therein), a so-called generalized averaged (GA) Gauss quadrature rules was introduced. This quadrature rule make use of the same information returned from a $k$-step Lanczos process, but gives an exact integral value for polynomials of degree $2k$. Hence it is more accurate at essentially the same cost.

When applying the GA Gauss quadrature rule to the matrix functional $v^*h(C)v$ in (8), we still use the expression (12), except that we have $(2k - 1)$ nodes $\theta_1, \theta_2, \ldots, \theta_{2k-1}$ which are the eigenvalues of the matrix

$$\tilde{J}_{2k-1} = \text{tridiag}\,\{(\alpha_1, \ldots, \alpha_k, \alpha_{k-1}, \ldots \alpha_1), (\beta_2, \ldots, \beta_k, \beta_{k+1}, \beta_{k-1} \ldots \beta_2)\} \quad (16)$$

obtained from $J_{k+1,k}$ in (11) by extending its tridiagonal part in a "reverse" order. The set $(\alpha_i)$ of numbers in (16) gives the diagonal entries of $J_{k+1,k}$, whereas $(\beta_i)$ define the upper and lower diagonals. Similarly, the associated weights $w_i$ are determined by squares of the first components of the properly normalized eigenvectors $z_i$ of $\tilde{J}_{2k-1}$ associated with the eigenvalues $\theta_i$; see [17] for more details.

Thus, we can expect to increase accuracy of the estimator by a minor modification of Algorithm 2. This modification will only affect step 5 of the algorithm, where $J_k$ must be replaced by the extended tridiagonal matrix (16).

# 4 Preconditioned Arnoldi

Sometimes, the preconditioner $T$ is not available in a factored form $T = MM^*$. In this case, it may be necessary to work with $T(A - \tau I)$ or $(A - \tau I)T$ directly. One possibility is to make use of the fact that $(A - \tau I)T$ is self adjoint with respect to an inner product induced by $T$. This property allows us to carry out a $T$-inner product Lanczos procedure that produces

$$(A - \tau I)TX_k = X_{k+1}J_{k+1,k}, \quad X_{k+1}^* TX_{k+1} = I. \tag{17}$$

Similarly, we can use a $T^{-1}$-inner product based Lanczos procedure to obtain

$$T(A - \tau I)Y_k = Y_{k+1}J_{k+1,k}, \quad Y_{k+1}^* T^{-1} Y_{k+1} = I, \tag{18}$$

where $Y_k = M^* Q_k$. Even though it may appear that we do not need $T$ in a factored form in either (17) or (18), the starting vectors we use to generate (17) and (18) are related to $M$. In particular, (17) must be generated from $x_1 = M^{-1} q_1$ and (18) must be generated from $y_1 = M^* q_1$, where $q_1$ is a random vector with i.i.d entries.

Another approach is to construct an estimator based on (5), where $C = T(A - \tau I)$. This will require evaluating the bilinear form $v^* h(T(A - \tau I))v$, where $h$ is a function of a matrix $T(A - \tau I)$ that has real spectrum but is non-Hermitian in standard inner product. Similar to the Hermitian case, the matrix functional $v^* h(T(A - \tau I))v$ can be viewed as an integral, such that

$$v^* h(T(A - \tau I))v = \frac{1}{4\pi^2} \int_\Gamma \int_\Gamma h(t) v^* (\bar{\omega} I - (A - \tau I)T)^{-1} (tI - T(A - \tau I))^{-1} v \overline{d\omega} dt, \tag{19}$$

where $\Gamma$ is a contour that encloses the spectrum of $T(A - \tau I)$ and the bar denotes complex conjugation; see, e.g., [9]. This integral can be approximated by a quadrature rule based on a few steps of the Arnoldi process (Algorithm 3) applied to the preconditioned operator $T(A - \tau I)$ with a starting vector $v$ [4, 6].

**Algorithm 3** The Arnoldi procedure for $T(A-\tau I)$

**Input**: Matrix $A-\tau I$, HPD preconditioner $T$, starting vector $v$, and number of steps $k$.
**Output**: Hessenberg matrix $H_{k+1,k}$ and the Arnoldi basis $Q_{k+1}=[q_1,q_2,\ldots,q_{k+1}]$.

1: $q_1 \leftarrow v/\|v\|$; $Q_1 \leftarrow q_1$;
2: **for** $j=1\rightarrow k$ **do:**
3: $\quad w \leftarrow T(A-\tau I)q_j$;
4: $\quad$ **for** $i=1\rightarrow j$ **do:**
5: $\qquad h_{i,j} \leftarrow q_i^* w$; $w \leftarrow w - h_{i,j}q_i$;
6: $\quad$ **end for**
7: $\quad h_{j+1,j} \leftarrow \|w\|$; $q_{j+1} \leftarrow w/h_{j+1,j}$; $Q_{j+1} \leftarrow [Q_j,\ q_{j+1}]$;
8: **end for**

Given $q_1 = v/\|v\|$, Algorithm 3 produces an orthonormal Arnoldi basis $Q_{k+1}$ and an extended upper Hessenberg matrix

$$H_{k+1,k} = \begin{bmatrix} h_{1,1} & h_{1,2} & \cdots & h_{1,k} \\ h_{2,1} & h_{2,2} & \ddots & h_{2,k} \\ & \ddots & \ddots & \vdots \\ & & h_{k,k-1} & h_{k,k} \\ & & & h_{k+1,k} \end{bmatrix} \in \mathbf{R}^{(k+1)\times k}, \tag{20}$$

such that $T(A-\tau I)Q_k = Q_{k+1}H_{k+1,k}$, $Q_{k+1}^* Q_{k+1} = I$. An Arnoldi quadrature rule for the integral (19) is fully determined by the $k$-by-$k$ leading submatrix $H_k$ of (20). Similar to (12), it gives

$$v^* h(T(A-\tau I))v \approx \|v\|^2 e_1^* h(H_k)e_1 \equiv \sum_{i=1}^{k\_} w_i t_i, \quad w_i = \|v\|^2 z_i(1),\ t_i = s_1(i), \tag{21}$$

where $w_i$ are determined by the first components of the (right) eigenvectors $z_1,\ldots,z_{k\_}$ of $H_k$ associated with its $k\_$ eigenvalues that have negative real parts, and $t_i$ is the $i$th entry of the first column of $S=Z^{-1}$. Similar to Proposition 1, it can be shown that if $T(A-\tau I)$ has $p$ distinct eigenvalues, then (21) is exact with at most $p$ nodes.

Let $H_k^{(j)}$ be the upper Hessenberg matrix produced by the Arnoldi process applied to $C=T(A-\tau I)$ with the starting vector $v_j$. Then (21) and (5) yield the estimator

$$A_\tau(k,m) = \frac{1}{m}\sum_{j=1}^{m}\sum_{i=1}^{k_j} w_i^{(j)} t_i^{(j)},\ w_i^{(j)} = \|v_j\|^2 z_i^{(j)}(1),\ t_i^{(j)} = s_1^{(j)}(i),\ v_j \in \mathcal{N}(0,I), \tag{22}$$

**Algorithm 4** The preconditioned Arnoldi-type estimator for $n\_(A - \tau I)$

**Input**: Matrix $A$, shift $\tau$, HPD preconditioner $T$ for $A - \tau I$, number of steps $k$, and parameter $m$.
**Output**: approximate number $A_\tau$ of eigenvalues of $A$ that are less than $\tau$;

1: $A_\tau \leftarrow 0$.
2: **for** $j = 1 \rightarrow m$ **do:**
3: Generate $v \sim \mathcal{N}(0, I)$.
4: Run $k$ steps of Arnoldi process in Algorithm 3 with the starting vector $v$ to obtain upper Hessenberg matrix $H_k$.
5: Find the eigendecomposition $(\Theta, Z)$ of $H_k$. Let $z_1, \ldots, z_{k\_}$ be unit eigenvectors associated with negative eigenvalues.
6: Compute $S = Z^{-1}$. Set $s \leftarrow S(1,:)$. Set $A_\tau \leftarrow A_\tau + \|v\|^2 \mathrm{Re}\left(\sum_{i=1}^{k\_} w_i s_i\right)$, where $w_i = z_i(1)$, $s_i = s(i)$.
7: **end for**
8: Return $A_\tau \leftarrow \lfloor A_\tau / m \rfloor$.

where $z_i^{(j)}(1)$ denotes the first component of the $k_j$ unit eigenvectors $z_i^{(j)}$ of $H_k^{(j)}$. and $s_i^{(j)}$ is the $i$th entries of the first column of the inverted matrix of eigenvectors of $H_k^{(j)}$. Similar to (13), we expect that, for a sufficiently large $m$, the real part of $A_\tau(k, m)$ approximates $n\_(A - \tau I)$. The computation of Re $(A_\tau(k, m))$ is described in Algorithm 4.

The cost of Algorithm 4 is comparable to that of Algorithm 2, and is slightly higher mainly due to the need to invert the eigenvector matrix of $H_k$. In contrast to Algorithm 2, the above described scheme assumes complex arithmetic, because the upper Hessenberg matrix $H_k$ is non-Hermitian and can have complex eigenpairs. However, for good choices of $T$, the imaginary parts tend to be small in practice as, for a sufficiently large $k$, the eigenpairs of $H_k$ converge rapidly to those of $T(A - \tau I)$, which are real. Finally, note that the derivation of the estimator (22) assumes an extension of the definition of the step function (1), such that $h(x)$ has the value of one on the left half of the complex plane, and is zero elsewhere.

## 5 Polynomial Viewpoint

Let $C = M(A - \tau I)M^*$ or $C = T(A - \tau I)$. Then, we can replace $h(C)$ in (5) by a polynomial approximation $p_l(C)$ of degree $l$. There are several ways to choose this polynomial. One option is to take $p_l(C)$ as formal truncated expansion of $h(x)$ in the basis of Chebyshev polynomials. This choice is related the approach described in [14].

The quality of a polynomial approximation $p_l(x)$ of $h(x)$ can be measured by the difference between $p_l(x)$ and $h(x)$ on the set of eigenvalues of $C$. When the spectrum of $C$ has an arbitrary distribution, constructing a polynomial that provides the best least squares fit on the entire interval containing all eigenvalues, as is done in [14], is well justified.

When a good preconditioner is used, the spectrum of $C$ tends to cluster around several points on the real line. Thus, a natural approach would be to choose $p_l$ such that it is only close to $h$ in regions that contain eigenvalue clusters. It can be quite different from $h$ elsewhere. An example of such an approach is an interpolating polynomial, e.g., [16], that interpolates $h$ at eigenvalue clusters. A practical construction of such a polynomial is given by the following theorem, which relates the the interpolation procedure to the Lanczos or Arnoldi process.

**Theorem 1 (See [8, 18])** *Let $Q_k$, $T_k$ be the orthonormal basis and the projection of the matrix $C$ generated from a k-step Lanczos (Arnoldi) process, with the starting vector $v$. Then*

$$\|v\|Q_k f(T_k)e_1 = p_{k-1,v}(C)v, \tag{23}$$

*where $p_{k-1,v}$ is the unique polynomial of degree at most $k-1$ that interpolates $f$ in the Hermite sense on the spectrum of $T_k$.*

The subscript "$v$" in $p_{k-1,v}$ is used to emphasize the dependence of the polynomial on the staring vector $v$. Note that $T_k$ is a symmetric tridiagonal matrix if $C$ is Hermitian. It is upper Hessenberg otherwise.

Using formula (23), it is easy to verify that if $C = M(A - \tau I)M^*$, then the bilinear form $v^* p_{k-1,v}(C)v$ is exactly the same as the Gauss quadrature rule on the right-hand side of (12). Similarly, if $C = T(A - \tau I)$, then $v^* p_{k-1,v}(C)v$ is given by the Arnoldi quadrature on the right-hand side of (21). Hence, both estimators (13) and (22) can be viewed as a stochastic approximation of trace $\{p_{k-1,v}(C)\}$, where $p_{k-1,v}(x)$ is an interpolating polynomial of degree $k-1$ for the step function $h$.

## 6 Preconditioning

The iterative scheme we presented earlier rely on the assumption that the operator $T$ is HPD, as this property guarantees that the inertia of the original matrix $A - \tau I$ is preserved after preconditioning. Furthermore, a good choice of $T$ should cluster spectrum of the preconditioned matrix $C$ around several points in the real axis.

An ideal HPD preconditioner will result in the preconditioned matrix with only two distinct eigenvalues. In this case, by Proposition 1, the Lanczos procedure should terminate in two steps. An example of such an ideal preconditioner is the matrix $T = |A - \tau I|^{-1}$, where the absolute value is understood in the matrix function sense.

Clearly, the choice $T = |A - \tau I|^{-1}$ is prohibitively costly in practice. However, it is possible to construct HPD preconditioners that only approximate $|A - \tau I|^{-1}$. Such a preconditioning strategy was proposed in [20] and is referred to as the absolute value (AV) preconditioning. It was shown in [20] that, e.g., for discrete Laplacian operators, AV preconditioners can be efficiently constructed using multigrid (MG).

Another possible option is to employ the incomplete $LDL^T$ (ILDL) factorization. Given a matrix $A - \tau I$ and a drop tolerance $t$, an ILDL($t$) preconditioner is of the form $T = L^{-*}D^{-1}L^{-1}$, where $L$ is lower triangular and $D$ is block-diagonal with diagonal blocks of size 1 and 2, such that $T \approx (A - \tau I)^{-1}$.

Clearly, since $A - \tau I$ is indefinite, the ILDL($t$) procedure will generally result in an indefinite $T$, which cannot be applied within the preconditioned estimators of this paper. Therefore, we suggest to modify it by taking the absolute value of diagonal blocks of $D$, so that $T = L^{-*}|D|^{-1}L^{-1}$. Such a preconditioner is HPD, and the cost of the proposed modification is marginal. This idea has been motivated by [5], where a similar approach was used in the context of full (complete) $LDL^T$ factorization.

Finally, in certain applications, HPD operators are readily available and traditionally used for preconditioning indefinite matrices. For example, this is the case in Density Functional Theory (DFT) based electronic structure calculations in which the solutions are expressed in terms of a linear combination of planewaves. A widely used preconditioner, often referred to as the Teter preconditioner [19], is diagonal in the planewave basis.

## 7 Numerical Experiments

We now study the numerical behavior of the proposed methods for three test problems listed in Table 1. The matrix "Laplace" represents a standard five-point finite differences (FD) discretization of the 2D Laplacian on a unit square with mesh size $h = 2^{-7}$. The problems "Benzene" and "H2" originates from the DFT based electronic structure calculations. The former is a FD discretization of a

**Table 1** Estimates of $n\_(A - \tau I)$ produced by Algorithm 2 with different preconditioners and the corresponding numbers $k$ of Lanczos iterations for three test problems

| Problem | $n$ | $\tau$ | $n\_(A - \tau I)$ | Preconditioner | Estimated $n\_(A - \tau I)$ | $k$ |
|---|---|---|---|---|---|---|
| Laplace | 16,129 | 3000 | 226 | No prec. | 232 | 134 |
| | | | | ILDL(1e-3) | 216 | 34 |
| | | | | ILDL(1e-5) | 229 | 6 |
| Benzene | 8219 | 5 | 344 | No prec. | 338 | 85 |
| | | | | ILDL(1e-5) | 350 | 18 |
| | | | | ILDL(1e-6) | 341 | 2 |
| H2 | 11,019 | 0.5 | 19 | No prec. | 20 | 50 |
| | | | | Teter | 20 | 11 |

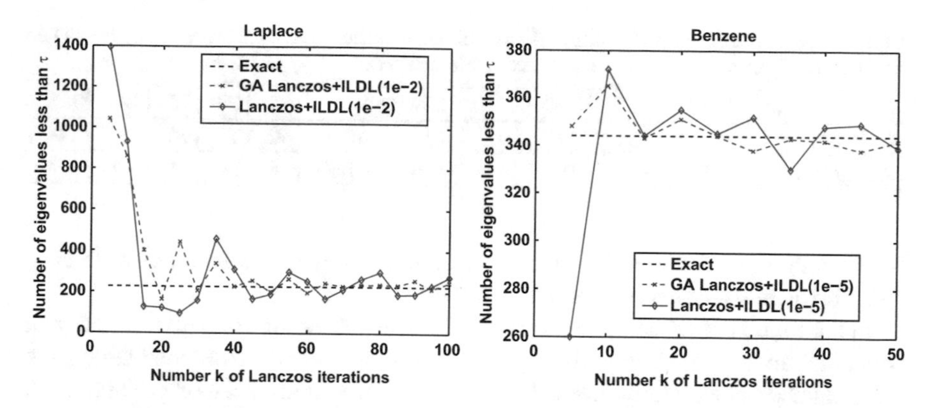


**Fig. 1** Effects of the GA Gauss quadrature of Sect. 3.4 on the accuracy of the estimator

Hamiltonian operator associated with a ground state benzene molecule,[2] whereas the latter corresponds to a Hamiltonian associated with the hydrogen molecule generated by the KSSOLV package [21]. Throughout, our goal is to estimate the quantity $n\_(A - \tau I)$ for a given value of the shift $\tau$.

Table 1 presents the results of applying the Lanczos-type estimator given in Algorithm 2 to the test problems with different preconditioner choices. For the "Laplace" and "Benzene" matrices, we use the positive definite ILDL($t$) based preconditioning with different drop tolerance $t$, discussed in the previous section. The ILDL factorizations of $A - \tau I$ are obtained using the `sym-ildl` package [7]. In the "H2" test, we employ the diagonal Teter preconditioner available in KSSOLV [19]. In both cases, the preconditioner is accessible in the factorized form $T = M^*M$. The number of random samples $m$ is set to 50 in all tests.

In the table, we report estimates of $n\_(A - \tau I)$ produced by Algorithm 2 along with the corresponding numbers of Lanczos iterations ($k$) performed at each sampling step. The reported values of $k$ correspond to the smallest numbers of Lanczos iterations that result in a sufficiently accurate estimate. The error associated with these approximations have been observed to be within 5%.

Table 1 demonstrates that the use of preconditioning significantly reduces the number of Lanczos iterations. Furthermore, $k$ becomes smaller as the quality of the preconditioner, which is controlled by the drop tolerance $t$ in the ILDL($t$) based preconditioners, improves for the "Laplace" and "Benzene" tests.

Figure 1 shows that the quality of the estimates can be further improved by using the GA Gauss quadrature rules discussed in Sect. 3.4. In both plots, the horizontal axis corresponds to the number of Lanczos iterations ($k$) per sampling step, and the vertical axis is the corresponding estimate of $n\_(A - \tau I)$. It can be seen that the estimator based on the GA Gauss quadrature (referred to as "GA Lanczos")

[2]Available in the PARSEC group of the University of Florida Sparse Matrix Collection at https://www.cise.ufl.edu/research/sparse/matrices/.

**Table 2** Independence of preconditioned Arnoldi- and Lanczos-type estimators for $n\_(A - \tau I)$ on the discretization parameter for the "Laplace" (left) and "H2" (right) problems

| $h$ | $2^{-6}$ | $2^{-7}$ | $2^{-8}$ | $2^{-9}$ | $2^{-10}$ |
|---|---|---|---|---|---|
| Chebyshev | 8 | 14 | 34 | 62 | 80 |
| Arnoldi+AV | 16 | 16 | 18 | 19 | 16 |

| ecut (Ry) | 25 | 50 | 75 | 100 | 125 |
|---|---|---|---|---|---|
| Chebyshev | 52 | 78 | 76 | 99 | 124 |
| Lanczos+Teter | 8 | 8 | 11 | 8 | 8 |

is generally more accurate for the two test problems, with the accuracy difference being especially evident for smaller values of $k$.

In the context of linear systems arising from discretizations of partial differential equations, an important property of preconditioning is that it allows maintaining the same number of iterations needed to obtain solution regardless of problem size. A similar phenomenon can be observed when estimating $n\_(A - \tau I)$ using the preconditioned methods of this paper.

In Table 2 (left) we consider a family of discrete Laplacians, whose size and condition numbers increase as the mesh parameter $h$ is refined. For each of the matrices, we apply the Arnoldi-type estimator of Algorithm 4 with the MG AV preconditioner from [20] and, similar to above, report the smallest numbers $k$ of Arnoldi iterations per sampling step needed to obtain a sufficiently accurate estimate (within 5% error) of $n\_(A - \tau I)$. The results are compared against those of an unpreconditioned estimator based on (5), where $C = A$ and the step function $h(A)$ is replaced by its least-squares polynomial approximation of degree $k$ constructed using the basis of Chebyshev polynomials. The latter (referred to as "Chebyshev") is essentially the approach proposed in [14].

It can be seen from the table, that Algorithm 4 with the AV preconditioner exhibits behavior that is independent of $h$. Regardless of the problem size and conditioning, the number of Arnoldi steps stays (roughly) the same (between 16 and 19).

In Table 2 (right) we report a similar test for a sequence of "H2" problems obtained by increasing the kinetic energy cutoff (ecut) from 25 to 125 Ry in the plane wave discretization. This gives Hamiltonian matrices with sizes ranging from 1024 to 23,583. Again, we observe that the behavior of the Lanczos-type estimator in Algorithm 2 with the Teter preconditioner [19] is essentially independent of the discretization parameter, whereas the "Chebyshev" approach tends to require higher polynomial degrees as the problem size grows.

**Acknowledgements** Support for this work was provided through Scientific Discovery through Advanced Computing (SciDAC) program funded by U.S. Department of Energy, Office of Science, Advanced Scientific Computing Research.

## References

1. Aktulga, H.M., Lin, L., Haine, C., Ng, E.G., Yang, C.: Parallel eigenvalue calculation based on multiple shift-invert Lanczos and contour integral based spectral projection method. Parallel Comput. **40**(7), 195–212 (2014)
2. Avron, H., Toledo, S.: Randomized algorithms for estimating the trace of an implicit symmetric positive semi-definite matrix. J. ACM **58**(2), 8:1–8:34 (2011). doi:10.1145/1944345.1944349
3. Calvetti, D., Golub, G., Reichel, L.: A computable error bound for matrix functionals. J. Comput. Appl. Math. **103**(2), 301–306 (1999)
4. Calvetti, D., Kim, S.M., Reichel, L.: Quadrature rules based on the Arnoldi process. SIAM J. Matrix Anal. Appl. **26**(3), 765–781 (2005)
5. Gill, P.E., Murray, W., Ponceleón, D.B., Saunders, M.A.: Preconditioners for indefinite systems arising in optimization. SIAM J. Matrix Anal. Appl. **13**(1), 292–311 (1992). doi:10.1137/0613022
6. Golub, G.H., Meurant, G.: Matrices, Moments and Quadrature with Applications. Princeton University Press, Princeton (2010)
7. Greif, C., He, S., Liu, P.: SYM-ILDL: incomplete ldlt factorization of symmetric indefinite and skew-symmetric matrices. CoRR **abs/1505.07589** (2015). http://arxiv.org/abs/1505.07589
8. Higham, N.J.: Functions of Matrices: Theory and Computation. Society for Industrial and Applied Mathematics (SIAM), Philadelphia, PA (2008)
9. Hochbruck, M., Lubich, C.: On Krylov subspace approximations to the matrix exponential operator. SIAM J. Sci. Comput. **34**(5), 1911–1925 (1997)
10. Hutchinson, M.F.: A stochastic estimator of the trace of the influence matrix for laplacian smoothing splines. Commun. Stat. Simul. Comput. **18**, 1059–1076 (1989)
11. Li, R., Xi, Y., Vecharynski, E., Yang, C., Saad, Y.: A thick-restart Lanczos algorithm with polynomial filtering for hermitian eigenvalue problems. Tech. rep. (2015). Http://arxiv.org/abs/1512.08135
12. Lin, L., Saad, Y., Yang, C.: Approximating spectral densities of large matrices. SIAM Rev. **58**, 34–654 (2016)
13. Maeda, Y., Futamura, Y., Imakura, A., Sakurai, T.: Filter analysis for the stochastic estimation of eigenvalue counts. JSIAM Lett. **7**, 53–56 (2015)
14. Napoli, E.D., Polizzi, E., Saad, Y.: Efficient estimation of eigenvalue counts in an interval. Tech. rep. (2015). Http://arxiv.org/abs/1308.4275
15. Parlett, B.N.: The Symmetric Eigenvalue Problem. Classics in Applied Mathematics, vol. 20. Society for Industrial and Applied Mathematics (SIAM), Philadelphia, PA (1998). Corrected reprint of the 1980 original
16. Powell, M.J.D.: Approximation Theory and Methods. Cambridge University Press, Cambridge (1981)
17. Reichel, L., Spalević, M.M., Tang, T.: Generalized averaged Gauss quadrature rules for the approximation of matrix functionals. BIT Numer. Math. **56**, 1045–1067 (2015)
18. Saad, Y.: Analysis of some Krylov subspace approximations to the matrix exponential operator. SIAM J. Numer. Anal. **29**(1), 209–228 (1992)
19. Teter, M.P., Payne, M.C., Allan, D.C.: Solution of Schrödinger's equation for large systems. Phys. Rev. B **40**(18), 12255–12263 (1989)
20. Vecharynski, E., Knyazev, A.V.: Absolute value preconditioning for symmetric indefinite linear systems. SIAM J. Sci. Comput. **35**(2), A696–A718 (2013)
21. Yang, C., Meza, J., Lee, B., Wang, L.W.: KSSOLV—a MATLAB toolbox for solving the Kohn-Sham equations. ACM Trans. Math. Softw. **36**(2), 10:1–10:35 (2009)
22. Zhang, Y., Wainwright, M.J., Jordan, M.I.: Distributed estimation of generalized matrix rank: efficient algorithms and lower bounds. Tech. rep. (2015). Http://arxiv.org/abs/1502.01403

# Comparison of Tridiagonalization Methods Using High-Precision Arithmetic with MuPAT

**Ryoya Ino, Kohei Asami, Emiko Ishiwata, and Hidehiko Hasegawa**

**Abstract** In general, when computing the eigenvalues of symmetric matrices, a matrix is tridiagonalized using some orthogonal transformation. The Householder transformation, which is a tridiagonalization method, is accurate and stable for dense matrices, but is not applicable to sparse matrices because of the required memory space. The Lanczos and Arnoldi methods are also used for tridiagonalization and are applicable to sparse matrices, but these methods are sensitive to computational errors. In order to obtain a stable algorithm, it is necessary to apply numerous techniques to the original algorithm, or to simply use accurate arithmetic in the original algorithm. In floating-point arithmetic, computation errors are unavoidable, but can be reduced by using high-precision arithmetic, such as double-double (DD) arithmetic or quad-double (QD) arithmetic. In the present study, we compare double, double-double, and quad-double arithmetic for three tridiagonalization methods; the Householder method, the Lanczos method, and the Arnoldi method. To evaluate the robustness of these methods, we applied them to dense matrices that are appropriate for the Householder method. It was found that using high-precision arithmetic, the Arnoldi method can produce good tridiagonal matrices for some problems whereas the Lanczos method cannot.

## 1 Introduction

Recently, eigenvalue computation has become very important in several applications. For a real symmetric dense matrix, the target matrix is usually reduced to symmetric tridiagonal form by orthogonal similarity transformations, and the

R. Ino • K. Asami • E. Ishiwata (✉)
Department of Mathematical Information Science, Tokyo University of Science, 1-3 Kagurazaka, Shinjuku-ku, Tokyo, Japan
e-mail: ishiwata@rs.kagu.tus.ac.jp

H. Hasegawa
Faculty of Library, Information and Media Science, University of Tsukuba, Kasuga 1-2, Tsukuba, Japan
e-mail: hasegawa@slis.tsukuba.ac.jp



T. Sakurai et al. (eds.), *Eigenvalue Problems: Algorithms, Software and Applications in Petascale Computing*, Lecture Notes in Computational Science and Engineering 117, https://doi.org/10.1007/978-3-319-62426-6_9

eigenvalues of the obtained symmetric tridiagonal matrix are then computed by, for example, the QR method or bisection and inverse iteration algorithms. On the other hand, for sparse matrices other than band matrices, tridiagonalization by the Householder transformation is so difficult because of requiring a great deal of memory. The Lanczos method involves simple matrix-vector multiplication and vector operations, and does not require modification of the given matrix. The Lanczos and Arnoldi methods are simple algorithms, but the roundoff error causes the Lanczos vectors to lose orthogonality [1]. However, they may require less memory.

Mathematically simple algorithms are often unstable because of computation errors. In order to obtain a stable algorithm, we can apply several techniques to the original algorithm, or simply use accurate arithmetic. In floating-point arithmetic, computation errors are unavoidable, but can be reduced through the use of high-precision arithmetic, such as double-double (DD) arithmetic or quad-double (QD) arithmetic.

Kikkawa et al. and Saito et al. [2, 3] developed the Multiple Precision Arithmetic Toolbox (MuPAT), a high-precision arithmetic software package, on Scilab (http://www.scilab.org/). The MuPAT uses double-double arithmetic and quad-double arithmetic in order to work on conventional computers. The computation time for double-double-precision arithmetic is approximately 20 times greater than that for ordinary double-precision arithmetic, but this cost can be reduced through the use of parallel processing.

In the present paper, we compare double, double-double, and quad-double arithmetic for the Lanczos method, the Arnoldi method, and the Householder method [1] for obtaining symmetric tridiagonal matrices from symmetric matrices, and the QR method for finding all eigenvalues thereof. We use a sparse storage format of MuPAT in order to reduce the memory requirement, but did not use parallel processing.

## 2 Multiple-Precision Arithmetic on MuPAT

### *2.1 Double-Double and Quad-Double Arithmetic*

Double-double and quad-double arithmetic were proposed as quasi-quadruple-precision and quasi-octuple-precision arithmetic by Hida et al. [4] and Dekker [5]. A double-double number is represented by two double-precision numbers, and a quad-double number is represented by four double-precision numbers. A double number $x_{(D)}$, a double-double number $x_{(DD)}$ and a quad-double number $x_{(QD)}$ are represented by an unevaluated sum of double-precision numbers $x_0, x_1, x_2, x_3$ as follows:

$$x_{(D)} = x_0, \quad x_{(DD)} = x_0 + x_1, \quad x_{(QD)} = x_0 + x_1 + x_2 + x_3,$$

**Table 1** Number of double-precision arithmetic operations

| Type | | Add & sub | Mul | div | Total |
|---|---|---|---|---|---|
| DD | Add & sub | 11 | 0 | 0 | 11 |
| | Mul | 15 | 9 | 0 | 24 |
| | Div | 17 | 8 | 2 | 27 |
| QD | Add & sub | 91 | 0 | 0 | 91 |
| | Mul | 171 | 46 | 0 | 217 |
| | Div | 579 | 66 | 0 | 649 |

where $x_0, x_1, x_2$ and $x_3$ satisfy the following inequalities:

$$|x_{i+1}| \leq \frac{1}{2}\text{ulp}(x_i), \quad i = 0, 1, 2,$$

where ulp stands for 'units in the last place'. For a given decimal input data $x$, we can also denote that

$$x_{(D)} = (x_0)_{(D)}, \quad x_{(DD)} = (x_0, x_1)_{(DD)}, \quad x_{(QD)} = (x_0, x_1, x_2, x_3)_{(QD)}.$$

The lower portion is ignored or truncated from the longer format data to the shorter format data, and is assumed to be zeros from the shorter format data to the longer format data. A double-double (quad-double) number has 31 (63) significant decimal digits.

In this paper, we abbreviate double-double and quad-double on DD and QD. Both DD and QD arithmetic are performed using error-free floating point arithmetic algorithms that use only double-precision arithmetic and so require only double-precision arithmetic operations. Both DD and QD arithmetic are described in detail in [4] and [5]. Table 1 shows the number of double-precision arithmetic operations for DD and QD arithmetic.

### 2.2 *Extended MuPAT with a Sparse Data Structure*

A quadruple- and octuple-precision arithmetic toolbox, i.e., the Multiple Precision Arithmetic Toolbox (MuPAT) and variants thereof [2, 3], allow the use of double-, quadruple-, and octuple-precision arithmetic with the same operators or functions, and mixed-precision arithmetic and partial use of different precision arithmetic becomes possible. The MuPAT is independent of hardware and operating system.

We developed an accelerated MuPAT for sparse matrices in [3] in order to reduce the amount of memory and computation time, and using the developed MuPAT, large matrices can easily be handled. We define two data types for a sparse matrix: `DDSP` for double-double numbers and `QDSP` for quad-double numbers.

These data types are based on the compressed column storage (CCS) format, which contains vectors in the form of row indices, column pointers, and values. Note that DDSP uses two value vectors and QDSP uses four value vectors to represent double-double and quad-double numbers, respectively. As such, it is possible to use a combination of double, double-double, and quad-double arithmetic for both dense and sparse data structures. Based on the definitions of these data types, MuPAT has six data types: `constant`, DD, and QD for dense data, and `sparse`, DDSP, and QDSP for sparse data of double, double-double and quad-double numbers, respectively.

Quad-double arithmetic requires a tremendous number of double-precision operations. In particular, one QD division requires 649 double-precision operations, so the required computation time is hundreds of times greater than that for double-precision arithmetic on Scilab. In order to accelerate QD and DD arithmetic operations, external routines written in the C language are prepared. These MuPAT functions achieve high-speed processing but depend on the hardware and operating system used. Currently, this code is not parallelized but can be accelerated through the use of parallel processing.

# 3 Eigenvalue Computation

In order to compute the eigenvalues of a real symmetric matrix $A$, the matrix $A$ is usually tridiagonalized to an similarity tridiagonal matrix $T$ by similarity transformations, and the eigenvalues of the matrix $T$ are then computed. The Lanczos, Arnoldi, and Householder methods can be used for this purpose.

The Lanczos and Arnoldi methods involve matrix-vector multiplication and some vector operations. Since, unlike in the Householder method, updating the original matrix $A$ is not necessary, the Lanczos and Arnoldi methods can be easily applied to sparse matrices.

The QR method and the bisection algorithm are used for computing the eigenvalues of a tridiagonal matrix $T$. For computing the eigenvectors of $T$, the QR method and inverse iteration are used. The quality of eigenvalues and eigenvectors depends only on the tridiagonal matrix $T$ and not on the tridiagonalization method. If $T$ is an inexact approximation of $A$, even if the eigenvalues and eigenvectors of $T$ are correctly calculated, they do not correspond to those of $A$.

In the present paper, we used the implicit single shift QR algorithm based on [6] for computing all eigenvalues. The QR method generates eigenvalues as diagonal elements in descending order.

In particular, for a sparse matrix, the transformations used for tridiagonalizing $A$ to $T$ are not used to compute eigenvectors of $A$, which would require tremendous computation and memory. If eigenvalues $\lambda_T$ of $T$ are accurately computed, an inverse iteration method can be applied to compute the eigenvectors of $(T - \lambda_T I)$ or $(A - \lambda_T I)$. The inverse iteration method for sparse matrices uses a direct solver or an iterative solver, such as the conjugate gradient method.

## 3.1 Tridiagonalization

For a given symmetric matrix $A$, it is possible to find an orthogonal $Q$ such that $Q^TAQ = T$ is tridiagonal. In the present paper, we consider three tridiagonalization methods for symmetric matrices: the Lanczos method, the Arnoldi method, and the Householder method. These methods are described in detail in [1].

### 3.1.1 The Lanczos Method

The Lanczos method can construct an equivalent tridiagonal matrix by generating orthogonal bases one after another. However, roundoff errors cause the Lanczos vectors to lose orthogonality [1].

Let $A$ be an $n \times n$ symmetric matrix, and let $Q$ be an $n \times n$ orthogonal matrix. Then, we generate $T = Q^TAQ$. We set the column of $Q$ by

$$Q = [\boldsymbol{q}_1|\boldsymbol{q}_2|\cdots|\boldsymbol{q}_n]$$

and the components of $T$ by

$$T = \begin{bmatrix} \alpha_1 & \beta_1 & & \cdots & 0 \\ \beta_1 & \alpha_2 & \beta_2 & & \vdots \\ & \ddots & \ddots & \ddots & \\ \vdots & & \ddots & \ddots & \beta_{n-1} \\ 0 & \cdots & & \beta_{n-1} & \alpha_n \end{bmatrix}.$$

Equating columns as $AQ = QT$, we conclude that

$$A\boldsymbol{q}_k = \beta_{k-1}\boldsymbol{q}_{k-1} + \alpha_k\boldsymbol{q}_k + \beta_k\boldsymbol{q}_{k+1} \qquad (\beta_0\boldsymbol{q}_0 \equiv \boldsymbol{0}),$$

for $k = 1, 2, \ldots, n-1$. The orthonormality of the vector $\boldsymbol{q}_k$ implies

$$\alpha_k = \boldsymbol{q}_k^T A\boldsymbol{q}_k.$$

If we define the vector $\boldsymbol{r}_k$ as

$$\boldsymbol{r}_k = (A - \alpha_k I)\boldsymbol{q}_k - \beta_{k-1}\boldsymbol{q}_{k-1},$$

and if it is nonzero, then

$$\boldsymbol{q}_{k+1} = \frac{\boldsymbol{r}_k}{\beta_k},$$

where $\beta_k = \pm\|\boldsymbol{r}_k\|_2$.

For a given symmetric matrix $A \in R^{n\times n}$ and an initial vector $\boldsymbol{q}_0 \in R^n$, Algorithm 1 computes a matrix $Q = [\boldsymbol{q}_1, \cdots, \boldsymbol{q}_n]$ with orthonormal columns and a tridiagonal matrix $T \in R^{n\times n}$ so that $AQ = QT$. The diagonal and superdiagonal entries of $T$ are $\alpha_1, \cdots, \alpha_n$ and $\beta_1, \cdots, \beta_{n-1}$, respectively.

**Algorithm 1** The Lanczos method [1]

1: $k = 0, \boldsymbol{r}_0 = \boldsymbol{q}_0, \beta_0 = \|\boldsymbol{q}_0\|_2$
2: **while** $\beta_k \neq 0$ **do:**
3: $\quad \boldsymbol{q}_{k+1} = \frac{\boldsymbol{r}_k}{\beta_k}$
4: $\quad k = k + 1$
5: $\quad \alpha_k = \boldsymbol{q}_k^T A \boldsymbol{q}_k$
6: $\quad \boldsymbol{r}_k = (A - \alpha_k I)\boldsymbol{q}_k - \beta_{k-1}\boldsymbol{q}_{k-1}$
7: $\quad \beta_k = \|\boldsymbol{r}_k\|_2$
8: **end while**

### 3.1.2 The Arnoldi Method

The Arnoldi method is a way to extend the Lanczos method to non-symmetric matrices and generate the Hessenberg matrix $Q^T AQ = H$. However, for a symmetric matrix $A$, this process produces a tridiagonal matrix $T = H$.

In the same manner as the Lanczos iteration, we set $Q = [\boldsymbol{q}_1, \boldsymbol{q}_2, \cdots, \boldsymbol{q}_n]$ and compare columns in $AQ = QH$. Then,

$$A\boldsymbol{q}_k = \sum_{i=1}^{k+1} h_{ik}\boldsymbol{q}_i, \qquad 1 \leq k \leq n - 1.$$

Isolating the last term in the summation gives

$$\boldsymbol{r}_k \equiv A\boldsymbol{q}_k - \sum_{i=1}^{k} h_{ik}\boldsymbol{q}_i,$$

where $h_{ik} = \boldsymbol{q}_i^T A \boldsymbol{q}_k$ for $i = 1, 2, \ldots, k$. It follows that if $\boldsymbol{r}_k \neq \boldsymbol{0}$, then $\boldsymbol{q}_{k+1}$ is specified by

$$\boldsymbol{q}_{k+1} = \frac{\boldsymbol{r}_k}{h_{k+1,k}},$$

where $h_{k+1,k} = \|\boldsymbol{r}_k\|_2$. These equations define the Arnoldi method.

For a given matrix $A \in R^{n \times n}$ and an initial vector $\boldsymbol{q}_0 \in R^n$, Algorithm 2 computes a matrix $Q = [\boldsymbol{q}_1, \cdots, \boldsymbol{q}_n] \in R^{n \times n}$ with orthonormal columns and an upper Hessenberg matrix $H \in R^{n \times n}$ so that $AQ = QH$. Especially for a symmetric matrix, this algorithm generates an orthogonal matrix $Q$ and a tridiagonal matrix $T$.

### 3.1.3 The Householder Method

The Householder method for a symmetric matrix can generate an tridiagonal matrix $Q^T AQ = T$ using the Householder matrix [1]. Suppose that the Householder matrices $P_1, \cdots, P_{k-1}$ have been determined such that if

$$A_{k-1} = (P_1 \cdots P_{k-1})^T A (P_1 \cdots P_{k-1}),$$

**Algorithm 2** The Arnoldi method [1]

1: $k = 0, \boldsymbol{r}_0 = \boldsymbol{q}_0, h_{1,0} = \|\boldsymbol{q}_0\|_2$
2: **while** $h_{k+1,k} \neq 0$ **do:**
3: $\quad \boldsymbol{q}_{k+1} = \frac{\boldsymbol{r}_k}{h_{k+1,k}}$
4: $\quad k = k + 1$
5: $\quad \boldsymbol{r}_k = A\boldsymbol{q}_k$
6: $\quad$ **for** $i = 1, 2, \cdots, k$ **do:**
7: $\quad\quad h_{ik} = \boldsymbol{q}_i^T \boldsymbol{r}_k$
8: $\quad\quad \boldsymbol{r}_k = \boldsymbol{r}_k - h_{ik}\boldsymbol{q}_i$
9: $\quad$ **end for**
10: $\quad h_{k+1,k} = \|\boldsymbol{r}_k\|_2$
11: **end while**

then

$$A_{k-1} = \begin{bmatrix} B_{11} & B_{12} & 0 \\ B_{21} & B_{22} & B_{23} \\ 0 & B_{32} & B_{22} \end{bmatrix}$$

is tridiagonal through its first $k-1$ columns. If $\tilde{P}_k$ is an order-$(n-k)$ Householder matrix such that $\tilde{P}_k B_{32}$ is a multiple of $I_{n-1}$ and if $P_k = diag(I_k, \tilde{P}_k)$, then the leading $k$-by-$k$ principal submatrix of

$$A_k = P_k A_{k-1} P_k = \begin{bmatrix} B_{11} & B_{12} & 0 \\ B_{21} & B_{22} & B_{23}\tilde{P}_k \\ 0 & \tilde{P}_k B_{32} & \tilde{P}_k B_{33} \tilde{P}_k \end{bmatrix}$$

is tridiagonal. Clearly, if $U = P_1 \cdots P_{n-2}$, then $U^T A U = T$ is tridiagonal. In the calculation of $A_k$, it is important to exploit symmetry during the formation of the matrix $\tilde{P}_k B_{33} \tilde{P}_k$. More specifically, suppose that $\tilde{P}_k$ has the form

$$\tilde{P}_k = I - \beta \boldsymbol{v}\boldsymbol{v}^T, \qquad \beta = \frac{2}{\boldsymbol{v}^T\boldsymbol{v}}, \qquad 0 \neq \boldsymbol{v} \in R^{n-k}.$$

Note that if $\boldsymbol{p} = \beta B_{33}\boldsymbol{v}$ and $\boldsymbol{w} = \boldsymbol{p} - (\frac{\beta \boldsymbol{p}^T \boldsymbol{v}}{2})\boldsymbol{v}$, then

$$\tilde{P}_k B_{33} \tilde{P}_k = B_{33} - \boldsymbol{v}\boldsymbol{w}^T - \boldsymbol{w}\boldsymbol{v}^T.$$

We used the Householder algorithm written in [1].

Since only the upper triangular portion of this matrix needs to be calculated, we see that the transition from $A_{k-1}$ to $A_k$ can be accomplished in only $4(n-k)^2$ flops for a dense matrix.

## 4 Numerical Experiments

In this section, we analyze the accuracy, numerical stability, and computing cost for three tridiagonalization methods and the computed eigenvalue by the implicit single shift QR algorithm [6] for the tridiagonal matrix $T$. For tridiagonalization, we compare three arithmetic precisions: double (D), DD, and QD.

The QR method can be applied to non-symmetric matrices (not only tridiagonal matrices), in which case complex eigenvalues would appear. Therefore, we use only tridiagonal factors in the Arnoldi method.

For the Lanczos and Arnoldi methods, the initial vector $\boldsymbol{q}_0$ is a uniformly distributed random vector between 0 and 1 using the 'rand' function of Scilab.

All experiments were carried out on an Intel Core i5-4200U, 1.60 GHz, 8 GB memory and Scilab 5.5.0 on Windows 7 Professional. We assumed the 'true eigenvalue' to be the computation result produced by the 'Eigenvalues' function of Mathematica with 200 decimal digits.

## 4.1 Example 1: Nos4 (Small Problem)

We demonstrate the results of the three tridiagonalization methods for a small matrix 'nos4' in MatrixMarket (http://math.nist.gov/MatrixMarket/). The dimension of this matrix was 100, the number of the nonzero elements was 594, the condition number on the matrix was $2.7 \times 10^3$, and the matrix originated from a structure problem. The eigenvalues of nos4 are distributed between 0.00053795... and 0.84913778... without any clustered eigenvalues.

Table 2 lists the accuracy of the eigenvalues, the loss of orthogonality, and the computation times for the three tridiagonalization methods and three precisions. Here, $max|\lambda_i - \bar{\lambda}_i|$ and $avg|\lambda_i - \bar{\lambda}_i|$ denote the maximum absolute error and the average of absolute errors, where $\lambda_i$ and $\bar{\lambda}_i$ represent the $i$th computed eigenvalue and the true eigenvalue, respectively. We checked the loss of orthogonality by $\frac{\|Q^TQ-I\|_F}{\|I\|_F}$, where $I$ and $Q$ are a unit matrix and an orthogonal matrix, respectively, and $\|\cdot\|_F$ denotes the Frobenius norm. '*Avg* time' for dense and sparse implies the computation time only for tridiagonalization part.

For the QR algorithm, the accuracy of the eigenvalues in D, DD, and QD are approximately the same for all tridiagonalization methods. This means that the accuracy of the QR method with double-precision arithmetic is sufficient, and the accuracy of tridiagonalization is important in eigenvalue computation. Therefore, we hereinafter apply the QR method with only double-precision arithmetic and focus on the difference in accuracy and computation time among the tridiagonalization methods and their arithmetic precisions.

Concerning the tridiagonalization methods, there is little difference between the maximum and average errors for Lanczos-QD, Arnoldi-DD, -QD, and Householder-D, -DD, -QD (where, for example, Lanczos-QD indicates the Lanczos method with QD precision).

The orthogonalities of Lanczos-QD, Arnoldi-DD, and Householder-D are approximately the same and can be improved by using DD and QD. The relationship between method and accuracy depends on the given matrix. In the case of nos4, however, Householder-D, Arnoldi-DD and Lanczos-QD are sufficient.

Table 2 Accuracy and tridiagonalization time [s] for nos4

| Tridiagonalization | | QR | $max\|\lambda_i - \bar{\lambda}_i\|$ | $avg\|\lambda_i - \bar{\lambda}_i\|$ | $avg\ \frac{\\|Q^TQ-I\\|_F}{\\|I\\|_F}$ | $avg$ time dense | $avg$ time sparse |
|---|---|---|---|---|---|---|---|
| Lanczos | D | D | $1.43 \times 10^{-1}$ | $4.56 \times 10^{-2}$ | $4.21 \times 10^{-1}$ | 0.016 | 0.006 |
| | | DD | $1.43 \times 10^{-1}$ | $4.56 \times 10^{-2}$ | | | |
| | | QD | $1.43 \times 10^{-1}$ | $4.56 \times 10^{-2}$ | | | |
| | DD | D | $7.96 \times 10^{-2}$ | $1.91 \times 10^{-2}$ | $2.45 \times 10^{-1}$ | 0.113 | 0.095 |
| | | DD | $7.96 \times 10^{-2}$ | $1.91 \times 10^{-2}$ | | | |
| | | QD | $7.96 \times 10^{-2}$ | $1.91 \times 10^{-2}$ | | | |
| | QD | D | $2.00 \times 10^{-15}$ | $2.31 \times 10^{-16}$ | $9.65 \times 10^{-21}$ | 0.554 | 0.157 |
| | | DD | $1.97 \times 10^{-16}$ | $3.86 \times 10^{-17}$ | | | |
| | | QD | $1.97 \times 10^{-16}$ | $3.86 \times 10^{-17}$ | | | |
| Arnoldi | D | D | $1.40 \times 10^{-2}$ | $1.15 \times 10^{-3}$ | $1.41 \times 10^{-1}$ | 0.031 | 0.028 |
| | | DD | $1.40 \times 10^{-2}$ | $1.15 \times 10^{-3}$ | | | |
| | | QD | $1.40 \times 10^{-2}$ | $1.15 \times 10^{-3}$ | | | |
| | DD | D | $1.55 \times 10^{-15}$ | $1.82 \times 10^{-16}$ | $7.25 \times 10^{-13}$ | 0.938 | 0.903 |
| | | DD | $4.94 \times 10^{-16}$ | $5.18 \times 10^{-16}$ | | | |
| | | QD | $4.94 \times 10^{-16}$ | $5.18 \times 10^{-17}$ | | | |
| | QD | D | $2.25 \times 10^{-15}$ | $2.75 \times 10^{-16}$ | $6.17 \times 10^{-46}$ | 2.172 | 1.744 |
| | | DD | $3.50 \times 10^{-16}$ | $3.88 \times 10^{-17}$ | | | |
| | | QD | $3.50 \times 10^{-16}$ | $3.88 \times 10^{-17}$ | | | |
| Householder | D | D | $1.44 \times 10^{-15}$ | $2.53 \times 10^{-16}$ | $5.99 \times 10^{-15}$ | 0.038 | – |
| | | DD | $4.54 \times 10^{-16}$ | $1.04 \times 10^{-16}$ | | | |
| | | QD | $4.54 \times 10^{-16}$ | $1.04 \times 10^{-16}$ | | | |
| | DD | D | $9.99 \times 10^{-16}$ | $1.90 \times 10^{-16}$ | $5.49 \times 10^{-31}$ | 5.054 | – |
| | | DD | $2.87 \times 10^{-16}$ | $4.85 \times 10^{-17}$ | | | |
| | | QD | $2.87 \times 10^{-16}$ | $4.85 \times 10^{-17}$ | | | |
| | QD | D | $9.99 \times 10^{-16}$ | $1.90 \times 10^{-16}$ | $1.61 \times 10^{-63}$ | 41.697 | – |
| | | DD | $2.87 \times 10^{-16}$ | $4.85 \times 10^{-17}$ | | | |
| | | QD | $2.87 \times 10^{-16}$ | $4.85 \times 10^{-17}$ | | | |

In Fig. 1, the horizontal axis indicates the index of the eigenvalues in descending order, and the vertical axis indicates the absolute error of eigenvalues $|\lambda_i - \bar{\lambda}_i|$.

Figure 1 shows that the absolute errors of Lanczos-D and -DD are large in general but become small for smaller eigenvalues, the absolute error of Arnoldi-D increases for smaller eigenvalues (from approximately the 50th eigenvalue), and Lanczos-QD, Arnoldi-DD, -QD, and Householder-D provide sufficient accuracy.

In Fig. 2, the horizontal axis again indicates the index of the eigenvalues in descending order, and the vertical axis indicates the value of the computed eigenvalues. The results for 'Mathematica', which represents the true eigenvalues, Lanczos-QD, Arnoldi-DD, and Householder-D are approximately the same. Both Lanczos-D and -DD have duplicative eigenvalues. Using higher-precision arithmetic, a plot is gradually brought closer to the true eigenvalue.

Table 3 lists the numbers of elements outside the tridiagonal part (upper triangular) for the Arnoldi method. The Arnoldi method is based on similarity transformation of non-symmetric matrices to Hessenberg matrices, and elements outside the tridiagonal part should be zero in the case of symmetric matrices. However, in our numerical experiments, nonzero elements appeared outside the tridiagonal part because of rounding errors. The relationship between nonzero elements and the accuracy of tridiagonalization is an area for future study.

In the case of using dense data, the ratio of the computation time for the Lanczos, Arnoldi, and Householder methods with double-precision arithmetic is approximately 1:2:2. For DD, the number of double-precision computations is 7 for the Lanczos method, 30 for the Arnoldi method, and 133 for the Householder method. For QD, the number of double-precision computations is 35 for the Lanczos method, 70 for the Arnoldi method, and 1,100 for the Householder method.

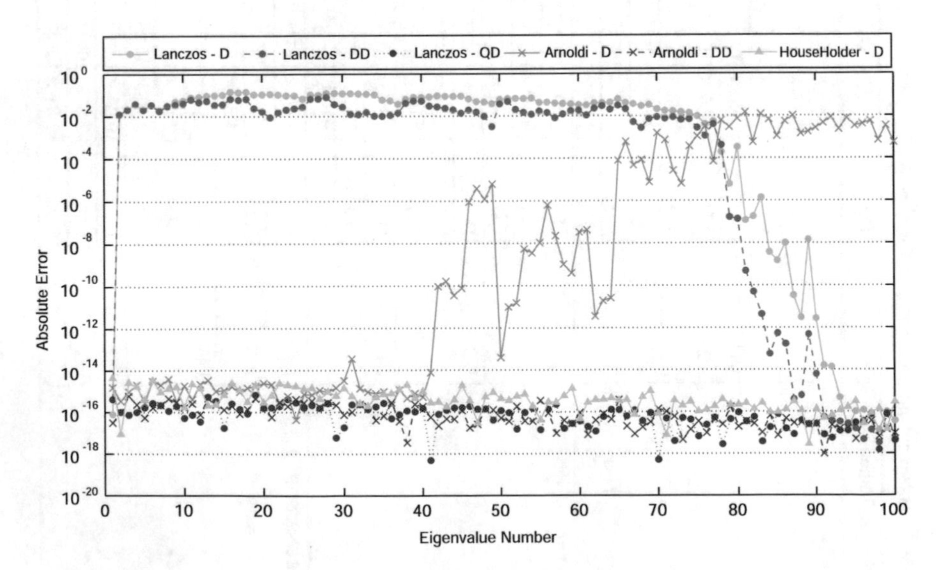


**Fig. 1** Absolute error in eigenvalues for nos4 in descending order

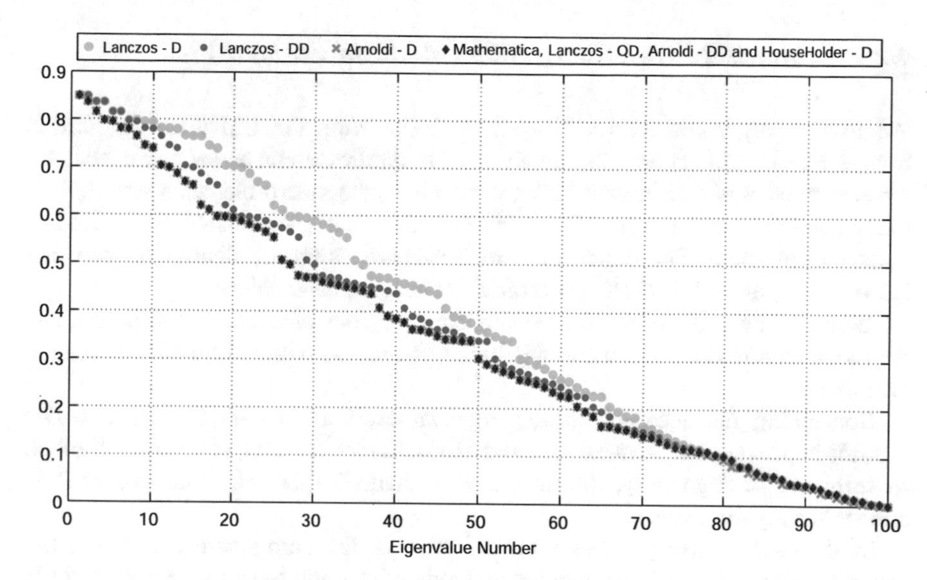


**Fig. 2** Eigenvalues for nos4 in descending order

**Table 3** Elements outside the tridiagonal part for the Arnoldi method for nos4

| | D | DD | QD |
|---|---|---|---|
| $10^{-5} \leq x$ | 598 | 0 | 0 |
| $10^{-10} \leq x < 10^{-5}$ | 1224 | 0 | 0 |
| $10^{-15} \leq x < 10^{-10}$ | 2716 | 103 | 0 |
| $10^{-20} \leq x < 10^{-15}$ | 313 | 453 | 0 |
| $10^{-30} \leq x < 10^{-20}$ | 0 | 3171 | 0 |
| $10^{-40} \leq x < 10^{-30}$ | 0 | 1124 | 0 |
| $10^{-50} \leq x < 10^{-40}$ | 0 | 0 | 263 |
| $x < 10^{-50}$ | 0 | 0 | 4588 |
| Maximum | $1.83 \times 10^{-1}$ | $3.97 \times 10^{-12}$ | $7.42 \times 10^{-45}$ |

The computation times for Lanczos-DD and Lanczos-QD for sparse data are 84% and 28%, respectively, of those for dense data, and the computation times for Arnoldi-DD and Arnoldi-QD for sparse data are 96% and 80%, respectively, of those for dense data. Costs of high-precision arithmetic and saving of computation time in sparse data type are depending on used algorithms and their implementations. For small matrices, Householder-D is the best method, but the computation time for Lanczos-QD using sparse data is only four times greater than that for Householder-D.

### 4.2 Example 2: Trefethen_200b (Medium Problem)

We used a larger test matrix 'Trefethen_200b' from the University of Florida Sparse Matrix Collection (http://www.cise.ufl.edu/research/sparse/matrices/). The dimension of this matrix was 199, the number of nonzero elements was 2,873, the condition number was $5.2 \times 10^2$, and the matrix originated from a combinatorial problem. The eigenvalues of Trefethen_200b are distributed between 2.3443911... and 1223.3718... without any clustered eigenvalues.

Table 4 shows the results for various combinations of methods and precisions. As mentioned in Sect. 4.1, we applied the QR method with only double-precision arithmetic.

Concerning the accuracy of the eigenvalues, Lanczos-QD and Arnoldi-DD are not improved, but Arnoldi-QD and Householder-D, -DD, -QD are sufficient. In terms of orthogonality, the accuracy of Arnoldi-QD and Householder-D is approximately the same.

In the case of using dense data, the ratio of the computation times for the Lanczos, Arnoldi, and Householder methods with double-precision arithmetic is approximately 1:14:30. For DD, the number of double-precision computation is 114 for the Lanczos method, 52 for the Arnoldi method, and 244 for the Householder method. For QD, the number of double-precision computations is 634 for the Lanczos method, 147 for the Arnoldi method, and 2,314 for the Householder method.

The computation times for Lanczos-DD and for Lanczos-QD for sparse data are 26% and 13%, respectively, of those for dense data, and the computation times for Arnoldi-DD and Arnoldi-QD are both 80% of those for dense data. For Trefethen_200b, the computation time was greatly reduced by the use of sparse data of MuPAT.

In Fig. 3, the horizontal and vertical axes are the same as in Fig. 1. Figure 3 reveals the following: The absolute errors for Lanczos-QD are large, but become small for smaller eigenvalues from approximately half of dimension. In contrast, the absolute error for Arnoldi-DD increases for the smaller eigenvalues (from approximately the 80th eigenvalue). Arnoldi-QD and Householder-D are sufficient.

Table 5 shows the upper triangular factors outside the tridiagonal part, which should be zero using the Arnoldi method. For Arnoldi-D and -DD, there are numerous nonzero elements within the range of double precision, and these elements affect the accuracy of the eigenvalues. For Arnoldi-QD, the number of nonzero elements is sufficiently small and does not affect the accuracy of eigenvalues in double precision.

### 4.3 Example 3: Nos5 (Slightly Large Problem)

We used the 'nos5' test matrix in Matrix Market (http://math.nist.gov/MatrixMarket/). The dimension of this matrix was 468, the number of nonzero

**Table 4** Accuracy and tridiagonalization time [s] and memory space [KB] for Trefethen_ 200b

| Tridiagonal -ization | | $max\lvert\lambda_i - \bar{\lambda}_i\rvert$ | $avg\lvert\lambda_i - \bar{\lambda}_i\rvert$ | $\frac{\lVert Q^T Q - I\rVert_F}{\lVert I\rVert_F}$ | Dense | | Sparse | |
|---|---|---|---|---|---|---|---|---|
| | | | | | time | Space | time | Space |
| Lanczos | D | $1.10 \times 10^2$ | $3.44 \times 10^1$ | $6.26 \times 10^4$ | 0.01 | 77.95 | 0.02 | 40.26 |
| | DD | $7.94 \times 10^1$ | $2.81 \times 10^1$ | $4.50 \times 10^{-4}$ | 1.14 | 158.23 | 0.30 | 82.75 |
| | QD | $2.14 \times 10^1$ | $3.75 \times 10^0$ | $1.48 \times 10^{-4}$ | 6.34 | 314.79 | 0.83 | 163.88 |
| Arnoldi | D | $2.32 \times 10^1$ | $4.61 \times 10^0$ | $1.42 \times 10^{-1}$ | 0.14 | 78.34 | 0.10 | 39.67 |
| | DD | $1.37 \times 10^1$ | $2.34 \times 10^0$ | $1.00 \times 10^{-1}$ | 7.32 | 158.90 | 5.86 | 81.70 |
| | QD | $7.84 \times 10^{-12}$ | $1.53 \times 10^{-12}$ | $1.00 \times 10^{-22}$ | 20.66 | 316.21 | 16.59 | 161.71 |
| Householder | D | $9.78 \times 10^{-12}$ | $6.30 \times 10^{-13}$ | $5.46 \times 10^{-15}$ | 0.30 | 154.72 | – | – |
| | DD | $1.82 \times 10^{-12}$ | $1.03 \times 10^{-14}$ | $4.51 \times 10^{-31}$ | 73.41 | 311.94 | – | – |
| | QD | $1.82 \times 10^{-12}$ | $1.03 \times 10^{-14}$ | $9.68 \times 10^{-64}$ | 694.42 | 621.71 | – | – |

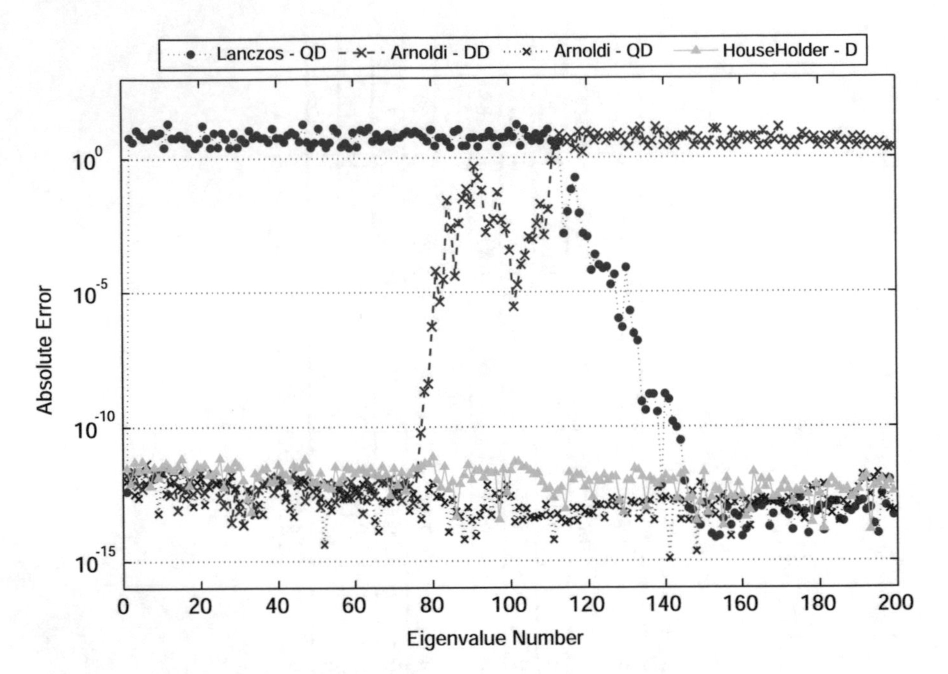


**Fig. 3** Absolute error in eigenvalues for Trefethen_ 200b in descending order

**Table 5** Elements outside the tridiagonal part for the Arnoldi method for Trefethen_ 200b

| | D | DD | QD |
|---|---|---|---|
| $10^0 \le x$ | 1474 | 269 | 0 |
| $10^{-5} \le x < 10^0$ | 6718 | 1543 | 0 |
| $10^{-10} \le x < 10^{-5}$ | 7465 | 2096 | 0 |
| $10^{-15} \le x < 10^{-10}$ | 3846 | 2517 | 0 |
| $10^{-20} \le x < 10^{-15}$ | 0 | 3168 | 56 |
| $10^{-30} \le x < 10^{-20}$ | 0 | 9906 | 1104 |
| $10^{-40} \le x < 10^{-30}$ | 0 | 4 | 2515 |
| $10^{-50} \le x < 10^{-40}$ | 0 | 0 | 4548 |
| $10^{-60} \le x < 10^{-50}$ | 0 | 0 | 9967 |
| $x < 10^{-60}$ | 0 | 0 | 1313 |
| Maximum | $2.91 \times 10^2$ | $3.70 \times 10^2$ | $2.19 \times 10^{-18}$ |

elements was 5, 172, the condition number was $1.1 \times 10^3$, and the matrix originated from a structure problem. The eigenvalues of nos5 are distributed between 52.899482… and 582029.11… without any clustered eigenvalues.

Table 6 shows the results for various combinations of methods and precisions. Concerning the accuracy of the eigenvalues, the accuracy of Arnoldi-QD was not improved, but the accuracy of Householder-D was sufficient. With respect to the orthogonality, only Householder-D was sufficient. Although the condition number of nos5 was not so large, the Arnoldi method with QD cannot generate an accurate

**Table 6** Accuracy and tridiagonalization time [s] and memory space [KB] for nos5

| Tridiagonal -ization | | $max\lvert\lambda_i - \bar{\lambda}_i\rvert$ | $avg\lvert\lambda_i - \bar{\lambda}_i\rvert$ | $\frac{\lVert Q^T Q - I \rVert_F}{\lVert I \rVert_F}$ | Dense time | Space | Sparse time | Space |
|---|---|---|---|---|---|---|---|---|
| Lanczos | DD | $1.31 \times 10^{5}$ | $5.27 \times 10^{4}$ | $7.59 \times 10^{-1}$ | 11.34 | 862.40 | 3.23 | 437.39 |
| | QD | $8.92 \times 10^{4}$ | $3.70 \times 10^{4}$ | $5.35 \times 10^{-1}$ | 56.03 | 1721.69 | 6.86 | 873.16 |
| Arnoldi | DD | $2.18 \times 10^{4}$ | $5.74 \times 10^{2}$ | $9.25 \times 10^{-2}$ | 199.91 | 860.62 | 188.57 | 434.76 |
| | QD | $2.15 \times 10^{4}$ | $2.87 \times 10^{2}$ | $6.54 \times 10^{-2}$ | 446.47 | 1718.11 | 409.35 | 867.83 |
| Householder | D | $6.64 \times 10^{-9}$ | $5.95 \times 10^{-10}$ | $1.04 \times 10^{-14}$ | 5.38 | 855.59 | – | – |

tridiagonal matrix. The modification of the implementation of the Lanczos method and the Arnoldi method and the choice of the initial value remain as areas for future research.

Arnoldi and Lanczos methods in high-precision arithmetic can not produce accurate eigenvalues in current implementation, for example, Lanczos-QD with sparse data structure can consume approximately the same as the computation time and the memory space with Householder-D. There are some possibility to improve the computation for high-precision arithmetics. By the compiled code and parallel processing, the computation will be improved, however their codes depend on computing environment and loose ease of use.

In the Arnoldi method, the ratio of matrix-vector operations becomes smaller as the dimension of matrix becomes larger, but in the Lanczos method, the ratio is not changed regardless of the dimension. Thus, in the Arnoldi method, the speed up by the sparse data is small.

## 5 Concluding Remarks

Although the authors believe that simple algorithms are good, floating-point number operations can break simple algorithms due to rounding errors. In the present paper, we attempted to stabilize the Lanczos and Arnoldi methods for tridiagonalization of symmetric matrices by using high-precision arithmetics; DD and QD. Since the Lanczos and Arnoldi methods are based on matrix-vector multiplication and do not change the given matrix, they have a possibility to be used for tridiagonalizing large sparse matrices.

We analyzed accuracy, numerical stability, and computing cost for tridiagonalization using dense and sparse matrix operations. We compared double (D), double-double (DD), and quad-double (QD) arithmetic for tridiagonalization by the Lanczos, Arnoldi, and Householder methods, and eigenvalue computation using the shifted QR method in only double-precision arithmetic.

The Lanczos method was stabilized by QD for only a small problem and required more precision. The Arnoldi method was also stabilized, although there were some problems in the case of relatively large test problems. A large matrix had some elements outside the tridiagonal part, resulting in an non-symmetric matrix. The Householder method was sufficient in double-precision arithmetic, but was not fit for large sparse matrices.

We conclude that a high-precision arithmetic is effective for tridiagonalization and no special technique is necessary for some problem. Lanczos and Arnoldi methods can work well with high-precision arithmetic. However, some improvement is necessary for other problems. The best combination of algorithm and computing precision depends on the problem to be solved. The controlling precision in automatic is one of our future issues.

The sparse data type in MuPAT could reduce the required memory space and computation time for sparse matrices in high-precision arithmetic. For accelerating computation, parallel computing for these operations will be necessary. The analysis of the numerical stability and additional improvement of algorithms and implementation are our future issues.

**Acknowledgements** The authors would like to thank the reviewers for their careful reading and much helpful suggestions, and Mr. Takeru Shiiba in Tokyo University of Science for his kind support in numerical experiments. The present study was supported by the Grant-in-Aid for Scientific Research (C) No. 25330141 from the Japan Society for the Promotion of Science.

## References

1. Golub, G.H., Van Loan, C.F.: Matrix Computations, 4th edn. The Johns Hopkins University Press, Baltimore (2013)
2. Kikkawa, S., Saito, T., Ishiwata, E., Hasegawa, H.: Development and acceleration of multiple precision arithmetic toolbox MuPAT for Scilab. J. SIAM Lett. **5**, 9–12 (2013)
3. Saito, T., Kikkawa, S., Ishiwata, E., Hasegawa, H.: Effectiveness of sparse data structure for double-double and quad-double arithmetics. In: Wyrzykowski, R., et al. (eds.) Parallel Processing and Applied Mathematics, Part I. Lecture Notes in Computer Science, vol.8384, pp. 1–9. Springer, Berlin/Heidelberg (2014)
4. Hida, Y., Li, X. S., Bailey, D.H.: Quad-double arithmetic: algorithms, implementation, and application. Technical Report LBNL-46996 (2000)
5. Dekker, T.J.: A floating-point technique for extending the available precision. Numer. Math. **18**, 224–242 (1971)
6. Demmel, J.W.: Applied Numerical Linear Algebra. SIAM, Philadelphia (1997)

# Computation of Eigenvectors for a Specially Structured Banded Matrix

**Hiroshi Takeuchi, Kensuke Aihara, Akiko Fukuda, and Emiko Ishiwata**

**Abstract** For a specially structured nonsymmetric banded matrix, which is related to a discrete integrable system, we propose a novel method to compute all the eigenvectors. We show that the eigenvector entries are arranged radiating out from the origin on the complex plane. This property enables us to efficiently compute all the eigenvectors. Although the intended matrix has complex eigenvalues, the proposed method can compute all the complex eigenvectors using only arithmetic of real numbers.

## 1 Introduction

Some eigenvalue algorithms are known to be related to integrable systems. There is an interesting analogy between the QR algorithm and the Toda flow [1, 2]. The recursion formula of the quotient difference (qd) algorithm is equivalent to the discrete Toda equation [3]. In addition to the discovery of the above relationships, some new algorithms have been formulated for computing eigenvalues or singular values based on the asymptotic properties of discrete integrable systems. A famous

H. Takeuchi
Graduate School of Science, Tokyo University of Science, 1-3 Kagurazaka, Shinjuku-ku, 162-8601 Tokyo, Japan
e-mail: 1414614@ed.tus.ac.jp

K. Aihara
Department of Computer Science, Tokyo City University, 1-28-1 Tamazutsumi, Setagaya-ku, 158-8557 Tokyo, Japan
e-mail: aiharak@tcu.ac.jp

A. Fukuda (✉)
Department of Mathematical Sciences, Shibaura Institute of Technology, 307 Fukasaku, Minuma-ku, Saitama-shi, 337-8570 Saitama, Japan
e-mail: afukuda@shibaura-it.ac.jp

E. Ishiwata
Department of Mathematical Information Science, Tokyo University of Science, 1-3 Kagurazaka, Shinjuku-ku, 162-8601 Tokyo, Japan
e-mail: ishiwata@rs.tus.ac.jp



T. Sakurai et al. (eds.), *Eigenvalue Problems: Algorithms, Software and Applications in Petascale Computing*, Lecture Notes in Computational Science and Engineering 117, https://doi.org/10.1007/978-3-319-62426-6_10

example is the dLV algorithm for computing singular values of bidiagonal matrices [4], which was designed based on the integrable discrete Lotka–Volterra (dLV) system, a prey–predator model in mathematical biology.

The discrete hungry Lotka–Volterra (dhLV) system:

$$\begin{cases} u_k^{(n+1)} = u_k^{(n)} \prod_{j=1}^{M} \dfrac{1+\delta^{(n)} u_{k+j}^{(n)}}{1+\delta^{(n+1)} u_{k-j}^{(n+1)}}, \quad k = 1, 2, \ldots, M_m, \quad n = 0, 1, \ldots, \\ u_{1-M}^{(n)} \equiv 0, \quad u_{2-M}^{(n)} \equiv 0, \ldots, u_0^{(n)} \equiv 0, \quad u_{M_m+1}^{(n)} \equiv 0, \ldots, u_{M_m+M}^{(n)} \equiv 0 \end{cases} \tag{1}$$

is a generalization of the dLV system, where $u_k^{(n)}$ is the population of the species and $\delta^{(n)}$ is the discrete step-size at a discrete time $n$. Note that $M_k := (M+1)k - M$, where the parameter $M$ denotes the number of species on which a species preys, and the case $M = 1$ corresponds to the dLV system. Based on the dhLV system (1), a new algorithm for computing the eigenvalues of nonsymmetric banded matrices was formulated, referred to as the dhLV algorithm [5]. Although the intended matrices of the dhLV algorithm are real and nonsymmetric and generally have complex eigenvalues, the algorithm can compute all the eigenvalues using only the arithmetic of real numbers [5].

However, the computation of eigenvectors for the intended matrices of the dhLV algorithm was not covered [5]. In general, the inverse iteration method can be used to compute eigenvectors when approximate eigenvalues are obtained. However, since the intended matrices have complex eigenvalues, the inverse iteration method requires the arithmetic of complex numbers. In this paper, we derive a recursion formula that satisfies in the characteristic equation, and show that eigenvector entries are arranged radiating out from the origin. Then, using this distinct distribution, we propose a new method for computing all the eigenvectors of the intended matrices without complex arithmetic.

The remainder of this paper is organized as follows. In Sect. 2, we present a brief review of the dhLV algorithm. In Sect. 3, we investigate the distribution of the eigenvector entries, and propose a new method for computing all the eigenvectors efficiently. In Sect. 4, some numerical examples are presented to show the effectiveness of the proposed method. Finally, Sect. 5 gives concluding remarks.

## 2 Computation of Eigenvalues Based on the dhLV System

In this section, we briefly review the dhLV algorithm [5], which is based on the asymptotic property of the dhLV system (1), for computing all the complex eigenvalues.

The intended matrix of the dhLV algorithm is a nonsymmetric banded matrix with size $(M_m + M) \times (M_m + M)$ for the given parameters $m$ and $M$, as follows.

$$S^{(n)} = \begin{pmatrix} \overbrace{0 \ldots 0}^{M} & U_1^{(n)} & & & & \\ 1 & 0 \ldots 0 & U_2^{(n)} & & & \\ & 1 \ddots & & \ddots & \ddots & \\ & & \ddots & \ddots & \ddots & U_{M_m}^{(n)} \\ & & & \ddots & \ddots & 0 \\ & & & & \ddots \ \ddots & \vdots \\ & & & & 1 & 0 \end{pmatrix}, \tag{2}$$

where $U_k^{(n)}$ is defined using the dhLV variables $u_k^{(n)}$ by

$$U_k^{(n)} := u_k^{(n)} \prod_{j=1}^{M} (1 + \delta^{(n)} u_{k-j}^{(n)}). \tag{3}$$

The matrix $S^{(n)}$ appears in the context of an integrable system and, in particular, is one of the Lax matrices. Let us introduce the following matrix

$$T^{(n)} = \begin{pmatrix} V_1^{(n)} & & & & \\ 0 & V_2^{(n)} & & & \\ \vdots & 0 & \ddots & & \\ 0 & \vdots & \ddots & \ddots & \\ \delta^{(n)} & 0 & & \ddots & \ddots \\ & \ddots & \ddots & & \ddots \ \ddots \\ & & \delta^{(n)} & \underbrace{0 \ \ldots \ 0}_{M} & V_{M_m+M}^{(n)} \end{pmatrix},$$

where

$$V_k^{(n)} := \prod_{j=0}^{M} (1 + \delta^{(n)} u_{k-j}^{(n)}).$$

Then, $S^{(n)}$ and $T^{(n)}$ satisfy

$$T^{(n)} S^{(n+1)} = S^{(n)} T^{(n)}, \tag{4}$$

which is called the Lax form for the dhLV system (1). Assume $u_k^{(0)} > 0$. Then, there exists the inverse $(T^{(n)})^{-1}$, and (4) can be transformed into

$$S^{(n+1)} = (T^{(n)})^{-1} S^{(n)} T^{(n)}. \tag{5}$$

This implies that the eigenvalues of $S^{(n)}$ and $S^{(n+1)}$ are equal to each other under the time evolution from $n$ to $n+1$ of the dhLV system (1). The positivity of the dhLV variables yields the following theorem concerning the asymptotic convergence of the dhLV system (1).

**Theorem 1 (Fukuda et al. [5, 6])** *Let* $0 < u_k^{(0)} < K_0,\ \ k = 1, 2, \ldots, M_m$ *for a positive real number* $K_0$. *Then, it holds that*

$$\lim_{n\to\infty} u_{M_k}^{(n)} = c_k, \quad k = 1, 2, \ldots, m, \tag{6}$$

$$\lim_{n\to\infty} u_{M_k+p}^{(n)} = 0, \quad k = 1, 2, \ldots, m-1, \quad p = 1, 2, \ldots, M, \tag{7}$$

*where* $c_1 > c_2 > \cdots > c_m > 0$.

From (6) and (7), the convergence of $U_k^{(n)}$ (3) in (2) is easily given by

$$\lim_{n\to\infty} U_{M_k}^{(n)} = c_k, \quad k = 1, 2, \ldots, m,$$

$$\lim_{n\to\infty} U_{M_k+p}^{(n)} = 0, \quad k = 1, 2, \ldots, m-1, \quad p = 1, 2, \ldots, M.$$

Therefore, the convergence of $S^{(n)}$ as $n \to \infty$ can be expressed by

$$S^* := \lim_{n\to\infty} S^{(n)} = \begin{pmatrix} S_1 & & & \\ E & S_2 & & \\ & \ddots & \ddots & \\ & & E & S_m \end{pmatrix},$$

where $S_k$ for $k = 1, 2, \ldots, m$, and $E$ are square matrices with size $(M+1)\times(M+1)$ defined by

$$S_k := \begin{pmatrix} 0 & \cdots & 0 & c_k \\ 1 & 0 & \cdots & 0 \\ & \ddots & \ddots & \vdots \\ & & 1 & 0 \end{pmatrix}, \quad E := \begin{pmatrix} 0 & \cdots & 0 & 1 \\ & \ddots & & 0 \\ & & \ddots & \vdots \\ & & & 0 \end{pmatrix}.$$

Since $\det(\lambda I - S_k) = \lambda^{M+1} - c_k$, the characteristic polynomial of $S^*$ can be written as

$$\det(\lambda I - S^*) = \prod_{k=1}^{m} [\det(\lambda I - S_k)] = \prod_{k=1}^{m} \left[\lambda^{M+1} - c_k\right], \tag{8}$$

where $I$ is the identity matrix. From (5) and (8), the eigenvalues of $S^{(n)}$ are given by the $(M+1)$th root of $c_k$. Since $S^{(0)}$ is similar to $S^{(1)}, S^{(2)}, \ldots$, all the eigenvalues of $S := S^{(0)}$ are given by

$$\lambda_{k,\ell} = r_k \exp\left(\frac{2\pi \ell i}{M+1}\right), \quad k = 1, 2, \ldots, m, \quad \ell = 1, 2, \ldots, M+1, \tag{9}$$

where $r_k = \lim_{n\to\infty} \sqrt[M+1]{u_{M_k}^{(n)}}$. Therefore, the eigenvalues of $S$ can be obtained using the time evolution of the dhLV system (1).

Based on $U_k^{(0)}$ for $k = 1, 2, \ldots, M_m$ in $S$, the initial values $u_k^{(0)}$ of the dhLV system (1) are set to

$$u_k^{(0)} = \frac{U_k^{(0)}}{\prod_{j=1}^{M}(1 + \delta^{(0)} u_{k-j}^{(0)})}, \quad k = 1, 2, \ldots, M_m.$$

Starting from $u_k^{(0)}$, the values $u_k^{(1)}, u_k^{(2)}, \ldots$ are recursively computed using the dhLV system (1). For sufficiently large $n$, $u_{M_k}^{(n)}$ becomes an approximation for $c_k$. Therefore, by using (9), we can obtain approximations for the eigenvalues of $S$. Although $S$ has complex eigenvalues, the recurrence (1) is performed only with the arithmetic of real numbers. Since the eigenvalues (9) can be rewritten as

$$\lambda_{k,\ell} = r_k \left[\cos\left(\frac{2\ell\pi}{M+1}\right) + i \sin\left(\frac{2\ell\pi}{M+1}\right)\right], \tag{10}$$

the right hand side of (10) can be obtained by computing real and imaginary parts individually with the arithmetic of real numbers.

## 3 Distribution and Computation of Eigenvector Entries

In this section, we propose a new method to compute all the eigenvectors of $S$ using only the arithmetic of real numbers. Let $\lambda$ be an eigenvalue of $S$. Then, its corresponding eigenvector $\boldsymbol{x}$ satisfies

$$(S - \lambda I)\boldsymbol{x} = \boldsymbol{0}, \tag{11}$$

where $\boldsymbol{x} = (x_1, x_2, \ldots, x_{M_m+M})^\top$. Let $U_k := U_k^{(0)}$ for $k = 1, 2, \ldots, M_m$ for simplicity, and by writing down each entry of (11), it holds that

$$x_j = \lambda x_{j+1}, \quad j = M_m, M_m + 1, \ldots, M_m + M - 1, \tag{12}$$

$$x_j = \lambda x_{j+1} - U_{j+1} \cdot x_{j+M+1}, \quad j = 1, 2, \ldots, M_m - 1. \tag{13}$$

Since, from Theorem 1, all the eigenvalues of $S$ are distinct, we have $\mathrm{rank}(S - \lambda I) = M_m + M - 1$. Therefore, by setting $x_{M_m+M} = 1$, the entries $x_{M_m+M-1}, x_{M_m+M-2}, \ldots, x_1$ can be computed recursively in this order using (12) and (13). Note here that, in the case $\lambda$ is a complex number, complex arithmetic is necessary when using the recursion formulas (12) and (13). For the purpose of avoiding complex arithmetic, we separate each entry of $\boldsymbol{x}$ into the absolute value and the argument.

**Lemma 1** *Let $\lambda$ and $\boldsymbol{x} = (x_1, x_2, \ldots, x_{M_m+M})^\top$ be the eigenvalue and the corresponding eigenvector of $S$. We assume that $\lambda$ is written as*

$$\lambda = r \exp\left(\frac{2\pi \ell i}{M+1}\right), \tag{14}$$

*where $r$ is a positive real number and $\ell \in \{1, 2, \ldots, M+1\}$. Then, $x_j$ can be expressed as*

$$x_j = y_j \exp\left(-\frac{2\pi \ell j i}{M+1}\right), \quad j = 1, 2, \ldots, M_m + M, \tag{15}$$

*where $y_j$ are real numbers that satisfy*

$$y_{M_m+M} = 1, \tag{16}$$

$$y_j = r y_{j+1}, \quad j = M_m, M_m + 1, \ldots, M_m + M - 1, \tag{17}$$

$$y_j = r y_{j+1} - U_{j+1} \cdot y_{j+M+1}, \quad j = 1, 2, \ldots, M_m - 1. \tag{18}$$

*Proof* We show that $y_j$ satisfying (15)–(18) exist for $j = 1, 2, \ldots, M_m + M$. The proof is by induction for $j = M_m + M, M_m + M - 1, \ldots, 1$ in this order. Let $x_{M_m+M} = 1$. By setting $y_{M_m+M} = 1$, it holds that

$$x_{M_m+M} = y_{M_m+M} \exp\left(-\frac{2\pi \ell (M_m + M) i}{M+1}\right).$$

Hence, (15) and (16) hold for $j = M_m + M$. We next consider the case of $j = M_m + M - 1, M_m + M - 2, \ldots, M_m$. From (12) and (14), we have

$$\begin{aligned} x_j &= \lambda x_{j+1} \\ &= \lambda^{M_m+M-j} \end{aligned}$$

$$= r^{M_m+M-j} \exp\left(\frac{2\pi\ell(M_m + M - j)i}{M+1}\right)$$
$$= r^{M_m+M-j} \exp\left(\frac{2\pi\ell((M+1)m - j)i}{M+1}\right)$$
$$= r^{M_m+M-j} \exp\left(-\frac{2\pi\ell ji}{M+1}\right).$$

By setting $y_j = r^{M_m+M-j}$, (15) and (17) are satisfied.

For the case $j = M_m - 1, M_m - 2, \dots, 1$, assume that (15)–(18) hold for $j = p+1$ and $j = p + M + 1$ with an arbitrary $p \in \{1, 2, \dots, M_m - 1\}$, that is,

$$x_{p+1} = y_{p+1} \exp\left(-\frac{2\pi\ell(p+1)i}{M+1}\right),$$
$$x_{p+M+1} = y_{p+M+1} \exp\left(-\frac{2\pi\ell(p+M+1)i}{M+1}\right)$$
$$= y_{p+M+1} \exp\left(-\frac{2\pi\ell pi}{M+1}\right).$$

From (13), it follows that

$$x_p = \lambda x_{p+1} - U_{p+1}\, x_{p+M+1}$$
$$= r\exp\left(\frac{2\pi\ell i}{M+1}\right) y_{p+1} \exp\left(-\frac{2\pi\ell(p+1)i}{M+1}\right) - U_{p+1}\, y_{p+M+1} \exp\left(-\frac{2\pi\ell pi}{M+1}\right)$$
$$= r y_{p+1} \exp\left(-\frac{2\pi\ell pi}{M+1}\right) - U_{p+1}\, y_{p+M+1} \exp\left(-\frac{2\pi\ell pi}{M+1}\right)$$
$$= (r y_{p+1} - U_{p+1}\, y_{p+M+1}) \exp\left(-\frac{2\pi\ell pi}{M+1}\right).$$

By letting $y_p = r y_{p+1} - U_{p+1} y_{p+M+1}$, it holds that $x_p = y_p \exp(-2\pi\ell pi/(M+1))$. Therefore, there exist $y_j$ for $j = 1, 2, \dots, M_m + M$, that satisfy (15)–(18).

From Lemma 1, the difference between $\arg x_j$ and $\arg x_{j+1}$ is $2\pi\ell/(M+1)$, which coincides with that for the corresponding eigenvalues, where $\arg X$ denotes the argument of $X \in \mathbb{C}$. We note that the *entries* of the eigenvectors are arranged radiating out from the origin on the complex plane. For example, Fig. 1 shows the distribution of the eigenvector entries on the complex plane in the case $M = 6$ and $m = 2$ with $U_k = 3/2$ for $k = 1, 2, \dots, M_m$. Here, the arguments of the eigenvalues are $2\pi/7$ and $4\pi/7$ on the left and right sides of Fig. 1, respectively. We can see that the difference between $\arg x_j$ and $\arg x_{j+1}$ coincides with the argument of the corresponding eigenvalue, and that the entries of the eigenvectors are distributed in a radial fashion.

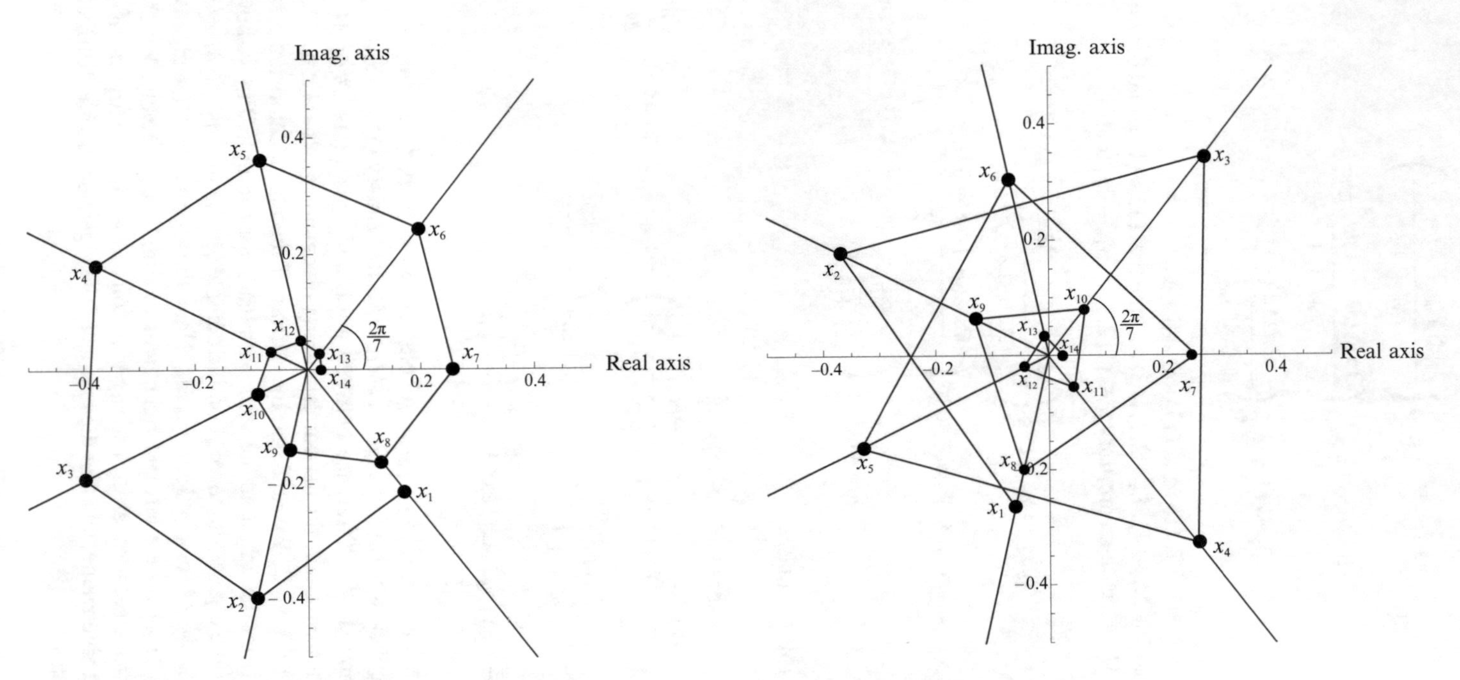


**Fig. 1** Distribution of the eigenvector entries for eigenvalues whose arguments are $2\pi/7$ (on the left) and $4\pi/7$ (on the right)

To compute an eigenvector $\boldsymbol{x}$, precisely $x_j$ for $j = 1, 2, \ldots, M_m + M$, we first compute $y_j$ for $j = M_m + M, M_m + M - 1, \ldots, 2, 1$ in this order by (16)–(18), and then give the quantities concerning the argument to $y_j$ using (15). It is noted here that (16)–(18) compute real eigenvectors that satisfy $S\boldsymbol{y} = r\boldsymbol{y}$, where $\boldsymbol{y} = (y_1, y_2, \ldots, y_{M_m+M})^\top$.

On the other hand, where approximate eigenvalues have been obtained, the inverse iteration method is a standard way to compute the corresponding eigenvectors. Specifically, for an approximate complex eigenvalue $\tilde{\lambda}$ of a matrix $A$, the iterations

$$\boldsymbol{x}^{(k+1)} = (A - \tilde{\lambda} I)^{-1} \boldsymbol{x}^{(k)}, \quad k = 0, 1, 2, \ldots \tag{19}$$

give a good approximation for an eigenvector within a few iterations. For the purpose of obtaining eigenvectors of $S$, the inverse iteration method might produce the desired eigenvectors. If we use a straightforward implementation, then complex arithmetic is required. However, using (15) of Lemma 1, it is not necessary to use complex arithmetic, even if we employ the inverse iteration method. Two alternative methods of computing all the eigenvectors of $S$ are summarized as follows:

(I) Compute only $m$ real eigenvectors, which correspond to $m$ real eigenvalues, using (16)–(18). Then, compute the remaining $mM$ complex eigenvectors by applying (15).
(II) Compute only $m$ real eigenvectors by the inverse iteration method, and then compute the remaining $mM$ complex eigenvectors by (15).

We here discuss the computational costs for the proposed methods. Let $N$ be the matrix size $M_m + M$. In the case of (I), from (16)–(18), the number of operations for computing $m$ real eigenvectors is $m(3N - 2M - 3)$. Moreover, computing a complex eigenvector from a real eigenvector using (15) requires $\xi N$ operations, where $\xi$ is the costs for computing $y_j \exp\left(-\dfrac{2\pi \ell j i}{M+1}\right)$, which we may assume to be $O(1)$. We need $mM\xi N$ operations to obtain the remaining $mM$ complex eigenvectors. Therefore, the total number of operations for (I) is $m(3N - 2M - 3) + mM\xi N$. On the other hand, when using the inverse iteration method, we need to solve a linear system (19) per iteration, and its computational costs are $O(N^3)$ in general. If we use the structure of the coefficient matrix, the linear system can be solved with operations of $O(N)$. Therefore, the total number of operations for (II) is $O(mN) + mM\xi N$, which is essentially the same cost as those in (I). Note here that the cost increases in proportion to the number of iterations of the inverse iteration method.

## 4 Numerical Examples

In this section, we present numerical examples and investigate the accuracy and efficiency of the proposed method. Numerical calculations are conducted in double-precision arithmetic in Matlab R2015a on a PC (Intel Core i7-3635QM CPU

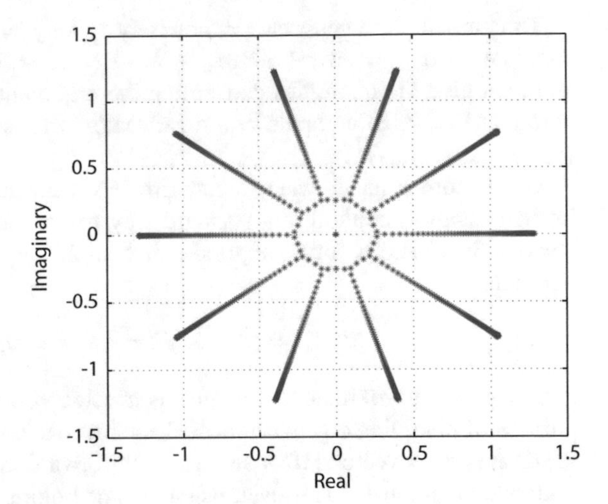


**Fig. 2** Distribution of eigenvalues computed by `eig()` for $S$

and 16 GB of RAM). The test matrix $S$ is given by setting $U_k = 0.5$ for $k = 1, 2, \ldots, M_m$, $M = 9$, and $m = 100$, that is, the size of $S$ is $1000 \times 1000$.

In Figs. 2 and 3, we present the distributions of eigenvalues and all the entries of some selected eigenvectors computed using the Matlab function `eig()`. Theoretically, all the eigenvalues and eigenvector entries are distributed radially from the origin $(0, 0)$. However, from Fig. 2, we can observe that the eigenvalues located near the origin are not arranged on the line

$$z = r \exp\left(\frac{2\pi \ell i}{10}\right), \quad \ell = 1, 2, \ldots, 10. \tag{20}$$

From Fig. 3, the entries of the eigenvectors corresponding to the eigenvalues (a) $\lambda = -0.39905 + 1.2281i$ and (b) $\lambda = 1.0004 - 0.72809i$ appear to be distributed on line defined by (20). However, those corresponding to the eigenvalues (c) $\lambda = 0.24683 - 0.069711i$ and (d) $\lambda = 0.13358 - 0.21754i$, which are near the origin, are clearly not distributed on line. This shows that the test matrix $S$ is an example of a matrix that the Matlab function `eig()` cannot accurately compute all the eigenpairs of.

Next, we present the CPU times for computing all the eigenvectors. The eigenvalues are computed using the dhLV algorithm [5], and we compute all the eigenvectors in the following four ways:

A. All the real and complex eigenvectors are computed by:

   (i) the recursion formulas (12) and (13), where real or complex arithmetic is used for real or complex eigenvectors, respectively,

   or

   (ii) the inverse iteration method, where real or complex arithmetic is used for real or complex eigenvectors, respectively.

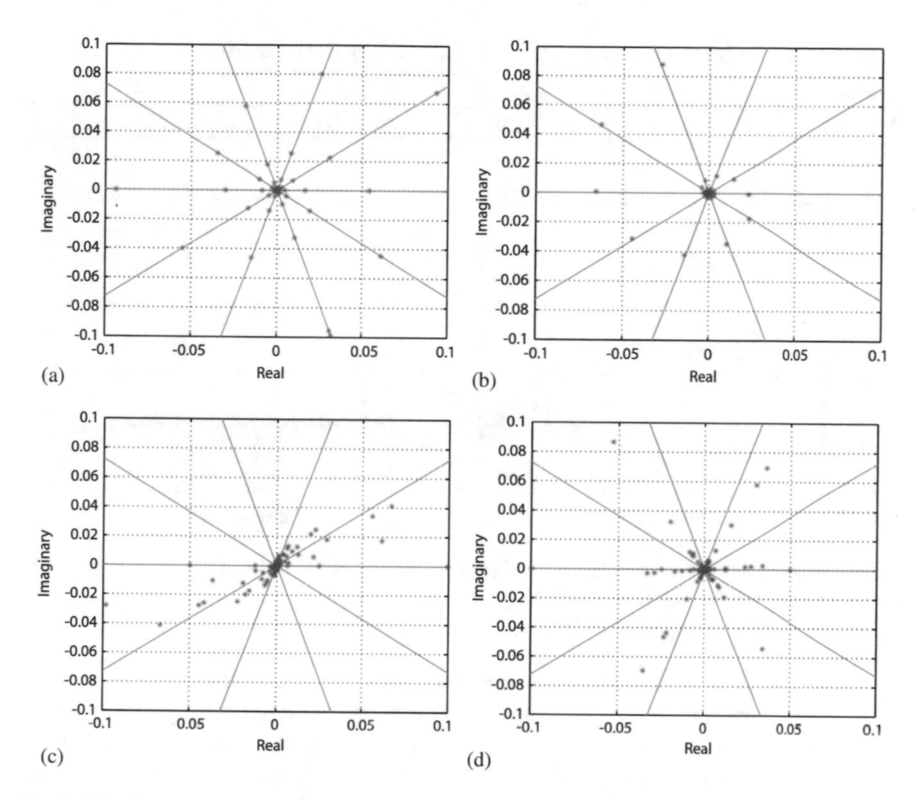


**Fig. 3** Distribution of the entries of the eigenvectors computed by `eig()`. (**a**) $\lambda = -0.39905 + 1.2281i$. (**b**) $\lambda = 1.0004 - 0.72809i$. (**c**) $\lambda = 0.24683 - 0.069711i$. (**d**) $\lambda = 0.13358 - 0.21754i$

B. The $m$ real eigenvectors corresponding to the $m$ real eigenvalues are computed by:

(i) the recursion formulas (16)–(18) with real arithmetic

or

(ii) the inverse iteration method with real arithmetic,

then, the remaining $mM$ complex eigenvectors are computed by multiplying the real eigenvector entries by the arguments (see methods (I) and (II) in Sect. 3).

The inverse iteration method was started with a random vector on $(0, 1)$, and was stopped after two iterations. The linear equations (19) were solved by the \ (`mldivide()`) function of Matlab.

The computation times for computing all the eigenvectors are shown in Table 1. Since the methods B–(i) and B–(ii) are carried out using only arithmetic of real numbers, their computation times are significantly shorter than those of A–(i) and A–(ii). Moreover, the computation time of B–(i) is 47 % of the computation time of B–(ii).

**Table 1** Computation times for the methods A–(i), A–(ii), B–(i) and B–(ii)

| Method | A–(i) | A–(ii) | B–(i) | B–(ii) |
|---|---|---|---|---|
| Time [s] | 2.45 | 5.22 | 0.33 | 0.70 |

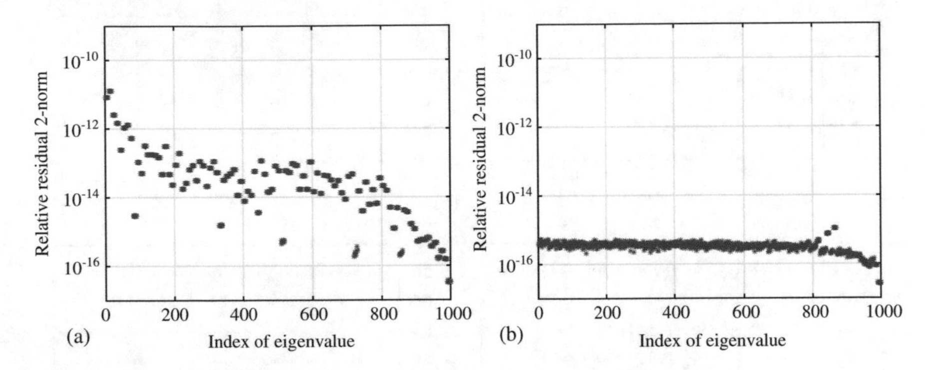


**Fig. 4** Relative residual norms for each eigenpair. (**a**) B–(i): Recursion formula (16)–(18). (**b**) B–(ii): Inverse iteration method

Next, we show the accuracy of the computed eigenpairs. Figure 4 shows the relative residual norms for the eigenpairs obtained by B–(i) and B–(ii) after the eigenvalues are computed using the dhLV algorithm. The horizontal axis is the index of eigenvalues in descending order of the absolute value, and the vertical axis is the relative residual 2-norms ($\|S\boldsymbol{x} - \lambda\boldsymbol{x}\|_2/\|\boldsymbol{x}\|_2$) corresponding to each eigenpair.

From Fig. 4, we can observe the following. The relative residual norms of the eigenpairs computed with B–(i) are at most $10^{-11}$, and more accurate eigenpairs are obtained for smaller absolute eigenvalues. The eigenpairs computed with B–(ii) are more accurate than those computed with B–(i); the relative residual norms of all the eigenpairs are at most $10^{-15}$.

The above examples demonstrate that the proposed method is more efficient than the recursion formula or the inverse iteration method with complex arithmetic. The methods B–(i) and B–(ii) are effective in terms of CPU time and accuracy, respectively. A possible future method may be a hybrid of B–(i) and B–(ii). In particular, the inverse iteration method, in which the initial vector is given by an approximate eigenvector obtained by B–(i), may give accurate results efficiently.

## 5 Concluding Remarks

In this paper, we showed that eigenvector entries for the structured nonsymmetric banded matrix are distributed radially from the origin. Using this property, we proposed a new method for computing all the eigenvectors using only arithmetic of real numbers. If all the eigenvalues are obtained, then, absolute values of complex eigenvectors can be computed using the recursion formula or the inverse iteration method with real arithmetic, and multiplying these by the arguments gives all the complex eigenvectors. Numerical examples show that the computation time can be reduced by computing the absolute value and the argument separately and avoiding complex arithmetic. The proposed recursion formula can compute eigenvectors more efficiently than does the inverse iteration method in terms of computation time. Eigenvectors computed by the inverse iteration method are more accurate than those computed by the recursion formula. Rounding error analysis for the recursion formula is still remaining as a future work.

**Acknowledgements** This work was partially supported by Grant-in-Aid for Encouragement of Young Scientists (B) No. 16K21368 and Scientific Research (C) No. 26400212 from the Japan Society for the Promotion of Science.

## References

1. Symes, W.W.: The QR algorithm and scattering for the finite nonperiodic Toda Lattice. Physica **4D**, 275–280 (1982)
2. Demmel, J.: Applied Numerical Linear Algebra. SIAM, Philadelphia (1997)
3. Hirota, R., Tsujimoto, S., Imai, T.: Difference scheme of soliton equations. In: Christiansen, P.L., Eilbeck, J.C., Parmentier, R.D. (eds.) Future Directions of Nonlinear Dynamics in Physics and Biological Systems. Series B: Physics, vol. 312, pp. 7–15. Plenum Press, New York (1993)
4. Iwasaki, M., Nakamura, Y.: On the convergence of a solution of the discrete Lotka-Volterra system. Inverse Probl. **18**, 1569–1578 (2002)
5. Fukuda, A., Ishiwata, E., Iwasaki, M., Nakamura, Y.: The discrete hungry Lotka-Volterra system and a new algorithm for computing matrix eigenvalues. Inverse Probl. **25**, 015007 (2009)
6. Fukuda, A., Ishiwata, E., Yamamoto, Y., Iwasaki, M., Nakamura, Y.: Integrable discrete hungry systems and their related matrix eigenvalues. Ann. Math. Pura Appl. **192**, 423–445 (2013)

# Monotonic Convergence to Eigenvalues of Totally Nonnegative Matrices in an Integrable Variant of the Discrete Lotka-Volterra System

**Akihiko Tobita, Akiko Fukuda, Emiko Ishiwata, Masashi Iwasaki, and Yoshimasa Nakamura**

**Abstract** The Lotka-Volterra (LV) system describes a simple predator-prey model in mathematical biology. The hungry Lotka-Volterra (hLV) system assumed that each predator preys on two or more species is an extension; those involving summations and products of nonlinear terms are referred to as summation-type and product-type hLV systems, respectively. Time-discretizations of these systems are considered in the study of integrable systems, and have been shown to be applicable to computing eigenvalues of totally nonnegative (TN) matrices. Monotonic convergence to eigenvalues of TN matrices, however, has not yet been observed in the time-discretization of the product-type hLV system. In this paper, we show the existence of a center manifold associated with the time-discretization of the product-type hLV system, and then clarify how the solutions approach an equilibrium corresponding to the eigenvalues of TN matrices in the final phase of convergence.

---

A. Tobita
Department of Mathematical Science for Information Sciences, Graduate School of Science, Tokyo University of Science, 162-8601 Tokyo, Japan
e-mail: ak.tobita@gmail.com

A. Fukuda (✉)
Department of Mathematical Sciences, Shibaura Institute of Technology, 337-8570 Saitama, Japan
e-mail: afukuda@shibaura-it.ac.jp

E. Ishiwata
Department of Mathematical Information Science, Tokyo University of Science, 162-8601 Tokyo, Japan
e-mail: ishiwata@rs.tus.ac.jp

M. Iwasaki
Faculty of Life and Environmental Sciences, Kyoto Prefectural University, 606-8522 Kyoto, Japan
e-mail: imasa@kpu.ac.jp

Y. Nakamura
Graduate School of Informatics, Kyoto University, 606-8501 Kyoto, Japan
e-mail: ynaka@i.kyoto-u.ac.jp



T. Sakurai et al. (eds.), *Eigenvalue Problems: Algorithms, Software and Applications in Petascale Computing*, Lecture Notes in Computational Science and Engineering 117, https://doi.org/10.1007/978-3-319-62426-6_11

# 1 Introduction

Predator-prey models in mathematical biology are often investigated through the analysis of dynamical systems known as Lotka-Volterra (LV) systems. One of the simplest LV systems is an integrable LV system, whose solution can be explicitly expressed using Hankel determinants [24]. The existence of explicit determinant solutions is sometimes said to be evidence of integrable systems. In this paper, we refer to the integrable LV system as "the LV system" for simplicity. The LV system describes the predator-prey relationships of $2m-1$ species, with the assumption that species $k$ preys on species $k+1$ for $k = 1, 2, \ldots, 2m-2$. Species 1 has no predator and species $2m-1$ has no prey in the LV system.

Two types of extension of the LV system have been derived in the study of integrable systems. Extended LV systems where each predator is assumed to prey on two or more species are called hungry LV (hLV) systems. One hLV system with positive integer $M$ involves summations [2, 10, 15],

$$\begin{cases} \dfrac{du_k(t)}{dt} = u_k(t)\left(\displaystyle\sum_{p=1}^{M} u_{k+p}(t) - \sum_{p=1}^{M} u_{k-p}(t)\right), \quad k = 1, 2, \ldots, M_m, \\ u_{1-M}(t) := 0, \ \ldots, \ u_0(t) := 0, \quad u_{M_m+1}(t) := 0, \ \ldots, \ u_{M_m+M}(t) := 0, \quad t \geq 0, \end{cases} \tag{1}$$

where $u_k(t)$ corresponds to the number of the $k$th species at continuous-time $t$ and $M_k := (M+1)k - M$. The other hLV system with positive integer $M$ involves products, as follows [2, 10].

$$\begin{cases} \dfrac{dv_k(t)}{dt} = v_k(t)\left(\displaystyle\prod_{p=1}^{M} v_{k+p}(t) - \prod_{p=1}^{M} v_{k-p}(t)\right), \quad k = 1, 2, \ldots, M_m + M - 1, \\ v_{1-M}(t) := 0, \ \ldots, \ v_0(t) := 0, \quad v_{M_m+M}(t) := 0, \ \ldots, \ v_{M_m+2M-1}(t) := 0, \quad t \geq 0. \end{cases} \tag{2}$$

Here, we denote the summation-type hLV system (1) and the product-type hLV system (2) as the $\mathrm{hLV_I}$ system and the $\mathrm{hLV_{II}}$ system, respectively. Note that both of these are, in the case $M = 1$, simply the LV system.

A skillful time-discretization of the LV system leads to the discrete LV (dLV) system, which has a solution [9, 21] that can be explicitly expressed using the same type of Hankel determinants as in the continuous-time case. Time-discretizations of the hLV$_{\mathrm{I}}$ system (1) and the hLV$_{\mathrm{II}}$ system (2) are respectively given as

$$\begin{cases} u_k^{(n+1)} = u_k^{(n)} \prod_{j=1}^{M} \dfrac{1+\delta^{(n)} u_{k+j}^{(n)}}{1+\delta^{(n+1)} u_{k-j}^{(n+1)}}, \quad k = 1, 2, \ldots, M_m, \\ u_{1-M}^{(n)} := 0, \ldots, u_0^{(n)} := 0, \quad u_{M_m+1}^{(n)} := 0, \ldots, u_{M_m+M}^{(n)} := 0, \quad n = 0, 1, \ldots, \end{cases} \tag{3}$$

$$\begin{cases} v_k^{(n+1)} = v_k^{(n)} \dfrac{1+\delta^{(n)} \prod_{j=1}^{M} v_{k+j}^{(n)}}{1+\delta^{(n+1)} \prod_{j=1}^{M} v_{k-j}^{(n+1)}}, \quad k = 1, 2, \ldots, M_m + M - 1, \\ v_{1-M}^{(n)} := 0, \ldots, v_0^{(n)} := 0, \quad v_{M_m+M}^{(n)} := 0, \ldots, v_{M_m+M+(M-1)}^{(n)} := 0, \quad n = 0, 1, \ldots, \end{cases} \tag{4}$$

where superscripts with parentheses denote the discrete-time, and $\delta^{(n)}$ are discretization parameters [9, 19]. Note that the discrete hLV$_{\mathrm{I}}$ (dhLV$_{\mathrm{I}}$) system (3), and the discrete hLV$_{\mathrm{II}}$ (dhLV$_{\mathrm{II}}$) system (4) are both simply the dLV system in the case $M = 1$. The solution to the dhLV$_{\mathrm{I}}$ system (3) has been written using the Casorati determinants [9, 17]. In contrast, a solution to the dhLV$_{\mathrm{II}}$ system (4) has not yet been explicitly shown.

Some authors have presented asymptotic analysis for the dLV system and dhLV$_{\mathrm{I}}$ system (3) as $n \to \infty$ by examining their determinant solutions [11, 17], associating them with the *LR* transformations [5, 7, 11, 12], or by applying the center manifold theory to them [6, 13, 20]. Here, the existence of a center manifold enables us to investigate local convergence and stability around equilibria [3, 23]. The center manifold approach actually proved that the solutions to the dLV system and dhLV$_{\mathrm{I}}$ systems (3) monotonically converge as $n \to \infty$. However, we did not observe monotonicity in the global analysis of the dLV system and dhLV$_{\mathrm{I}}$ system (3) as $n \to \infty$.

These asymptotic analyses contributed to the design of numerical algorithms based on the dLV system and dhLV$_{\mathrm{I}}$ system (3) for computing singular values of bidiagonal matrices and eigenvalues of totally nonnegative (TN) matrices, respectively. The monotonic convergence is also a desirable property in designing numerical algorithms based on the dLV system and dhLV$_{\mathrm{I}}$ system (3). This is because the accuracy of computed eigenvalues is improved as the iteration number grows larger. TN matrices appear in several branches of mathematics, such as stochastic processes, probability, and combinatorics [1, 4, 14, 16]. Sun et al. showed that the dhLV$_{\mathrm{II}}$ system (4) can be applied for computing eigenvalues of band matrices, which are essentially equivalent to TN matrices [18], but the monotonic convergence to matrix eigenvalues has not yet been reported. Watkins [22] and Fukuda et al. [7] presented the relationship between band matrices and TN matrices.

In this paper, using the center manifold theory, we clarify the asymptotic behavior of the dhLV$_{\mathrm{II}}$ variables as $n \to \infty$. The center manifold theory is a classical analytic tool that requires neither explicit expressions of solutions nor relationships to the *LR* transformations. It enables us to examine how to approach the equilibria of the solutions in the final phase of the convergence, rather than global convergence of the solutions.

The remainder of this paper is organized as follows. In Sect. 2, we first clarify the existence of a center manifold associated with the dhLV$_{\mathrm{II}}$ system (4). In Sect. 3, by applying the center manifold theory to the dhLV$_{\mathrm{II}}$ system (4), we show the asymptotic behavior of the solutions to the dhLV$_{\mathrm{II}}$ system (4) around their equilibria. Finally, we give concluding remarks in Sect. 4.

## 2 Existence of the Center Manifold

In this section, we show the existence of a center manifold associated with the dhLV$_{\mathrm{II}}$ system (4), in particular, that for an auxiliary discrete system of the dhLV$_{\mathrm{II}}$ system (4).

For convenience of explanation, we employ the subscripts $M_k + \ell$ instead of $k$ in the dhLV$_{\mathrm{II}}$ system (4). The use of the subscripts $M_k + \ell$, of course, covers and distinguishes all of the dhLV$_{\mathrm{II}}$ variables. According to Sun et al. [18], the dhLV$_{\mathrm{II}}$ variables $v_{M_k}^{(n)}, v_{M_k+1}^{(n)}, \ldots, v_{M_k+M-1}^{(n)}$ and $v_{M_k+M}^{(n)}$ converge to positive constants $c_{M_k}, c_{M_k+1}, \ldots, c_{M_k+M-1}$ and $0$ as $n \to \infty$, respectively, where $c_{M_k+\ell} > c_{M_{k+1}+\ell}$ for $\ell = 0, 1, \ldots, M-1$. The equilibria $c_{M_k}, c_{M_k+1}, \ldots, c_{M_k+M-1}$ have also been shown to be related to eigenvalues of TN matrices. Now, we introduce auxiliary variables given by subtracting the equilibria $c_{M_k}, c_{M_k+1}, \ldots, c_{M_k+M-1}$ and $0$ from the dhLV$_{\mathrm{II}}$ variables $v_{M_k}^{(n)}, v_{M_k+1}^{(n)}, \ldots, v_{M_k+M-1}^{(n)}$ and $v_{M_k+M}^{(n)}$, respectively, as

$$\bar{v}_{M_k+\ell}^{(n)} = v_{M_k+\ell}^{(n)} - c_{M_k+\ell}, \quad k = 1, 2, \ldots, m, \quad \ell = 0, 1, \ldots, M-1, \tag{5}$$

$$\bar{v}_{M_k+M}^{(n)} = v_{M_k+M}^{(n)}, \quad k = 1, 2, \ldots, m-1. \tag{6}$$

We then divide the dhLV$_{\mathrm{II}}$ system (4) into linear terms and other terms with respect to $\bar{v}_{M_1}^{(n)}, \bar{v}_{M_1+1}^{(n)}, \ldots, \bar{v}_{M_m+M-1}^{(n)}$ and $\bar{v}_{M_1}^{(n+1)}, \bar{v}_{M_1+1}^{(n+1)}, \ldots, \bar{v}_{M_m+M-1}^{(n+1)}$.

**Lemma 1** *Let us assume that the discretization parameter $\delta^{(n+1)}$ in the* dhLV$_{\mathrm{II}}$ *system* (4) *satisfies*

$$|\delta^{(n+1)}| < \left[ \max\left( \max_{k,\ell} \left| \prod_{p=1}^{M} v_{M_k+\ell-p}^{(n+1)} \right|, \max_{k} \left| \prod_{p=0}^{M-1} v_{M_k+p}^{(n+1)} - 2\prod_{p=0}^{M-1} c_{M_k+p} \right| \right) \right]^{-1}. \tag{7}$$

*Let* $\alpha_{k,\ell} := \prod_{p=\ell+1}^{M-1} c_{M_k+p} \prod_{p=0}^{\ell-1} c_{M_{k+1}+p}$, $\beta_k := (1 + \delta^{(n)} \prod_{p=0}^{M-1} c_{M_{k+1}+p})/(1 + \delta^{(n+1)} \prod_{p=0}^{M-1} c_{M_k+p})$ *and* $\boldsymbol{v}^{(n)} := (\bar{v}_{M_1}^{(n)}, \bar{v}_{M_1+1}^{(n)}, \ldots, \bar{v}_{M_m+M-1}^{(n)})^\top \in \mathbb{R}^{M_m+M-1}$. *Then, it holds that*

$$\bar{v}_{M_k+\ell}^{(n+1)} = -\delta^{(n+1)} c_{M_k+\ell} \alpha_{k-1,\ell} \bar{v}_{M_{k-1}+M}^{(n+1)} + \bar{v}_{M_k+\ell}^{(n)} + \delta^{(n)} c_{M_k+\ell} \alpha_{k,\ell} \bar{v}_{M_k+M}^{(n)} + \hat{f}_{M_k+\ell}(\boldsymbol{v}^{(n)}, \boldsymbol{v}^{(n+1)}), \quad k = 1, 2, \ldots, m, \quad \ell = 0, 1, \ldots, M-1, \tag{8}$$

$$\bar{v}_{M_k+M}^{(n+1)} = \beta_k \bar{v}_{M_k+M}^{(n)} + \hat{g}_{M_k+M}(\boldsymbol{v}^{(n)}, \boldsymbol{v}^{(n+1)}), \quad k = 1, 2, \ldots, m-1, \tag{9}$$

*where* $\hat{f}_{M_k+\ell}(\boldsymbol{v}^{(n)}, \boldsymbol{v}^{(n+1)})$ *and* $\hat{g}_{M_k+M}(\boldsymbol{v}^{(n)}, \boldsymbol{v}^{(n+1)})$ *denote multivariate nonlinear polynomials with respect to entries of* $\boldsymbol{v}^{(n)}$ *and* $\boldsymbol{v}^{(n+1)}$.

*Proof* Under the assumption of $\delta^{(n+1)}$, noting that $1/(1+x) = 1 + \sum_{j=1}^{\infty}(-x)^j$ for $|x| < 1$, we can rewrite the equations, except for $\ell = M$, in the dhLV$_{\text{II}}$ system (4) as

$$v_{M_k+\ell}^{(n+1)} = v_{M_k+\ell}^{(n)} \left(1 + \delta^{(n)} \prod_{p=1}^{M} v_{M_k+\ell+p}^{(n)}\right) \left[1 + \sum_{j=1}^{\infty} \left(-\delta^{(n+1)} \prod_{p=1}^{M} v_{M_k+\ell-p}^{(n+1)}\right)^j\right],$$
$$k = 1, 2, \ldots, m, \quad \ell = 1, 2, \ldots, M-1. \tag{10}$$

Here, by considering $M_k = M_{k-1} + M + 1$, we can express two products, $\prod_{p=1}^{M} v_{M_k+\ell+p}^{(n)}$ and $\prod_{p=1}^{M} v_{M_k+\ell-p}^{(n+1)}$, using $\bar{v}_{M_k+\ell+1}^{(n)}, \bar{v}_{M_k+\ell+2}^{(n)}, \ldots, \bar{v}_{M_{k+1}+\ell-1}^{(n)}$ and $\bar{v}_{M_{k-1}+\ell+1}^{(n+1)}, \bar{v}_{M_{k-1}+\ell+2}^{(n+1)}, \ldots, \bar{v}_{M_k+\ell-1}^{(n+1)}$, respectively, as

$$\prod_{p=1}^{M} v_{M_k+\ell+p}^{(n)} = \left(\prod_{p=\ell+1}^{M-1} v_{M_k+p}^{(n)}\right) v_{M_k+M}^{(n)} \left(\prod_{p=0}^{\ell-1} v_{M_{k+1}+p}^{(n)}\right)$$
$$= \left[\prod_{p=\ell+1}^{M-1} (\bar{v}_{M_k+p}^{(n)} + c_{M_k+p})\right] \bar{v}_{M_k+M}^{(n)} \left[\prod_{p=0}^{\ell-1} (\bar{v}_{M_{k+1}+p}^{(n)} + c_{M_{k+1}+p})\right], \tag{11}$$

$$\prod_{p=1}^{M} v_{M_k+\ell-p}^{(n+1)} = \left(\prod_{p=\ell+1}^{M-1} v_{M_{k-1}+p}^{(n+1)}\right) v_{M_{k-1}+M}^{(n+1)} \left(\prod_{p=0}^{\ell-1} v_{M_k+p}^{(n+1)}\right)$$
$$= \left[\prod_{p=\ell+1}^{M-1} (\bar{v}_{M_{k-1}+p}^{(n+1)} + c_{M_{k-1}+p})\right] \bar{v}_{M_{k-1}+M}^{(n+1)} \left[\prod_{p=0}^{\ell-1} (\bar{v}_{M_k+p}^{(n+1)} + c_{M_k+p})\right], \tag{12}$$

where $\prod_{p=0}^{-1}(\bar{v}_{M_{k+1}+p}^{(n)} + c_{M_{k+1}+p}) := 1$, $\prod_{p=0}^{-1}(\bar{v}_{M_k+p}^{(n+1)} + c_{M_k+p}) := 1$, $\prod_{p=M}^{M-1}(\bar{v}_{M_k+p}^{(n)} + c_{M_k+p}) := 1$, and $\prod_{p=M}^{M-1}(\bar{v}_{M_{k-1}+p}^{(n+1)} + c_{M_{k-1}+p}) := 1$. Moreover, by expanding the right-hand sides of (11) and (12) and taking into account $\prod_{p=\ell+1}^{M-1} c_{M_k+p} \prod_{p=0}^{\ell-1} c_{M_{k+1}+p} = \alpha_{k,\ell}$, we derive

$$\prod_{p=1}^{M} v_{M_k+\ell+p}^{(n)} = \alpha_{k,\ell}\bar{v}_{M_k+M}^{(n)} + \bar{f}(\bar{v}_{M_k+\ell+1}^{(n)}, \bar{v}_{M_k+\ell+2}^{(n)}, \ldots, \bar{v}_{M_{k+1}+\ell-1}^{(n)}), \tag{13}$$

$$\prod_{p=1}^{M} v_{M_k+\ell-p}^{(n+1)} = \alpha_{k-1,\ell}\bar{v}_{M_{k-1}+M}^{(n+1)} + \bar{f}(\bar{v}_{M_{k-1}+\ell+1}^{(n+1)}, \bar{v}_{M_{k-1}+\ell+2}^{(n+1)}, \ldots, \bar{v}_{M_k+\ell-1}^{(n+1)}), \tag{14}$$

where $\bar{f}(\bar{v}_{M_k+\ell+1}^{(n)}, \bar{v}_{M_k+\ell+2}^{(n)}, \ldots, \bar{v}_{M_{k+1}+\ell-1}^{(n)})$ and $\bar{f}(\bar{v}_{M_{k-1}+\ell+1}^{(n+1)}, \bar{v}_{M_{k-1}+\ell+2}^{(n+1)}, \ldots, \bar{v}_{M_k+\ell-1}^{(n+1)})$ are multivariate polynomials with respect to $\bar{v}_{M_k+\ell+1}^{(n)}, \bar{v}_{M_k+\ell+2}^{(n)}, \ldots, \bar{v}_{M_{k+1}+\ell-1}^{(n)}$, and $\bar{v}_{M_{k-1}+\ell+1}^{(n+1)}, \bar{v}_{M_{k-1}+\ell+2}^{(n+1)}, \ldots, \bar{v}_{M_k+\ell-1}^{(n+1)}$ involving no terms of order 0 or 1, respectively. Using (13) and (14), we observe that the term $-\delta^{(n+1)} v_{M_k+\ell}^{(n)} \prod_{p=1}^{M} v_{M_k+\ell-p}^{(n+1)} + v_{M_k+\ell}^{(n)} + \delta^{(n)} v_{M_k+\ell}^{(n)} \prod_{p=1}^{M} v_{M_k+\ell+p}^{(n)}$ on the right-hand side of (10) coincides with the sum of the constant $c_{M_k+\ell}$, the multivariate linear polynomial $-\delta^{(n+1)} c_{M_k+\ell}\alpha_{k-1,\ell}\bar{v}_{M_{k-1}+M}^{(n+1)} + \bar{v}_{M_k+\ell}^{(n)} + \delta^{(n)} c_{M_k+\ell}\alpha_{k,\ell}\bar{v}_{M_k+M}^{(n)}$, and the multivariate nonlinear polynomial with respect to $\bar{v}_{M_k+\ell+1}^{(n)}, \bar{v}_{M_k+\ell+2}^{(n)}, \ldots, \bar{v}_{M_{k+1}+\ell-1}^{(n)}$, and $\bar{v}_{M_{k-1}+\ell+1}^{(n+1)}, \bar{v}_{M_{k-1}+\ell+2}^{(n+1)}, \ldots, \bar{v}_{M_k+\ell-1}^{(n+1)}$. We can simultaneously regard the other term on the right-hand side of (10) as the multivariate nonlinear polynomial with respect to $\bar{v}_{M_k+\ell+1}^{(n)}, \bar{v}_{M_k+\ell+2}^{(n)}, \ldots, \bar{v}_{M_{k+1}+\ell-1}^{(n)}$, and $\bar{v}_{M_{k-1}+\ell+1}^{(n+1)}, \bar{v}_{M_{k-1}+\ell+2}^{(n+1)}, \ldots, v_{M_k+\ell-1}^{(n+1)}$. Since it holds that $v_{M_k+\ell}^{(n+1)} = \bar{v}_{M_k+\ell}^{(n+1)} + c_{M_k+\ell}$ on the left-hand side of (10), we therefore have (8).

Furthermore, let $\bar{g}(\bar{v}_{M_k}^{(n)}, \bar{v}_{M_k+1}^{(n)}, \ldots, \bar{v}_{M_k+M-1}^{(n)}) := \prod_{p=0}^{M-1} v_{M_k+p}^{(n)} - \gamma_k$, where $\gamma_k := \prod_{p=0}^{M-1} c_{M_k+p}$. Then, it is obvious that $\bar{g}(\bar{v}_{M_k}^{(n)}, \bar{v}_{M_k+1}^{(n)}, \ldots, \bar{v}_{M_k+M-1}^{(n)})$ are multivariate polynomials with respect to $\bar{v}_{M_k}^{(n)}, \bar{v}_{M_k+1}^{(n)}, \ldots, \bar{v}_{M_k+M-1}^{(n)}$ and without constants. By using (5) and (6) in the dhLV$_{\rm II}$ system (4) with $\ell = M$, and $\delta^{(n+1)}$ satisfying (7), we derive

$$\begin{aligned}
\bar{v}_{M_k+M}^{(n+1)} &= \frac{1+\delta^{(n)}\gamma_{k+1}}{1+\delta^{(n+1)}\gamma_k}\bar{v}_{M_k+M}^{(n)} \frac{1 + \dfrac{\delta^{(n)}\bar{g}(\bar{v}_{M_{k+1}}^{(n)}, \bar{v}_{M_{k+1}+1}^{(n)}, \ldots, \bar{v}_{M_{k+1}+M-1}^{(n)})}{1+\delta^{(n)}\gamma_{k+1}}}{1 + \dfrac{\delta^{(n+1)}\bar{g}(\bar{v}_{M_k}^{(n+1)}, \bar{v}_{M_k+1}^{(n+1)}, \ldots, \bar{v}_{M_k+M-1}^{(n+1)})}{1+\delta^{(n+1)}\gamma_k}} \\
&= \beta_k \bar{v}_{M_k+M}^{(n)} \left(1 + \frac{\delta^{(n)}\bar{g}(\bar{v}_{M_{k+1}}^{(n)}, \bar{v}_{M_{k+1}+1}^{(n)}, \ldots, \bar{v}_{M_{k+1}+M-1}^{(n)})}{1+\delta^{(n)}\gamma_{k+1}}\right)
\end{aligned}$$

$$\times \left[ 1 + \sum_{j=1}^{\infty} \left( -\frac{\delta^{(n+1)} \bar{g}(\bar{v}_{M_k}^{(n+1)}, \bar{v}_{M_k+1}^{(n+1)}, \ldots, \bar{v}_{M_k+M-1}^{(n+1)})}{1 + \delta^{(n+1)} \gamma_k} \right)^j \right],$$

$$k = 1, 2, \ldots, m-1. \tag{15}$$

The right-hand side of (15) comprises multivariate nonlinear polynomials with respect to $\bar{v}_{M_k+M}^{(n)}$ and $\bar{v}_{M_{k+1}}^{(n)}, \bar{v}_{M_{k+1}+1}^{(n)}, \ldots, \bar{v}_{M_{k+1}+M-1}^{(n)}$ and $\bar{v}_{M_k}^{(n+1)}, \bar{v}_{M_k+1}^{(n+1)}, \ldots, \bar{v}_{M_k+M-1}^{(n+1)}$, except for $\beta_k \bar{v}_{M_k+M}^{(n)}$. This immediately leads to (9). □

Noting that $v_k^{(n+1)}$ are given using $v_1^{(n)}, v_2^{(n)}, \ldots, v_{k+M}^{(n)}$ in the dhLV$_{\text{II}}$ system (4), we immediately see that the entries of $\boldsymbol{v}^{(n+1)}$ can be written using those of $\boldsymbol{v}^{(n)}$. Thus, we derive a lemma concerning $\hat{f}_{M_k+\ell}(\boldsymbol{v}^{(n)}, \boldsymbol{v}^{(n+1)})$ and $\hat{g}_{M_k+M}(\boldsymbol{v}^{(n)}, \boldsymbol{v}^{(n+1)})$ involving $\boldsymbol{v}^{(n+1)}$.

**Lemma 2** *Let us assume that* $\delta^{(n+1)}$ *in the* dhLV$_{\text{II}}$ *system* (4) *satisfies* (7). *There exist multivariate nonlinear polynomials with respect to entries of* $\boldsymbol{v}^{(n)}$, *denoted by* $\check{f}_{M_k+\ell}(\boldsymbol{v}^{(n)})$ *and* $\check{g}_{M_k+M}(\boldsymbol{v}^{(n)})$, *such that*

$$\check{f}_{M_k+\ell}(\boldsymbol{v}^{(n)}) = \hat{f}_{M_k+\ell}(\boldsymbol{v}^{(n)}, \boldsymbol{v}^{(n+1)}), \quad k = 1, 2, \ldots, m, \quad \ell = 0, 1, \ldots, M-1, \tag{16}$$

$$\check{g}_{M_k+M}(\boldsymbol{v}^{(n)}) = \hat{g}_{M_k+M}(\boldsymbol{v}^{(n)}, \boldsymbol{v}^{(n+1)}), \quad k = 1, 2, \ldots, m-1. \tag{17}$$

*Proof* Since $v_{M_0+1}^{(n+1)} = 0, v_{M_0+2}^{(n+1)} = 0, \ldots, v_{M_0+M}^{(n+1)} = 0$ in (10) with (11) and (12), it is obvious that $\hat{f}_{M_1}(\boldsymbol{v}^{(n)}, \boldsymbol{v}^{(n+1)})$ is the multivariate nonlinear polynomial with respect to $\bar{v}_{M_1}^{(n)}, \bar{v}_{M_1+1}^{(n)}, \ldots, \bar{v}_{M_1+M}^{(n)}$. Thus, considering this in (8) with $k = 1$ and $\ell = 0$, we can regard $\bar{v}_{M_1}^{(n+1)}$ as a multivariate polynomial with respect to $\bar{v}_{M_1}^{(n)}, \bar{v}_{M_1+1}^{(n)}, \ldots, \bar{v}_{M_1+M}^{(n)}$ and without constants. For $\ell' = 1, 2, \ldots, \ell-1$, let us assume that $\bar{v}_{M_k+\ell'}^{(n+1)}$ are multivariate polynomials with respect to $\bar{v}_{M_1}^{(n)}, \bar{v}_{M_1+1}^{(n)}, \ldots, \bar{v}_{M_{k+1}+\ell'-1}^{(n)}$ and without constants. Let us recall that $\hat{f}_{M_k+\ell}(\boldsymbol{v}^{(n)}, \boldsymbol{v}^{(n+1)})$ is the multivariate nonlinear polynomial with respect to $\bar{v}_{M_k+\ell+1}^{(n)}, \bar{v}_{M_k+\ell+2}^{(n)}, \ldots, \bar{v}_{M_{k+1}+\ell-1}^{(n)}$ and $\bar{v}_{M_{k-1}+\ell+1}^{(n+1)}, \bar{v}_{M_{k-1}+\ell+2}^{(n+1)}, \ldots, \bar{v}_{M_k+\ell-1}^{(n+1)}$. Then, we find that $\hat{f}_{M_k+\ell}(\boldsymbol{v}^{(n)}, \boldsymbol{v}^{(n+1)})$ is also a multivariate nonlinear polynomial with respect to $\bar{v}_{M_1}^{(n)}, \bar{v}_{M_1+1}^{(n)}, \ldots, \bar{v}_{M_{k+1}+\ell-1}^{(n)}$. By combining these with (8), we see that $\bar{v}_{M_k+\ell}^{(n+1)}$ are multivariate polynomials with respect to $\bar{v}_{M_1}^{(n)}, \bar{v}_{M_1+1}^{(n)}, \ldots, \bar{v}_{M_{k+1}+\ell-1}^{(n)}$ and without constants. Simultaneously, we observe that $\hat{f}_{M_k+\ell+1}(\boldsymbol{v}^{(n)}, \boldsymbol{v}^{(n+1)})$ is also a multivariate nonlinear polynomial with respect to $\bar{v}_{M_1}^{(n)}, \bar{v}_{M_1+1}^{(n)}, \ldots, \bar{v}_{M_{k+1}+\ell}^{(n)}$. Therefore, by induction for $k$ and $\ell$, we obtain (16).

Similarly, by taking into account the fact that $\hat{g}_{M_k+M}(\boldsymbol{v}^{(n)}, \boldsymbol{v}^{(n+1)})$ is a multivariate polynomial with respect to $\bar{v}_{M_k+M}^{(n)}$ and $\bar{v}_{M_{k+1}}^{(n)}, \bar{v}_{M_{k+1}+1}^{(n)}, \ldots, \bar{v}_{M_{k+1}+M-1}^{(n)}$ and $\bar{v}_{M_k}^{(n+1)}, \bar{v}_{M_k+1}^{(n+1)}, \ldots, \bar{v}_{M_k+M-1}^{(n+1)}$, we have (17). □

For joining linear terms with respect to entries of $\boldsymbol{v}^{(n)}$ in (8), let us introduce new variables

$$w_{M_k+\ell}^{(n)} := -\xi_{k,\ell}\bar{v}_{M_{k-1}+M}^{(n)} + \bar{v}_{M_k+\ell}^{(n)} + \eta_{k,\ell}\bar{v}_{M_k+M}^{(n)},$$
$$k = 1, 2, \ldots, m, \quad \ell = 0, 1, \ldots, M-1, \tag{18}$$

where $\xi_{k,\ell} := (\delta^{(n+1)} c_{M_k+\ell}\, \alpha_{k-1,\ell}\, \beta_{k-1})/(1-\beta_{k-1})$ and $\eta_{k,\ell} := (\delta^{(n)} c_{M_k+\ell}\, \alpha_{k,\ell}(1 + \delta^{(n+1)}\gamma_k))/(\delta^{(n+1)}\gamma_k - \delta^{(n)}\gamma_{k+1})$. The following lemma then gives the evolution from $n$ to $n+1$ of $\boldsymbol{w}^{(n)} := (\boldsymbol{w}_1^{(n)}, \boldsymbol{w}_2^{(n)}, \ldots, \boldsymbol{w}_m^{(n)})^\top \in \mathbb{R}^{M_m+M-m}$ with $\boldsymbol{w}_k^{(n)} := (w_{M_k}^{(n)}, w_{M_k+1}^{(n)}, \ldots, w_{M_k+M-1}^{(n)})^\top \in \mathbb{R}^M$ and $\boldsymbol{v}_0^{(n)} := (\bar{v}_{M_1+M}^{(n)}, \bar{v}_{M_2+M}^{(n)}, \ldots, \bar{v}_{M_{m-1}+M}^{(n)})^\top \in \mathbb{R}^{m-1}$.

**Lemma 3** *The* dhLV$_{\text{II}}$ *system* (4) *with* $\delta^{(n+1)}$ *satisfying* (7) *can be rewritten using* $A := \text{diag}(1, 1, \ldots, 1) \in \mathbb{R}^{(M_m+M-m)\times(M_m+M-m)}$ *and* $B := \text{diag}(\beta_1, \beta_2, \ldots, \beta_{m-1}) \in \mathbb{R}^{(m-1)\times(m-1)}$ *as*

$$\boldsymbol{w}^{(n+1)} = A\boldsymbol{w}^{(n)} + \boldsymbol{f}(\boldsymbol{w}^{(n)}, \boldsymbol{v}_0^{(n)}), \tag{19}$$

$$\boldsymbol{v}_o^{(n+1)} = B\boldsymbol{v}_0^{(n)} + \boldsymbol{g}(\boldsymbol{w}^{(n)}, \boldsymbol{v}_0^{(n)}), \tag{20}$$

*where the functions* $\boldsymbol{f} : \mathbb{R}^{M_m+M-1} \to \mathbb{R}^{M_m+M-m}$ *and* $\boldsymbol{g} : \mathbb{R}^{M_m+M-1} \to \mathbb{R}^{m-1}$ *with respect to* $\boldsymbol{w}^{(n)}$ *and* $\boldsymbol{v}_0^{(n)}$ *and their first derivatives* $\nabla\boldsymbol{f}$ *and* $\nabla\boldsymbol{g}$ *satisfy*

$$\boldsymbol{f}(\boldsymbol{0}, \boldsymbol{0}) = \boldsymbol{0}, \quad \nabla\boldsymbol{f}(\boldsymbol{0}, \boldsymbol{0}) = \boldsymbol{O}, \tag{21}$$

$$\boldsymbol{g}(\boldsymbol{0}, \boldsymbol{0}) = \boldsymbol{0}, \quad \nabla\boldsymbol{g}(\boldsymbol{0}, \boldsymbol{0}) = \boldsymbol{O}. \tag{22}$$

*Proof* Using Lemmas 1 and 2, we can rewrite $w_{M_k+\ell}^{(n+1)}$ using $\bar{v}_{M_1}^{(n)}, \bar{v}_{M_1+1}^{(n)}, \ldots, \bar{v}_{M_k+M}^{(n)}$ as

$$\begin{aligned} w_{M_k+\ell}^{(n+1)} = &- \beta_{k-1}(\xi_{k,\ell} + \delta^{(n+1)} c_{M_k+\ell}\alpha_{k-1,\ell})\bar{v}_{M_{k-1}+M}^{(n)} + \bar{v}_{M_k+\ell}^{(n)} \\ &+ (\beta_k\eta_{k,\ell} + \delta^{(n)} c_{M_k+\ell}\alpha_{k,\ell})\bar{v}_{M_k+M}^{(n)} + \check{f}_{M_k+\ell}(\boldsymbol{v}^{(n)}) \\ &- (\xi_{k,\ell} + \delta^{(n+1)} c_{M_k+\ell}\alpha_{k-1,\ell})\check{g}_{M_{k-1}+M}(\boldsymbol{v}^{(n)}) + \eta_{k,\ell}\check{g}_{M_k+M}(\boldsymbol{v}^{(n)}). \end{aligned} \tag{23}$$

From the definitions of $\alpha_{k,\ell}$ and $\beta_k$ and $\xi_{k,\ell}$ and $\eta_{k,\ell}$ in terms of $c_{M_k+\ell}$, we can verify that $\xi_{k,\ell} + \delta^{(n+1)} c_{M_k+\ell}\alpha_{k-1,\ell} = \xi_{k,\ell}/\beta_{k-1}$ and $\beta_k\eta_{k,\ell} + \delta^{(n)} c_{M_k+\ell}\alpha_{k,\ell} = \eta_{k,\ell}$. We can then derive

$$w_{M_k+\ell}^{(n+1)} = w_{M_k+\ell}^{(n)} + \tilde{f}_{M_k+\ell}(\boldsymbol{v}^{(n)}), \tag{24}$$

where $\tilde{f}_{M_k+\ell}(\boldsymbol{v}^{(n)}) := \check{f}_{M_k+\ell}(\boldsymbol{v}^{(n)}) - (\xi_{k,\ell}/\beta_{k-1})\check{g}_{M_{k-1}+M}(\boldsymbol{v}^{(n)}) + \eta_{k,\ell}\check{g}_{M_k+M}(\boldsymbol{v}^{(n)})$. Since $\check{f}_{M_k+\ell}(\boldsymbol{v}^{(n)})$, $\check{g}_{M_{k-1}+M}(\boldsymbol{v}^{(n)})$, and $\check{g}_{M_k+M}(\boldsymbol{v}^{(n)})$ are multivariate nonlinear polynomials with respect to the entries of $\boldsymbol{v}^{(n)}$, their linear combinations $\tilde{f}_{M_k+\ell}(\boldsymbol{v}^{(n)})$ are also. Moreover, there exist multivariate nonlinear polynomials with respect to the entries of $\boldsymbol{w}^{(n)}$ and $\boldsymbol{v}_0^{(n)}$, denoted by $f_{M_k+\ell}(\boldsymbol{w}^{(n)}, \boldsymbol{v}_0^{(n)})$, such that $f_{M_k+\ell}(\boldsymbol{w}^{(n)}, \boldsymbol{v}_0^{(n)}) = \tilde{f}_{M_k+\ell}(\boldsymbol{v}^{(n)})$ for each $k$ and $\ell$. This is because some entries of $\boldsymbol{v}^{(n)}$ are equal to entries of $\boldsymbol{v}_0^{(n)}$ and the others can be rewritten using (18) as linear combinations of the entries of $\boldsymbol{w}^{(n)}$ and $\boldsymbol{v}_0^{(n)}$. Thus, we observe that $f_{M_k+\ell}(\boldsymbol{0}, \boldsymbol{0}) = 0$ and $\nabla f_{M_k+\ell}(\boldsymbol{0}, \boldsymbol{0}) = \boldsymbol{0}$. Consequently, by letting $\boldsymbol{f} := (\boldsymbol{f}_1, \boldsymbol{f}_2, \dots, \boldsymbol{f}_m)^\top \in \mathbb{R}^{M_m+M-m}$ with $\boldsymbol{f}_k := (f_{M_k}, f_{M_k+1}, \dots, f_{M_k+M-1}) \in \mathbb{R}^M$, we obtain (19) with (21).

Similarly, there exist multivariate nonlinear polynomials $g_k(\boldsymbol{w}^{(n)}, \boldsymbol{v}_0^{(n)})$ satisfying $g_{M_k+M}(\boldsymbol{w}^{(n)}, \boldsymbol{v}_0^{(n)}) = \check{g}_k(\boldsymbol{v}^{(n)})$, $g_{M_k+M}(\boldsymbol{0}, \boldsymbol{0}) = 0$, and $\nabla g_{M_k+M}(\boldsymbol{0}, \boldsymbol{0}) = \boldsymbol{0}$. Thus, we have (20) with (22), where $\boldsymbol{g} := (g_{M_1+M}, g_{M_2+M}, \dots, g_{M_{m-1}+M})^\top \in \mathbb{R}^{m-1}$. □

According to Carr [3] and Wiggins [23], a center manifold for the discrete dynamical system in (19) and (20) exists if the moduli of the eigenvalues of $A$ and $B$ are equal to and are less than 1, respectively. It is important to note that $0 < \beta_k < 1$ if $\delta^{(n+1)} > \delta^{(n)} \prod_{p=0}^{M-1}(c_{M_{k+1}+p}/c_{M_k+p})$. Lemma 3 therefore leads to the following theorem concerning the existence of a center manifold associated with the dhLV$_{\mathrm{II}}$ system (4).

**Theorem 1** *There exists a center manifold $\boldsymbol{h} : \mathbb{R}^{M_m+M-m} \to \mathbb{R}^{m-1}$ for the discrete dynamical system in* (19) *and* (20) *derived from the* dhLV$_{\mathrm{II}}$ *system* (4) *with $\delta^{(n)}$ and $\delta^{(n+1)}$ satisfying* (7) *and*

$$\delta^{(n)} \prod_{p=0}^{M-1} \frac{c_{M_{k+1}+p}}{c_{M_k+p}} < \delta^{(n+1)}. \tag{25}$$

For example, from $c_{M_k+p} < c_{M_{k+1}+p}$, we find that (25) always holds if $\delta^{(n)} = \delta^{(n+1)}$. In this case, the center manifold associated with the dhLV$_{\mathrm{II}}$ system (4) exists only if (7) holds. It is obvious that a sufficiently small $\delta^{(n+1)}$ always satisfies (7).

## 3 Monotonic Convergence

We first determine a center manifold for the discrete system defined in (19) and (20) with (21) and (22), and then show the asymptotic behavior of its solution with the help of theorems concerning the center manifold.

According to Carr [3] and Wiggins [23], the center manifold for the discrete system in (19) and (20) with (21) and (22) is the function $\boldsymbol{h} : \mathbb{R}^{M_m+M-m} \to \mathbb{R}^{m-1}$

with $\boldsymbol{h}(\mathbf{0}) = \mathbf{0}$ and $\nabla\boldsymbol{h}(\mathbf{0}) = \boldsymbol{O}$, such that $\boldsymbol{v}_0^{(n)} = \boldsymbol{h}(\boldsymbol{w}^{(n)})$ and $\boldsymbol{v}_0^{(n+1)} = \boldsymbol{h}(\boldsymbol{w}^{(n+1)})$. In other words, the center manifold $\boldsymbol{h}$ with $\boldsymbol{h}(\mathbf{0}) = \mathbf{0}$ and $\nabla\boldsymbol{h}(\mathbf{0}) = \boldsymbol{O}$ satisfies

$$\boldsymbol{h}(A\boldsymbol{w}^{(n)} + \boldsymbol{f}(\boldsymbol{w}^{(n)}, \boldsymbol{h}(\boldsymbol{w}^{(n)}))) - B\boldsymbol{h}(\boldsymbol{w}^{(n)}) - \boldsymbol{g}(\boldsymbol{w}^{(n)}, \boldsymbol{h}(\boldsymbol{w}^{(n)})) = \mathbf{0}. \tag{26}$$

Since it has been observed that $f_{M_k+\ell}(\boldsymbol{w}^{(n)}, \boldsymbol{v}_0^{(n)})$ and $g_{M_k+M}(\boldsymbol{w}^{(n)}, \boldsymbol{v}_0^{(n)})$ are multivariate nonlinear polynomials with a common factor $\bar{v}_{M_k+M}^{(n)}$ in the proofs of Lemmas 1–3, we obtain $f_{M_k+\ell}(\boldsymbol{w}^{(n)}, \mathbf{0}) = 0$ and $g_{M_k+M}(\boldsymbol{w}^{(n)}, \mathbf{0}) = 0$. This immediately leads to $\boldsymbol{f}(\boldsymbol{w}^{(n)}, \mathbf{0}) = \mathbf{0}$ and $\boldsymbol{g}(\boldsymbol{w}^{(n)}, \mathbf{0}) = \mathbf{0}$. Thus, by letting $\boldsymbol{h}(\boldsymbol{w}^{(n)}) = \mathbf{0}$ and taking into account the fact that $A\boldsymbol{w}^{(n)} = \boldsymbol{w}^{(n)}$ in (26), we derive

$$\boldsymbol{h}(\boldsymbol{w}^{(n)} + \boldsymbol{f}(\boldsymbol{w}^{(n)}, \mathbf{0})) - \boldsymbol{g}(\boldsymbol{w}^{(n)}, \mathbf{0}) = \boldsymbol{h}(\boldsymbol{w}^{(n)}) = \mathbf{0}. \tag{27}$$

Equation (27) implies that $\boldsymbol{h}(\boldsymbol{w}^{(n)}) = \mathbf{0}$ is a solution to (26). Combining this with theorems concerning the center manifold [3, 23], we see that the asymptotic behavior of a small solution to the discrete system in (19) and (20) with (21) and (22) is governed by the reduced discrete system

$$\hat{\boldsymbol{w}}^{(n+1)} = A\hat{\boldsymbol{w}}^{(n)} + \boldsymbol{f}(\hat{\boldsymbol{w}}^{(n)}, \mathbf{0}) = \hat{\boldsymbol{w}}^{(n)}. \tag{28}$$

Obviously, the zero solution $\hat{\boldsymbol{w}}^{(n)} = \mathbf{0}$ is stable. Thus the zero solution to the discrete system (19) and (20) with (21) and (22) is also so. Consequently, we have the following theorem for monotonic convergence in the discrete system in (19) and (20) with (21) and (22) derived from the dhLV$_{\text{II}}$ system (4).

**Theorem 2** *There exists some positive integer $n^*$ such that $|\boldsymbol{w}^{(n)}| < \kappa\omega^{n-n^*}$ and $|\boldsymbol{v}_0^{(n)}| < \kappa\omega^{n-n^*}$ for $n \geq n^*$ if all entries of $\boldsymbol{w}^{(n)}$ and $\boldsymbol{v}_0^{(n)}$ are sufficiently small, where $\kappa$ and $\omega$ are positive constants with $\omega < 1$.*

Theorems 1 and 2 also suggest that the dhLV$_{\text{II}}$ variables $v_{M_k}^{(n)}, v_{M_k+1}^{(n)}, \ldots, v_{M_k+M-1}^{(n)}$ and $v_{M_k+M}^{(n)}$ moderately approach their equilibria in the final phase of convergence if the discretization parameters $\delta^{(n)}$ and $\delta^{(n+1)}$ satisfy (7) and (25). The numerical algorithm based on the dhLV$_{\text{II}}$ system (4) can succeed to this convergence property. For example, in the practical algorithm, we may set $\delta^{(n_i)}, \delta^{(n_i+1)}, \ldots, \delta^{(n_i+n_p)}$ as $\delta^{(n_i)} = \delta^{(n_i+1)} = \cdots = \delta^{(n_i+n_p)} = \delta$ for some integers $n_i$ and $n_p$ where $n_{i+1} = n_i + n_p + 1$, and then we gradually make the value of $\delta$ smaller as $i$ grows larger.

We here give a numerical example concerning the monotonic convergence appearing in the dhLV$_{\text{II}}$ system (4) with $M = 3$, $m = 2$ and $\delta^{(n)} = 1$ for $n = 0, 1, \ldots$. Figure 1 shows differences between the dhLV$_{\text{II}}$ variables $v_1^{(n)}, v_2^{(n)}, \ldots, v_7^{(n)}$ and their equilibria for $n = 0, 1, \ldots, 13$ under the initial settings $v_1^{(0)} = v_2^{(0)} = \cdots = v_7^{(0)} = 1$ where the equilibria were computed beforehand using 200-digits precision arithmetic in Mathematica. We then observe that each $v_k^{(n)}$ approaches monotonically to its equilibrium as $n$ increases.

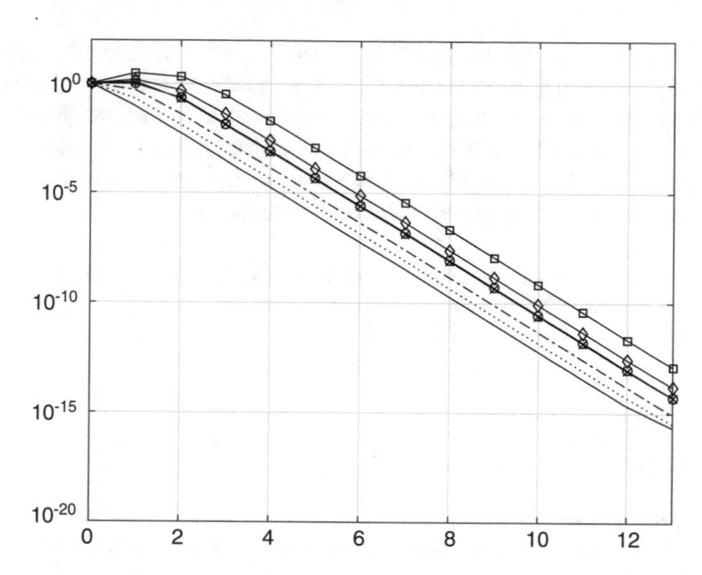


**Fig. 1** Monotonic convergence in the dhLV$_{\text{II}}$ system (4). The horizontal axis is the discrete time $n$ and the vertical axis corresponds to the differences between $v_1^{(n)}, v_2^{(n)}, \ldots, v_7^{(n)}$ and their equilibria for $n = 0, 1, \ldots, 13$. The solid line with *opensquare*: the values of $|v_1^{(n)} - c_1|$, the solid line with *opendiamond*: the values of $|v_2^{(n)} - c_2|$, the solid line with *cross*: the values of $|v_3^{(n)} - c_3|$, the solid line with *opencircle*: the values of $v_4^{(n)}$, and the dashed line: the values of $|v_5^{(n)} - c_5|$, the dotted line: the values of $|v_6^{(n)} - c_6|$, the solid line: the values of $|v_7^{(n)} - c_7|$

# 4 Concluding Remarks

In this paper, we demonstrated monotonic convergence in a time-discretization of the product-type hungry Lotka-Volterra system, referred to as the dhLV$_{\text{II}}$ system, around the equilibria of the dhLV$_{\text{II}}$ variables. This monotonic convergence, of course, also appears in a numerical algorithm based on the dhLV$_{\text{II}}$ system for computing eigenvalues of totally nonnegative (TN) matrices. The monotonic convergence is a valuable property in deciding the stopping criterion of a numerical algorithm in finite arithmetic.

In Sect. 2, we investigated the existence of a center manifold associated with the dhLV$_{\text{II}}$ system. We first considered an auxiliary discrete system given by shifting the equilibria of the dhLV$_{\text{II}}$ variables to the origin, and dividing the discrete system into multivariate linear and nonlinear polynomials with respect to the shifted dhLV$_{\text{II}}$ variables. Next, by simplifying the auxiliary discrete system, we proved that a center manifold associated with the dhLV$_{\text{II}}$ system exists if the discretization parameters are suitable. In Sect. 3, using theorems concerning the center manifold, we finally obtained the monotonic convergence in the dhLV$_{\text{II}}$ system around the solutions' equilibria. A future work for the dhLV$_{\text{II}}$ system is to clarify how to actually find the discretization parameters such that the associated center manifold always exists.

A quotient difference (qd)-type $\mathrm{dhLV_{II}}$ system, similar to the recursion formula of the qd algorithm, has been formulated as a variant of the $\mathrm{dhLV_{II}}$ system, and is potentially applicable to computing eigenvalues of TN matrices [8]. It has been established that there always exists a center manifold associated with the qd-type $\mathrm{dhLV_{I}}$ system [20]. Future work may examine local convergence in the qd-type $\mathrm{dhLV_{II}}$ system from the viewpoint of the center manifold.

**Acknowledgements** This work is supported by the Grants-in-Aid for Scientific Research (C) Nos. 23540163 and 26400208 and Encouragement of Young Scientists (B) No. 16K21368 from the Japan Society for the Promotion of Science.

## References

1. Ando, T.: Totally positive matrices. Linear Algebra Appl. **90**, 165–219 (1987)
2. Bogoyavlensky, O.I.: Some constructions of integrable dynamical systems. Math. USSR Izv **31**, 47–75 (1998)
3. Carr, J.: Applications of Centre Manifold Theory. Springer, New York (1981)
4. Fallat, S.M., Johnson, C.R.: Totally Nonnegative Matrices. Princeton University Press, Princeton (2011)
5. Fukuda, A., Ishiwata, E., Iwasaki, M., Nakamura, Y.: The discrete hungry Lotka-Volterra system and a new algorithm for computing matrix eigenvalues. Inverse Probl. **25**, 015007 (2009)
6. Fukuda, A., Ishiwata, E., Iwasaki, M., Nakamura, Y.: Discrete hungry integrable systems related to matrix eigenvalue and their local analysis by center manifold theory. RIMS Kôkyûroku Bessatsu **B13**, 1–17 (2009)
7. Fukuda, A., Ishiwata, E., Yamamoto, Y., Iwasaki, M., Nakamura, Y.: Integrable discrete hungry systems and their related matrix eigenvalues. Ann. Math. Pura Appl. **192**, 423–445 (2013)
8. Hama, Y., Fukuda, A., Yamamoto, Y., Iwasaki, M., Ishiwata, E., Nakamura, Y.: On some properties of a discrete hungry Lotka-Volterra system of multiplicative type, J. Math Ind. **5**, 5–15 (2012)
9. Hirota, R., Tsujimoto, S.: Conserved quantities of discrete Lotka-Volterra equation. RIMS Kôkyûroku **868**, 31–38 (1994)
10. Itoh, Y.: Integrals of a Lotka-Volterra system of odd number of variables. Prog. Theor. Phys. **78**, 507–510 (1987)
11. Iwasaki, M., Nakamura, Y.: On the convergence of a solution of the discrete Lotka–Volterra system. Inverse Probl. **18**, 1569–1578 (2002)
12. Iwasaki, M., Nakamura, Y.: An application of the discrete Lotka–Volterra system with variable step-size to singular value computation. Inverse Probl. **20**, 553–563 (2004)
13. Iwasaki, M., Nakamura, Y.: Center manifold approach to discrete integrable systems related to eigenvalues and singular values. Hokkaido Math. J. **36**, 759–775 (2007)
14. Karlin, S.: Total Positivity, vol.1. Stanford University Press, Stanford (1968)
15. Narita, K.: Soliton solution to extended Volterra equations. J. Phys. Soc. Jpn. **51**,1682–1685 (1982)
16. Pinkus, A.: Totally Positive Matrices. Cambridge University Press, New York (2009)
17. Shinjo, M., Iwasaki, M., Fukuda, A., Ishiwata, E., Yamamoto, Y., Nakamura, Y.: An asymptotic analysis for an integrable variant of the Lotka-Volterra prey-predator model via a determinant expansion technique. Cogent Math. **2**, 1046538 (2015)
18. Sun, J.-Q., Hu, X.-B., Tam, H.-W.: Short note: an integrable numerical algorithm for computing eigenvalues of a specially structured matrix. Numer. Linear Algebra Appl. **18**, 261–274 (2011)

19. Suris, Y.B.: Integrable discretizations of the Bogoyavlensky lattices. J. Math. Phys. **37**, 3982–3996 (1996)
20. Takahashi, Y., Iwasaki, M., Fukuda, A., Ishiwata, E., Nakamura, Y.: Asymptotic analysis for an extended discrete Lotka-Volterra system related to matrix eigenvalues. Appl. Anal. **92**, 586–594 (2013)
21. Tsujimoto, S., Nakamura, Y., Iwasaki, M.: The discrete Lotka–Volterra system computes singular values. Inverse Probl. **17**, 53–58 (2001)
22. Watkins, D.S.: Product eigenvalue problems. SIAM Rev. **47**, 3–40 (2005)
23. Wiggins, S.: Introduction to Applied Nonlinear Dynamical Systems and Chaos. Springer, Berlin (2003)
24. Yamazaki, S.: On the system of non-linear differential equations $\dot{y}_k = y_k(y_{k+1} - y_{k-1})$. J. Phys. A: Math. Gen. **20**, 6237–6241 (1987)

# Accuracy Improvement of the Shifted Block BiCGGR Method for Linear Systems with Multiple Shifts and Multiple Right-Hand Sides

**Hiroto Tadano, Shusaku Saito, and Akira Imakura**

**Abstract** We consider solving linear systems with multiple shifts and multiple right-hand sides. In order to solve these linear systems efficiently, we develop the Shifted Block BiCGGR method. This method is based on the shift-invariance property of Block Krylov subspaces. Using this property, the Shifted systems can be solved in the process of solving the Seed system without matrix-vector multiplications. The Shifted Block BiCGGR method can generate high accuracy approximate solutions of the Seed system. However, the accuracy of the approximate solutions of the Shifted systems may deteriorate due to the error of the matrix multiplications appearing in the algorithm. In this paper, we improve the accuracy of the approximate solutions of the Shifted systems generated by the Shifted Block BiCGGR method.

## 1 Introduction

We consider solving linear systems with multiple right-hand sides

$$AX = B, \tag{1}$$

$$(A + \sigma I)X^{(\sigma)} = B, \tag{2}$$

where $A \in \mathbb{C}^{n\times n}$ is a non-singular matrix, $X, X^{(\sigma)}, B \in \mathbb{C}^{n\times L}$ and $\sigma \in \mathbb{C}$. The linear systems (1) and (2) are called "Seed system" and "Shifted system", respectively. These linear systems appear in many scientific applications such as the eigensolver based on contour integral [8], lattice quantum chromodynamics (QCD) calculation, and so on.

As methods for solving linear systems with multiple shifts and single right-hand side, Shifted Krylov subspace methods have been proposed such as the Shifted BiCGSTAB($\ell$) method [5], the Shifted GMRES method [6]. Shifted Krylov

H. Tadano (✉) • S. Saito • A. Imakura
University of Tsukuba, Tennodai 1-1-1, Tsukuba, Ibaraki 305-8573, Japan
e-mail: tadano@cs.tsukuba.ac.jp; shusaku@hpcs.cs.tsukuba.ac.jp; imakura@cs.tsukuba.ac.jp


T. Sakurai et al. (eds.), *Eigenvalue Problems: Algorithms, Software and Applications in Petascale Computing*, Lecture Notes in Computational Science and Engineering 117, https://doi.org/10.1007/978-3-319-62426-6_12

subspace methods are based on the shift-invariance property of Krylov subspaces. Using the shift-invariance property, the Shifted systems can be solved in the process of solving the Seed system without matrix-vector multiplications.

On the other hand, Block Krylov subspace methods have been proposed for solving linear systems with multiple right-hand sides such as the Block BiCG method [7], the Block BiCGSTAB method [4], and the Block BiCGGR method [10]. These methods can solve the linear system with multiple right-hand sides efficiently in terms of the number of iterations and the computation time compared with the standard Krylov subspace methods for single right-hand side.

In this paper, we extend the Block BiCGGR method to the Shifted Block BiCGGR method for solving the Shifted systems (2) efficiently. The Shifted Block BiCGGR method can generate highly accurate approximate solutions of the Seed system. However, the accuracy of the approximate solutions of (2) deteriorate due to the error of the matrix multiplications appearing in the algorithm. In this paper, we also improve the accuracy of the approximate solutions of the Shifted systems.

This paper is organized as follows. In Sect. 2, the Block BiCGGR method is briefly described. In Sect. 3, the Shifted Block BiCGGR method is developed. Moreover, we improve numerical stability of the Shifted Block BiCGGR method. In Sect. 4, the modified Shifted Block BiCGGRrQ method is proposed for computing more accurate approximate solutions of the Shifted systems. Section 5 provides the results of numerical experiments to show the performance of the proposed method. This paper is concluded in Sect. 6.

## 2 Block BiCGGR Method

In this section, we describe the Block BiCGGR method [10] for solving the linear system (1). Let $X_{k+1} \in \mathbb{C}^{n \times L}$ be a $(k+1)$th approximate solution of (1). $X_{k+1}$ is computed so that the condition:

$$X_{k+1} = X_0 + Z_{k+1}, \quad Z_{k+1} \in \mathscr{B}^{\square}_{2k+2}(A; R_0)$$

is satisfied. Here, $R_0 = B - AX_0$ is an initial residual, and $\mathscr{B}^{\square}_{k+1}(A; R_0)$ is a Block Krylov subspace defined by

$$\mathscr{B}^{\square}_{k+1}(A; R_0) \equiv \left\{ \sum_{j=0}^{k} A^j R_0 \phi_j \mid \phi_j \in \mathbb{C}^{L \times L} \ (j = 0, 1, \ldots, k) \right\}.$$

The corresponding residual $R_{k+1}$ is written as

$$R_{k+1} = B - AX_{k+1} = R_0 - AZ_{k+1} \in \mathscr{B}^{\square}_{2k+3}(A; R_0).$$

We define the $(k+1)$th residual $R_{k+1}$ of the Block BiCGGR method as

$$R_{k+1} = B - AX_{k+1} \equiv H_{k+1}(A)R_{k+1}^{(\mathrm{BiCG})} \in \mathscr{B}_{2k+3}^{\square}(A; R_0). \tag{3}$$

Here, $R_{k+1}^{(\mathrm{BiCG})} \in \mathbb{C}^{n\times L}$ and $H_{k+1}(A)$ denote the $(k+1)$th residual generated by the Block BiCG method and the matrix polynomial of degree $k+1$, respectively. The matrix $R_{k+1}^{(\mathrm{BiCG})}$ is computed by the following recurrence relations.

$$R_0^{(\mathrm{BiCG})} = P_0^{(\mathrm{BiCG})} = R_0,$$

$$R_{k+1}^{(\mathrm{BiCG})} = R_k^{(\mathrm{BiCG})} - AP_k^{(\mathrm{BiCG})}\alpha_k \in \mathscr{B}_{k+2}^{\square}(A; R_0), \tag{4}$$

$$P_{k+1}^{(\mathrm{BiCG})} = R_{k+1}^{(\mathrm{BiCG})} + P_k^{(\mathrm{BiCG})}\beta_k \in \mathscr{B}_{k+2}^{\square}(A; R_0), \tag{5}$$

where $P_{k+1}^{(\mathrm{BiCG})} \in \mathbb{C}^{n\times L}, \alpha_k, \beta_k \in \mathbb{C}^{L\times L}$. From (4) and (5), the recurrence relation of $R_{k+1}^{(\mathrm{BiCG})}$ is rewritten as

$$R_{k+1}^{(\mathrm{BiCG})} = -AR_k^{(\mathrm{BiCG})}\alpha_k + R_k^{(\mathrm{BiCG})}\left[I_L + \gamma_{k-1}\alpha_k\right] - R_{k-1}^{(\mathrm{BiCG})}\gamma_{k-1}\alpha_k. \tag{6}$$

Here, the matrix $\gamma_{k-1} \in \mathbb{C}^{L\times L}$ is defined as $\gamma_{k-1} \equiv \alpha_{k-1}^{-1}\beta_{k-1}$, and $I_L$ denotes $L \times L$ identity matrix. The matrix polynomial $H_{k+1}(A)$ is defined as follows:

$$H_{k+1}(A) = (I - \zeta_k A)H_k(A), \;\; \zeta_k \in \mathbb{C}, \;\; H_0(A) = I. \tag{7}$$

Here, $I$ denotes the $n \times n$ identity matrix.

We construct the recurrence relations for computing the residual $R_{k+1} = H_{k+1}(A)R_{k+1}^{(\mathrm{BiCG})}$. From (4), (5) and (7), the recurrence relations for computing the matrices $H_{k+1}(A)R_{k+1}^{(\mathrm{BiCG})}$, $H_{k+1}(A)P_k^{(\mathrm{BiCG})}$, and $H_{k+1}(A)P_{k+1}^{(\mathrm{BiCG})}$ can be constructed as follows:

$$H_{k+1}(A)R_{k+1}^{(\mathrm{BiCG})} = H_k(A)R_k^{(\mathrm{BiCG})} - \zeta_k AH_k(A)R_k^{(\mathrm{BiCG})} - AH_{k+1}P_k^{(\mathrm{BiCG})}\alpha_k,$$

$$H_{k+1}(A)P_k^{(\mathrm{BiCG})} = H_k(A)P_k^{(\mathrm{BiCG})} - \zeta_k AH_k(A)P_k^{(\mathrm{BiCG})},$$

$$H_{k+1}(A)P_{k+1}^{(\mathrm{BiCG})} = H_{k+1}(A)R_{k+1}^{(\mathrm{BiCG})} + H_{k+1}(A)P_k^{(\mathrm{BiCG})}\beta_k.$$

Defining the matrices $P_k \equiv H_k(A)P_k^{(\mathrm{BiCG})} \in \mathbb{C}^{n\times L}$ and $U_k \equiv H_{k+1}(A)P_k^{(\mathrm{BiCG})}\alpha_k \in \mathbb{C}^{n\times L}$, the following recurrence relations can be obtained.

$$R_{k+1} = R_k - \zeta_k AR_k - AU_k, \tag{8}$$

$$U_k = (P_k - \zeta_k AP_k)\alpha_k, \tag{9}$$

$$P_{k+1} = R_{k+1} + U_k\gamma_k. \tag{10}$$

From (3) and (8), the approximate solution $X_{k+1}$ is computed as follows:

$$X_{k+1} = X_k + \zeta_k R_k + U_k. \tag{11}$$

The matrices $\alpha_k, \beta_k$ are determined so that the condition:

$$R_{k+1}^{(\mathrm{BiCG})} \perp \mathscr{B}_{k+1}^{\square}(A^{\mathrm{H}}; \tilde{R}_0)$$

is satisfied. Here, $\tilde{R}_0 \in \mathbb{C}^{n \times L}$ is an arbitrary matrix such that $\tilde{R}_0^{\mathrm{H}} R_0$ and $\tilde{R}_0^{\mathrm{H}} A P_0$ are non-singular. From this condition, the matrices $\alpha_k$, $\beta_k$ and $\gamma_k$ can be obtained as follows:

$$\alpha_k = (\tilde{R}_0^{\mathrm{H}} A P_k)^{-1} \tilde{R}_0^{\mathrm{H}} R_k,$$
$$\beta_k = (\tilde{R}_0^{\mathrm{H}} A P_k)^{-1} \tilde{R}_0^{\mathrm{H}} R_{k+1} / \zeta_k,$$
$$\gamma_k = (\tilde{R}_0^{\mathrm{H}} R_k)^{-1} \tilde{R}_0^{\mathrm{H}} R_{k+1} / \zeta_k.$$

The scalar $\zeta_k \in \mathbb{C}$ is determined so that $\|R_k - \zeta_k A R_k\|_{\mathrm{F}}$ is minimized.

# 3 Development of the Shifted Block BiCGGR Method

In this section, we develop the Shifted Block BiCGGR method for solving the Seed system (1) and the Shifted systems (2) simultaneously. First, the Shifted Block BiCG method is described, then the Shifted Block BiCGGR method is developed. After that, we improve numerical stability of the Shifted Block BiCGGR method by using the QR factorization of the residual matrix.

## *3.1 Shifted Block BiCG Method*

In this subsection, the Shifted Block BiCG method is described. Let $R_{k+1}^{(\sigma,\mathrm{BiCG})}$ be the $(k+1)$th residual of (2) generated by the Block BiCG method. The residual $R_{k+1}^{(\sigma,\mathrm{BiCG})}$ is computed by the following recurrence relations.

$$R_0^{(\sigma,\mathrm{BiCG})} = P_0^{(\sigma,\mathrm{BiCG})} = R_0^{(\sigma)},$$
$$R_{k+1}^{(\sigma,\mathrm{BiCG})} = R_k^{(\sigma,\mathrm{BiCG})} - (A + \sigma I) P_k^{(\sigma,\mathrm{BiCG})} \alpha_k^{(\sigma)} \in \mathscr{B}_{k+2}^{\square}(A + \sigma I; R_0^{(\sigma)}), \tag{12}$$
$$P_{k+1}^{(\sigma,\mathrm{BiCG})} = R_{k+1}^{(\sigma,\mathrm{BiCG})} + P_k^{(\sigma,\mathrm{BiCG})} \beta_k^{(\sigma)} \in \mathscr{B}_{k+2}^{\square}(A + \sigma I; R_0^{(\sigma)}), \tag{13}$$

where $R_0^{(\sigma)} = B - (A + \sigma I)X_0^{(\sigma)}$ is the initial residual of (2), and $\alpha_k^{(\sigma)}, \beta_k^{(\sigma)} \in \mathbb{C}^{L \times L}$. From (12) and (13), the recurrence relation of $R_{k+1}^{(\sigma,\mathrm{BiCG})}$ is rewritten as follows:

$$\begin{aligned} R_{k+1}^{(\sigma,\mathrm{BiCG})} &= -(A + \sigma I)R_k^{(\sigma,\mathrm{BiCG})}\alpha_k^{(\sigma)} + R_k^{(\sigma,\mathrm{BiCG})}\left[I_L + \gamma_{k-1}^{(\sigma)}\alpha_k^{(\sigma)}\right] \\ &\quad - R_{k-1}^{(\sigma,\mathrm{BiCG})}\gamma_{k-1}^{(\sigma)}\alpha_k^{(\sigma)}. \end{aligned} \tag{14}$$

Here, the matrix $\gamma_{k-1}^{(\sigma)}$ is defined as $\gamma_{k-1}^{(\sigma)} \equiv (\alpha_{k-1}^{(\sigma)})^{-1}\beta_{k-1}^{(\sigma)}$.

For the Block Krylov subspaces $\mathscr{B}_k^{\square}(A; B)$ and $\mathscr{B}_k^{\square}(A + \sigma I; B)$, the following shift-invariance property holds [9].

$$\mathscr{B}_k^{\square}(A; B) = \mathscr{B}_k^{\square}(A + \sigma I; B).$$

Hence the residual matrices $R_{k+1}^{(\mathrm{BiCG})} \in \mathscr{B}_{k+2}^{\square}(A; B)$ and $R_{k+1}^{(\sigma,\mathrm{BiCG})} \in \mathscr{B}_{k+2}^{\square}(A + \sigma I; B)$ belong to the same Block Krylov subspace $\mathscr{B}_{k+2}^{\square}(A; B)$. For the matrices $R_{k+1}^{(\mathrm{BiCG})}$ and $R_{k+1}^{(\sigma,\mathrm{BiCG})}$, we assume that the following relation holds:

$$R_{k+1}^{(\sigma,\mathrm{BiCG})} = R_{k+1}^{(\mathrm{BiCG})}(\pi_{k+1}^{(\sigma)})^{-1}. \tag{15}$$

Here, $\pi_{k+1}^{(\sigma)} \in \mathbb{C}^{L \times L}$.

Using the relation (15), the Eq. (6) can be transformed as follows:

$$\begin{aligned} R_{k+1}^{(\sigma,\mathrm{BiCG})} &= -(A + \sigma I)R_k^{(\sigma,\mathrm{BiCG})}\pi_k^{(\sigma)}\alpha_k(\pi_{k+1}^{(\sigma)})^{-1} \\ &\quad + R_k^{(\sigma,\mathrm{BiCG})}\pi_k^{(\sigma)}\left[I_L + \sigma\alpha_k + \gamma_{k-1}\alpha_k\right](\pi_{k+1}^{(\sigma)})^{-1} \\ &\quad - R_{k-1}^{(\sigma,\mathrm{BiCG})}\pi_{k-1}^{(\sigma)}\gamma_{k-1}\alpha_k(\pi_{k+1}^{(\sigma)})^{-1} \end{aligned} \tag{16}$$

By comparing the coefficients of (14) and (16), the matrices $\pi_{k+1}^{(\sigma)}$, $\alpha_k^{(\sigma)}$, and $\gamma_k^{(\sigma)}$ are determined as follows:

$$\pi_{k+1}^{(\sigma)} = \pi_k^{(\sigma)}\left[I_L + \sigma\alpha_k + \gamma_{k-1}\alpha_k\right] - \pi_{k-1}^{(\sigma)}\gamma_{k-1}\alpha_k, \tag{17}$$

$$\alpha_k^{(\sigma)} = \pi_k^{(\sigma)}\alpha_k(\pi_{k+1}^{(\sigma)})^{-1}, \tag{18}$$

$$\gamma_k^{(\sigma)} = \pi_k^{(\sigma)}\gamma_k(\pi_{k+1}^{(\sigma)})^{-1}. \tag{19}$$

## 3.2 *Shifted Block BiCGGR Method*

In this subsection, we develop the Shifted Block BiCGGR method. Let $X_{k+1}^{(\sigma)} \in \mathbb{C}^{n \times L}$ be a $(k+1)$th approximate solution of (2) generated by the Shifted Block BiCGGR method. The corresponding residual $R_{k+1}^{(\sigma)} \in \mathbb{C}^{n \times L}$ is defined as follows:

$$R_{k+1}^{(\sigma)} = B - (A + \sigma I)X_{k+1}^{(\sigma)} \equiv H_{k+1}^{(\sigma)}(A + \sigma I)R_{k+1}^{(\sigma,\mathrm{BiCG})} \in \mathscr{B}_{2k+3}^{\square}(A + \sigma I; R_0^{(\sigma)}).$$

Here, $H_{k+1}^{(\sigma)}(A + \sigma I)$ is a matrix polynomial of degree $k + 1$ defined as

$$H_{k+1}^{(\sigma)}(A + \sigma I) = \left[I - \zeta_k^{(\sigma)}(A + \sigma I)\right] H_k^{(\sigma)}(A + \sigma I),$$

where $\zeta_k^{(\sigma)} \in \mathbb{C}$, and $H_0^{(\sigma)}(A + \sigma I) = I$.

For the matrix polynomials $H_{k+1}(A)$ and $H_{k+1}^{(\sigma)}(A+\sigma I)$, we find a scalar $\theta_{k+1}^{(\sigma)} \in \mathbb{C}$ such that

$$H_{k+1}^{(\sigma)}(A + \sigma I) = H_{k+1}(A)/\theta_{k+1}^{(\sigma)}. \tag{20}$$

Using the relation (20), the matrix polynomial $H_{k+1}^{(\sigma)}(A + \sigma I)$ is expressed as

$$\begin{aligned} H_{k+1}^{(\sigma)}(A + \sigma I) &= \left[I - \zeta_k^{(\sigma)}(A + \sigma I)\right] H_k^{(\sigma)}(A + \sigma I) \\ &= \left[I - \zeta_k^{(\sigma)}(A + \sigma I)\right] H_k(A)/\theta_k^{(\sigma)} \\ &= (1 - \sigma\zeta_k^{(\sigma)})H_k(A)/\theta_k^{(\sigma)} - \zeta_k^{(\sigma)} A H_k(A)/\theta_k^{(\sigma)}. \end{aligned} \tag{21}$$

On the other hand, $H_{k+1}^{(\sigma)}(A + \sigma I)$ is also written as

$$H_{k+1}^{(\sigma)}(A + \sigma I) = H_k(A)/\theta_{k+1}^{(\sigma)} - \zeta_k A H_k(A)/\theta_{k+1}^{(\sigma)}. \tag{22}$$

Comparing the coefficients of (21) and (22), $\theta_{k+1}^{(\sigma)}$ and $\zeta_k^{(\sigma)}$ can be obtained as follows:

$$\begin{aligned} \theta_{k+1}^{(\sigma)} &= (1 + \sigma\zeta_k)\theta_k^{(\sigma)}, \theta_0^{(\sigma)} = 1, \\ \zeta_k^{(\sigma)} &= \zeta_k/(1 + \sigma\zeta_k). \end{aligned} \tag{23}$$

From the relations (15) and (20), the residual $R_{k+1}^{(\sigma)}$ is expressed as

$$R_{k+1}^{(\sigma)} = R_{k+1}(\pi_{k+1}^{(\sigma)})^{-1}/\theta_{k+1}^{(\sigma)}. \tag{24}$$

Using the relation (24), the approximate solution $X_{k+1}^{(\sigma)}$ and the matrices $U_k^{(\sigma)} \equiv H_{k+1}^{(\sigma)}(A + \sigma I)P_k^{(\sigma,\text{BiCG})}\alpha_k^{(\sigma)}$, $P_{k+1}^{(\sigma)} \equiv H_{k+1}^{(\sigma)}(A + \sigma I)P_{k+1}^{(\sigma,\text{BiCG})}$ can be computed as follows:

$$\begin{aligned} X_{k+1}^{(\sigma)} &= X_k^{(\sigma)} + \zeta_k^{(\sigma)} R_k(\pi_k^{(\sigma)})^{-1}/\theta_k^{(\sigma)} + U_k^{(\sigma)}, \\ U_k^{(\sigma)} &= P_k^{(\sigma)}\alpha_k^{(\sigma)} - \zeta_k^{(\sigma)}\left[R_k(\pi_k^{(\sigma)})^{-1} - (R_k - AP_k\alpha_k)(\pi_{k+1}^{(\sigma)})^{-1}\right]/\theta_{k+1}^{(\sigma)}, \\ P_{k+1}^{(\sigma)} &= R_{k+1}(\pi_{k+1}^{(\sigma)})^{-1}/\theta_{k+1}^{(\sigma)} + U_k^{(\sigma)}\gamma_k^{(\sigma)}. \end{aligned}$$

**Set** $X_0 = O, R_0 = P_0 = B, \gamma_{-1} = O_L$,

**Set** $X_0^{(\sigma)} = O, P_0^{(\sigma)} = B, \pi_{-1}^{(\sigma)} = \pi_0^{(\sigma)} = I_L, \theta_0^{(\sigma)} = 1$,

**Choose** $\tilde{R}_0 \in \mathbb{C}^{n \times L}$,

**For** $k = 0, 1, \ldots$,

  **Solve** $(\tilde{R}_0^{\mathrm{H}} A P_k)\alpha_k = \tilde{R}_0^{\mathrm{H}} R_k$ for $\alpha_k$,

  $\zeta_k = \mathrm{Tr}\left[(AR_k)^{\mathrm{H}} R_k\right] / \mathrm{Tr}\left[(AR_k)^{\mathrm{H}} AR_k\right]$,

  $U_k = (P_k - \zeta_k A P_k)\alpha_k$,

  $X_{k+1} = X_k + \zeta_k R_k + U_k$,

  `// --- Shift part --- //`

  $\pi_{k+1}^{(\sigma)} = \pi_k^{(\sigma)}(I_L + \sigma\alpha_k + \gamma_{k-1}\alpha_k) - \pi_{k-1}^{(\sigma)}\gamma_{k-1}\alpha_k$,

  $\alpha_k^{(\sigma)} = \pi_k^{(\sigma)}\alpha_k(\pi_{k+1}^{(\sigma)})^{-1}$,

  $\zeta_k^{(\sigma)} = \zeta_k/(1+\sigma\zeta_k)$,

  $U_k^{(\sigma)} = P_k^{(\sigma)}\alpha_k^{(\sigma)} - \zeta_k^{(\sigma)}\left[R_k(\pi_k^{(\sigma)})^{-1} - (R_k - AP_k\alpha_k)(\pi_{k+1}^{(\sigma)})^{-1}\right]/\theta_k^{(\sigma)}$,

  $X_{k+1}^{(\sigma)} = X_k^{(\sigma)} + \zeta_k^{(\sigma)} R_k(\pi_k^{(\sigma)})^{-1}/\theta_k^{(\sigma)} + U_k^{(\sigma)}$,

  $\theta_{k+1}^{(\sigma)} = (1+\sigma\zeta_k)\theta_k^{(\sigma)}$,

  `// ------------------ //`

  $R_{k+1} = R_k - \zeta_k A R_k - A U_k$,

  **Solve** $(\tilde{R}_0^{\mathrm{H}} R_k)\gamma_k = \tilde{R}_0^{\mathrm{H}} R_{k+1}/\zeta_k$ for $\gamma_k$,

  $P_{k+1} = R_{k+1} + U_k\gamma_k$,

  $AP_{k+1} = AR_{k+1} + AU_k\gamma_k$,

  `// --- Shift part --- //`

  $\gamma_k^{(\sigma)} = \pi_k^{(\sigma)}\gamma_k(\pi_{k+1}^{(\sigma)})^{-1}$,

  $P_{k+1}^{(\sigma)} = R_{k+1}(\pi_{k+1}^{(\sigma)})^{-1}/\theta_{k+1}^{(\sigma)} + U_k^{(\sigma)}\gamma_k^{(\sigma)}$,

  `// ------------------ //`

**End For**

**Fig. 1** Algorithm of the Shifted Block BiCGGR method

By summarizing the above equations, algorithm of the Shifted Block BiCGGR method shown in Fig. 1 can be obtained.

## 3.3 *Improvement of Numerical Stability*

When the number $L$ of right-hand sides is large, the residual norm may not converge due to numerical instability. This numerical instability comes from the loss of linear independence among column vectors of $n \times L$ matrices. In this subsection, numerical stability of the Shifted Block BiCGGR method is improved by orthonormalizing the column vectors of the residual $R_k$ [3].

The residual $R_k$ is factored as

$$R_k = Q_k \xi_k \tag{25}$$

by the QR factorization. Here, $Q_k$ is an $n \times L$ matrix such that $Q_k^{\mathrm{H}} Q_k = I_L$, and $\xi_k$ is an $L \times L$ upper triangular matrix.

Using (25), the Eqs. (8), (9), (10) and (11) are rewritten as follows:

$$\begin{aligned} Q_{k+1}\tau_{k+1} &= Q_k - \zeta_k A Q_k - AV_k, \\ V_k &= (S_k - \zeta_k AS_k)\alpha'_k, \\ S_{k+1} &= Q_{k+1} + V_k \gamma'_k, \\ X_{k+1} &= X_k + [\zeta_k Q_k + V_k]\,\xi_k. \end{aligned}$$

Here, $V_k \equiv U_k \xi_k^{-1} \in \mathbb{C}^{n\times L}$, $S_k \equiv P_k \xi_k^{-1} \in \mathbb{C}^{n\times L}$, $\tau_{k+1} = \xi_{k+1}\xi_k^{-1} \in \mathbb{C}^{L\times L}$, $\alpha'_k \equiv \xi_k \alpha_k \xi_k^{-1} \in \mathbb{C}^{L\times L}$, and $\gamma'_k \equiv \xi_k \gamma_k \xi_{k+1}^{-1} \in \mathbb{C}^{L\times L}$. The $L \times L$ matrices $\alpha'_k$ and $\gamma'_k$ are computed by

$$\begin{aligned} \alpha'_k &= (\tilde{R}_0^{\mathrm{H}} AS_k)^{-1} \tilde{R}_0^{\mathrm{H}} Q_k, \\ \gamma'_k &= (\tilde{R}_0^{\mathrm{H}} Q_k)^{-1} \tilde{R}_0^{\mathrm{H}} Q_{k+1}/\zeta_k. \end{aligned}$$

The scalar $\zeta_k$ is determined so that $\|Q_k - \zeta_k AQ_k\|_{\mathrm{F}}$ is minimized.

Then, the recurrence relations for the Shifted systems are modified. The matrix $\pi_k^{\prime(\sigma)} \in \mathbb{C}^{L\times L}$ is defined as $\pi_k^{\prime(\sigma)} \equiv \pi_k^{(\sigma)} \xi_k^{-1}$. Using the matrix $\pi_k^{\prime(\sigma)}$, the recurrence relations (17), (18), and (19) are rewritten as follows:

$$\pi_{k+1}^{\prime(\sigma)} = \left[\pi_k^{\prime(\sigma)}(I_L + \sigma\alpha'_k + \tau_k \gamma'_{k-1}\alpha'_k) - \pi_{k-1}^{\prime(\sigma)}\gamma'_{k-1}\alpha'_k\right]\tau_{k+1}^{-1}, \tag{26}$$

$$\alpha_k^{(\sigma)} = \pi_k^{\prime(\sigma)}\alpha'_k \tau_{k+1}^{-1}(\pi_{k+1}^{\prime(\sigma)})^{-1}, \tag{27}$$

$$\gamma_k^{(\sigma)} = \pi_k^{\prime(\sigma)}\gamma'_k(\pi_{k+1}^{\prime(\sigma)})^{-1}. \tag{28}$$

The approximate solution $X_{k+1}^{(\sigma)}$ and the matrices $U_k^{(\sigma)}$, $P_{k+1}^{(\sigma)}$ can be computed as follows:

$$X_{k+1}^{(\sigma)} = X_k^{(\sigma)} + \zeta_k^{(\sigma)} Q_k (\pi_k^{\prime(\sigma)})^{-1}/\theta_k^{(\sigma)} + U_k^{(\sigma)},$$

$$U_k^{(\sigma)} = P_k^{(\sigma)}\alpha_k^{(\sigma)} - \zeta_k^{(\sigma)}\left[Q_k(\pi_k^{\prime(\sigma)})^{-1} - (Q_k - AS_k\alpha'_k)\tau_{k+1}^{-1}(\pi_{k+1}^{\prime(\sigma)})^{-1}\right]/\theta_k^{(\sigma)}, \tag{29}$$

$$P_{k+1}^{(\sigma)} = Q_{k+1}(\pi_{k+1}^{\prime(\sigma)})^{-1}/\theta_{k+1}^{(\sigma)} + U_k^{(\sigma)}\gamma_k^{(\sigma)}. \tag{30}$$

By summarizing above equations, algorithm of the Shifted Block BiCGGRrQ method shown in Fig. 2 can be obtained.

**Perform** the QR factorization $Q_0\xi_0 = B$,
**Set** $X_0 = O, S_0 = Q_0, \gamma'_{-1} = O_L, \tau_0 = \xi_0$,
**Set** $X_0^{(\sigma)} = O, P_0^{(\sigma)} = B, \pi'^{(\sigma)}_{-1} = I_L, \pi'^{(\sigma)}_0 = \xi_0^{-1}, \theta_0^{(\sigma)} = 1$,
**Choose** $\tilde{R}_0 \in \mathbb{C}^{n\times L}$,
**For** $k = 0, 1, \ldots,$

  **Solve** $(\tilde{R}_0^{\mathrm{H}} A S_k)\alpha'_k = \tilde{R}_0^{\mathrm{H}} Q_k$ for $\alpha'_k$,
  $\zeta_k = \mathrm{Tr}\left[(AQ_k)^{\mathrm{H}} Q_k\right] / \mathrm{Tr}\left[(AQ_k)^{\mathrm{H}} AQ_k\right]$,
  $V_k = (S_k - \zeta_k AS_k)\alpha'_k$,
  $X_{k+1} = X_k + [\zeta_k Q_k + V_k]\,\xi_k$,
  $Q_{k+1}\tau_{k+1} = Q_k - \zeta_k AQ_k - AV_k$,
  $\xi_{k+1} = \tau_{k+1}\xi_k$,

```
// --- Shift part --- //
```

  $\pi'^{(\sigma)}_{k+1} = \left[\pi'^{(\sigma)}_k (I_L + \sigma\alpha'_k + \tau_k \gamma'_{k-1}\alpha'_k) - \pi'^{(\sigma)}_{k-1}\gamma'_{k-1}\alpha'_k\right]\tau_{k+1}^{-1}$,
  $\alpha_k^{(\sigma)} = \pi'^{(\sigma)}_k \alpha'_k \tau_{k+1}^{-1} (\pi'^{(\sigma)}_{k+1})^{-1}$,
  $\zeta_k^{(\sigma)} = \zeta_k/(1+\sigma\zeta_k)$,
  $U_k^{(\sigma)} = P_k^{(\sigma)}\alpha_k^{(\sigma)} - \zeta_k^{(\sigma)}\left[Q_k(\pi'^{(\sigma)}_k)^{-1} - (Q_k - AS_k\alpha'_k)\tau_{k+1}^{-1}(\pi'^{(\sigma)}_{k+1})^{-1}\right]/\theta_k^{(\sigma)}$,
  $X_{k+1}^{(\sigma)} = X_k^{(\sigma)} + \zeta_k^{(\sigma)} Q_k (\pi'^{(\sigma)}_k)^{-1}/\theta_k^{(\sigma)} + U_k^{(\sigma)}$,
  $\theta_{k+1}^{(\sigma)} = (1+\sigma\zeta_k)\theta_k^{(\sigma)}$,

```
// ----------------- //
```

  **Solve** $(\tilde{R}_0^{\mathrm{H}} Q_k)\gamma'_k = \tilde{R}_0^{\mathrm{H}} Q_{k+1}/\zeta_k$ for $\gamma'_k$,
  $S_{k+1} = Q_{k+1} + V_k\gamma'_k$,
  $AS_{k+1} = AQ_{k+1} + AV_k\gamma'_k$,

```
// --- Shift part --- //
```

  $\gamma_k^{(\sigma)} = \pi'^{(\sigma)}_k \gamma'_k (\pi'^{(\sigma)}_{k+1})^{-1}$,
  $P_{k+1}^{(\sigma)} = Q_{k+1}(\pi'^{(\sigma)}_{k+1})^{-1}/\theta_{k+1}^{(\sigma)} + U_k^{(\sigma)}\gamma_k^{(\sigma)}$,

```
// ----------------- //
```

**End For**

**Fig. 2** Algorithm of the Shifted Block BiCGGRrQ method

## 3.4 *Preliminary Experiments*

In this subsection, the performance of the Shifted Block BiCGGRrQ method is evaluated. The test matrix used in the preliminary experiments is `epb2` [2]. The structure of the matrix is real and nonsymmetric. The size of the matrix and the number of nonzero elements are 25,228 and 175,027, respectively. The right-hand sides $B$ and the matrix $\tilde{R}_0$ are given by the random number generator. The number $L$ of the right-hand sides is $1, 2, \ldots, 16$. The shift parameter $\sigma$ is 0.01. The iteration is stopped when the stopping conditions $\|R_k\|_{\mathrm{F}}/\|B\|_{\mathrm{F}} \le 10^{-12}$ and $\|R_k^{(\sigma)}\|_{\mathrm{F}}/\|B\|_{\mathrm{F}} \le 10^{-12}$ are satisfied. Preliminary experiments are carried out in double precision arithmetic on MATLAB R2015b.

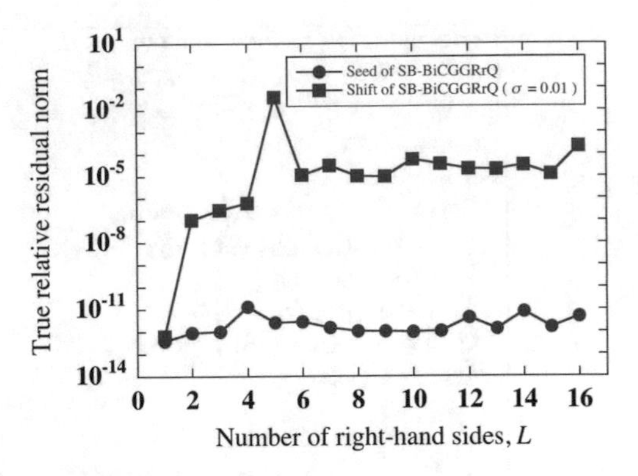


**Fig. 3** The true relative residual norm of the seed system and that of the shifted system generated by the Shifted Block BiCGGRrQ method

Figure 3 shows the true relative residual norm of the Seed system and that of the Shifted system generated by the Shifted Block BiCGGRrQ method. The true relative residual norm of the Seed system and that of the Shifted systems are calculated by $\|B - AX_k\|_F / \|B\|_F$ and $\|B - (A + \sigma I)X_k^{(\sigma)}\|_F / \|B\|_F$, respectively. The true relative residual norms of the Seed system are sufficiently small for all $L$. When $L = 1$, that of the Shifted system is also sufficiently small. However, those of the Shifted system are large in the case of $L \geq 2$. When $L \geq 2$, the main operation of the Shift part is the product of an $n \times L$ matrix and an $L \times L$ matrix.

In [10], it is reported that the error of the matrix multiplication causes the accuracy deterioration of the approximate solutions. In the next section, we try to improve the accuracy of the approximate solutions of the Shifted systems by reducing the number of matrix multiplications.

# 4 Accuracy Improvement of Approximate Solutions of the Shifted Systems

In the previous section, we developed the Shifted Block BiCGGR method and the Shifted Block BiCGGRrQ method. Through the preliminary experiments, it is shown that the accuracy of the approximate solutions of the Shifted system generated by the Shifted Block BiCGGRrQ method is not good in the case of $L \geq 2$. So it is expected that the accuracy of the approximate solutions of the Shifted systems can be improved by reducing the number of matrix multiplications.

In this section, we modify the Shifted Block BiCGGRrQ method in order to improve the accuracy of the approximate solutions of the Shifted systems.

## 4.1 Modification of the Shifted Block BiCGGRrQ Method

Using (27), the Eq. (29) can be transformed as follows:

$$\begin{aligned}\theta_k^{(\sigma)} U_k^{(\sigma)} \pi_k^{\prime(\sigma)} =& -\zeta_k^{(\sigma)} Q_k \left[ I_L - \tau_{k+1}^{-1} (\pi_{k+1}^{\prime(\sigma)})^{-1} \pi_k^{\prime(\sigma)} \right] \\ &+ (\theta_k^{(\sigma)} P_k^{(\sigma)} \pi_k^{\prime(\sigma)} - \zeta_k^{(\sigma)} A S_k) \alpha_k' \tau_{k+1}^{-1} (\pi_{k+1}^{\prime(\sigma)})^{-1} \pi_k^{\prime(\sigma)}.\end{aligned} \tag{31}$$

Similarly, using (23) and (28), the equation (30) can be transformed as follows:

$$\theta_{k+1}^{(\sigma)} P_{k+1}^{(\sigma)} \pi_{k+1}^{\prime(\sigma)} = Q_{k+1} + (1 + \sigma \zeta_k) \cdot (\theta_k^{(\sigma)} U_k^{(\sigma)} \pi_k^{\prime(\sigma)}) \gamma_k'. \tag{32}$$

We define the matrices $\hat{P}_k^{(\sigma)} \in \mathbb{C}^{n \times L}$, $\hat{U}_k^{(\sigma)} \in \mathbb{C}^{n \times L}$, and $\mu_k^{(\sigma)} \in \mathbb{C}^{L \times L}$ as $\hat{P}_k^{(\sigma)} \equiv \theta_k^{(\sigma)} P_k^{(\sigma)} \pi_k^{\prime(\sigma)}$, $\hat{U}_k^{(\sigma)} \equiv \theta_k^{(\sigma)} P_k^{(\sigma)} \pi_k^{\prime(\sigma)}$, and $\mu_k^{(\sigma)} \equiv (\pi_{k-1}^{\prime(\sigma)})^{-1} \pi_k^{\prime(\sigma)} \tau_k$, respectively. Using these matrices, the Eqs. (31) and (32) can be simplified as follows:

$$\begin{aligned}\hat{U}_k^{(\sigma)} &= -\zeta_k^{(\sigma)} Q_k \left[ I_L - (\mu_{k+1}^{(\sigma)})^{-1} \right] + (\hat{P}_k^{(\sigma)} - \zeta_k^{(\sigma)} A S_k) \alpha_k' (\mu_{k+1}^{(\sigma)})^{-1}, \\ \hat{P}_{k+1}^{(\sigma)} &= Q_{k+1} + (1 + \sigma \zeta_k) \hat{U}_k^{(\sigma)} \gamma_k'.\end{aligned}$$

From (26), the matrix $\mu_{k+1}^{(\sigma)}$ can be computed as:

$$\mu_{k+1}^{(\sigma)} = I_L + \sigma \alpha_k' + \tau_k \left[ I_L - (\mu_k^{(\sigma)})^{-1} \right] \gamma_{k-1}' \alpha_k'.$$

The approximate solution $X_{k+1}^{(\sigma)}$ is computed by the following equation.

$$X_{k+1}^{(\sigma)} = X_k^{(\sigma)} + (\zeta_k^{(\sigma)} Q_k + \hat{U}_k^{(\sigma)}) (\pi_k^{\prime(\sigma)})^{-1} / \theta_k^{(\sigma)}.$$

By summarizing above equations, algorithm of the modified Shifted Block BiCGGRrQ method shown in Fig. 4 can be obtained. In this algorithm, the matrices $\alpha_k^{(\sigma)}$ and $\gamma_k^{(\sigma)}$ are no longer required.

## 4.2 Computational Costs

The main computational costs per iteration of the Shifted Block BiCGGR method, the Shifted Block BiCGGRrQ method, and the modified Shifted Block BiCGGRrQ method are shown in Table 1. $s$ denotes the number of shift parameters. In this table, the operations "SPMM", "Block Dot", "Block AXPY", and "QR" are defined as follows:

- SPMM: The product of the $n \times n$ sparse matrix $A$ and an $n \times L$ matrix.
- Block Dot: The product of an $L \times n$ matrix and an $n \times L$ matrix.

**Perform** the QR factorization $Q_0\xi_0 = B$,
**Set** $X_0 = O, S_0 = Q_0, \gamma'_{-1} = O_L, \tau_0 = \xi_0$,
**Set** $X_0^{(\sigma)} = O, \hat{P}_0^{(\sigma)} = Q_0, \pi_0'^{(\sigma)} = \xi_0^{-1}, \mu_0^{(\sigma)} = I_L, \theta_0^{(\sigma)} = 1$,
**Choose** $\tilde{R}_0 \in \mathbb{C}^{n\times L}$,
**For** $k = 0, 1, \ldots,$
  **Solve** $(\tilde{R}_0^{\mathrm{H}} AS_k)\alpha'_k = \tilde{R}_0^{\mathrm{H}} Q_k$ for $\alpha'_k$,
  $\zeta_k = \mathrm{Tr}\left[(AQ_k)^{\mathrm{H}} Q_k\right] / \mathrm{Tr}\left[(AQ_k)^{\mathrm{H}} AQ_k\right]$,
  $V_k = (S_k - \zeta_k AS_k)\alpha'_k$,
  $X_{k+1} = X_k + [\zeta_k Q_k + V_k]\,\xi_k$,
  $Q_{k+1}\tau_{k+1} = Q_k - \zeta_k AQ_k - AV_k$,
  $\xi_{k+1} = \tau_{k+1}\xi_k$,

  `// --- Shift part --- //`
  $\mu_{k+1}^{(\sigma)} = I_L + \sigma\alpha'_k + \tau_k\left[I_L - (\mu_k^{(\sigma)})^{-1}\right]\gamma'_{k-1}\alpha'_k$,
  $\zeta_k^{(\sigma)} = \zeta_k/(1+\sigma\zeta_k)$,
  $\hat{U}_k^{(\sigma)} = -\zeta_k^{(\sigma)} Q_k\left[I_L - (\mu_{k+1}^{(\sigma)})^{-1}\right] + (\hat{P}_k^{(\sigma)} - \zeta_k^{(\sigma)} AS_k)\alpha'_k(\mu_{k+1}^{(\sigma)})^{-1}$,
  $X_{k+1}^{(\sigma)} = X_k^{(\sigma)} + (\zeta_k^{(\sigma)} Q_k + \hat{U}_k^{(\sigma)})(\pi_k'^{(\sigma)})^{-1}/\theta_k^{(\sigma)}$,
  $\theta_{k+1}^{(\sigma)} = (1+\sigma\zeta_k)\theta_k^{(\sigma)}$,
  $\pi_{k+1}'^{(\sigma)} = \pi_k'^{(\sigma)}\mu_{k+1}^{(\sigma)}\tau_{k+1}^{-1}$,
  `// ------------------ //`

  **Solve** $(\tilde{R}_0^{\mathrm{H}} Q_k)\gamma'_k = \tilde{R}_0^{\mathrm{H}} Q_{k+1}/\zeta_k$ for $\gamma'_k$,
  $S_{k+1} = Q_{k+1} + V_k\gamma'_k$,
  $AS_{k+1} = AQ_{k+1} + AV_k\gamma'_k$,

  `// --- Shift part --- //`
  $\hat{P}_{k+1}^{(\sigma)} = Q_{k+1} + (1+\sigma\zeta_k)\hat{U}_k^{(\sigma)}\gamma'_k$,
  `// ------------------ //`

**End For**

**Fig. 4** Algorithm of the modified Shifted Block BiCGGRrQ method

**Table 1** Main computational costs per iteration of the three Shifted Block Krylov subspace methods

| Method | SPMM | Block Dot | Block AXPY | QR |
|---|---|---|---|---|
| Shifted Block BiCGGR | 2 | 2 | 5*s*+4 | 0 |
| Shifted Block BiCGGRrQ | 2 | 2 | 5*s*+5 | 1 |
| Modified Shifted Block BiCGGRrQ | 2 | 2 | 4*s*+4 | 1 |

*s* denotes the number of shift parameters

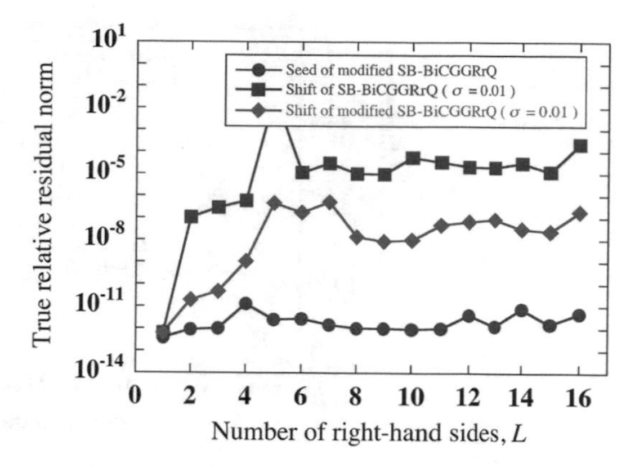


**Fig. 5** The true relative residual norm of the seed system and that of the shifted system generated by the modified Shifted Block BiCGGRrQ method

- Block AXPY: The operation of $Z = Y + X\alpha$, where $X$, $Y$ and $Z$ are $n \times L$ matrices, and $\alpha$ is an $L \times L$ matrix.
- QR: The QR factorization of an $n \times L$ matrix.

### 4.3 Preliminary Experiments

In this subsection, the performance of the modified Block BiCGGRrQ method is evaluated through the preliminary experiments. The experimental conditions and the experimental environment are the same as Sect. 3.4.

Figure 5 shows the true relative residual norm of the Seed system and that of the Shifted system generated by the modified Shifted Block BiCGGRrQ method. For $L \geq 2$, the true relative residual norms of the Shifted system calculated by the modified Block BiCGGRrQ method are smaller than that by calculating the Shifted Block BiCGGRrQ method.

## 5 Numerical Experiments

In this section, the performance of the Shifted Block BiCGGRrQ method and the modified Shifted Block BiCGGRrQ method is evaluated.

Test problem is a linear system with non-Hermitian matrix, which derived from lattice quantum chromodynamics (QCD) calculation [1]. The size $n$ of the matrix is 1,572,864, and the number of nonzero elements is 80,216,064. The number $L$ of right-hand sides is 12. The right-hand sides $B$ is set as $B = [\boldsymbol{e}_1, \boldsymbol{e}_2, \ldots, \boldsymbol{e}_L]$, where $\boldsymbol{e}_j$ is a $j$th unit vector. The elements of the matrix $\tilde{R}_0$ are given by a random number generator. The shift parameter $\sigma$ is $\sigma = 0.001, 0.002, \ldots, 0.01$. The iteration is stopped when the stopping conditions $\|R_k\|_{\mathrm{F}}/\|B\|_{\mathrm{F}} \leq 10^{-12}$ and $\|R_k^{(\sigma)}\|_{\mathrm{F}}/\|B\|_{\mathrm{F}} \leq 10^{-12}$ are satisfied.

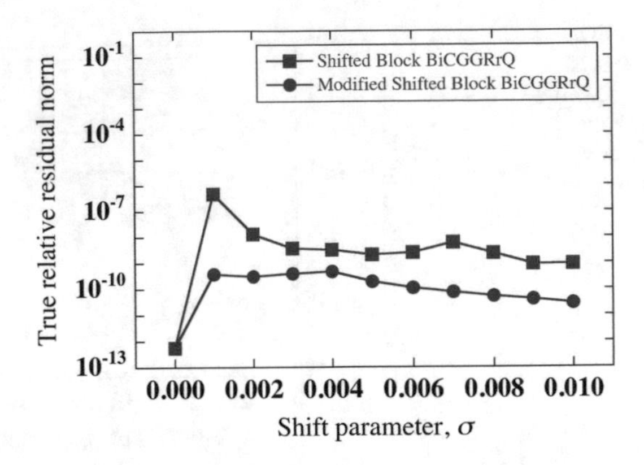


**Fig. 6** True relative residual norm for each shift point

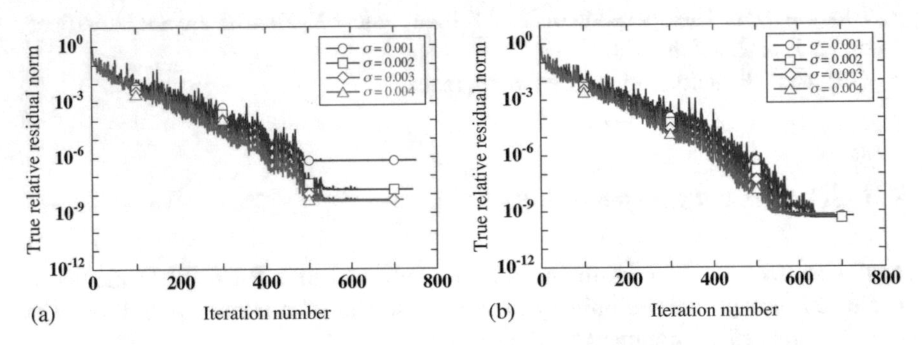


**Fig. 7** True relative residual history of the Shifted Block BiCGGRrQ method and the modified Shifted Block BiCGGRrQ method. (**a**) Shifted Block BiCGGRrQ. (**b**) Modified Shifted Block BiCGGRrQ

Numerical experiments are carried out in double precision arithmetic on CPU: Intel Xeon E5-2620v3 2.4 GHz (6 cores) $\times$ 2, Memory: 64 GB, Compiler: gfortran ver. 4.9.2, Compile options: `-O3 -fopenmp`. All calculations are parallelized by OpenMP with 24 threads.

Figure 6 shows the true relative residual norm for each shift point. For the Seed system, the Shifted Block BiCGGRrQ method and the modified Shifted Block BiCGGRrQ method can generate high accuracy approximate solution. However, the accuracy of the approximate solutions of the Shifted systems are not good even though the stopping condition is satisfied. By using the modified Shifted Block BiCGGRrQ method, we can generate more accurate approximate solutions.

Figure 7 shows the true relative residual history of the Shifted Block BiCGGRrQ method and the modified Shifted Block BiCGGRrQ method in the case of $\sigma = 0.001, 0.002, 0.003$ and $0.004$. When $\sigma = 0.001$, the true relative residual norm of the Shifted Block BiCGGRrQ method stagnates around $4.6 \times 10^{-7}$. On the other hand, that of the modified Shifted Block BiCGGRrQ method decreases around $3.8 \times 10^{-10}$.

The elapsed time of the Shifted Block BiCGGRrQ method and that of the modified Shifted Block BiCGGRrQ method are 2305.58 and 1814.38 s, respectively. In this numerical experiment, the relative residual norm of the Shifted Block BiCGGR diverged.

## 6 Conclusions

In this paper, we have developed the Shifted Block BiCGGR method and the Shifted Block BiCGGRrQ method for solving linear systems with multiple right-hand sides and multiple shifts. Then, we have proposed the modified Shifted Block BiCGGRrQ method in order to improve the accuracy of the approximate solutions of the Shifted systems. By using the modified Shifted Block BiCGGRrQ method, the accuracy of the obtained approximate solutions was improved.

Our future work is to precisely analyze the cause of the accuracy deterioration of the approximate solutions of the Shifted systems.

**Acknowledgements** This work was partly supported by JSPS KAKENHI Grant Numbers 15K15996, 25286097, 25870099, and University of Tsukuba Basic Research Support Program Type A.

## References

1. CCS Matrix Forum: http://ccsmf.ccs.tsukuba.ac.jp/
2. Davis, T.: The University of Florida Sparse Matrix Collection. http://www.cise.ufl.edu/research/sparse/matrices/
3. Dubrulle, A.A.: Retooling the method of block conjugate gradients. Electron. Trans. Numer. Anal. **12**, 216–233 (2001)
4. El Guennouni, A., Jbilou, K., Sadok, H.: A block version of BiCGSTAB for linear systems with multiple right-hand sides. Electron. Trans. Numer. Anal. **16**, 129–142 (2003)
5. Frommer, A.: BiCGSTAB($\ell$) for families of shifted linear systems. Computing **70**, 87–109 (2003)
6. Frommer, A., Glassner, U.: Restarted GMRES for shifted linear systems. SIAM J. Sci. Comput. **19**(1), 15–26 (1998)
7. O'Leary, D.P.: The block conjugate gradient algorithm and related methods. Linear Algebra Appl. **29**, 293–322 (1980)
8. Sakurai, T., Sugiura, H.: A projection method for generalized eigenvalue problems using numerical integration. J. Comput. Appl. Math. **159**, 119–128 (2003)
9. Soodhalter, K.M.: Block Krylov subspace recycling for shifted systems with unrelated right-hand sides. SIAM J. Sci. Comput. **38**(1), A302–A324 (2016)
10. Tadano, H., Sakurai, T., Kuramashi, Y.: Block BiCGGR: a new block Krylov subspace method for computing high accuracy solutions. JSIAM Lett. **1**, 44–47 (2009)

# Memory-Saving Technique for the Sakurai–Sugiura Eigenvalue Solver Using the Shifted Block Conjugate Gradient Method

**Yasunori Futamura and Tetsuya Sakurai**

**Abstract** In recent years, a numerical quadrature-based sparse eigensolver—the so-called Sakurai–Sugiura method—and its variants have attracted attention because of their highly coarse-grained parallelism. In this paper, we propose a memory-saving technique for a variant of the Sakurai–Sugiura method. The proposed technique can be utilized when inner linear systems are solved with the shifted block conjugate gradient method. Using our technique, eigenvalues and residual norms can be obtained without the explicit need to compute the eigenvector. This technique saves a considerable amount of memory space when eigenvectors are unnecessary. Our technique is also beneficial in cases where eigenvectors are necessary, because the residual norms of the target eigenpairs can be cheaply computed and monitored during each iteration step of the inner linear solver.

## 1 Introduction

In this study, we consider a standard eigenvalue problem:

$$Au = \lambda u \tag{1}$$

where $A \in \mathbb{C}^{n\times n}$ is a Hermitian matrix, $\lambda \in \mathbb{R}$ is an eigenvalue, and $\boldsymbol{u} \in \mathbb{C}^n\backslash\{\boldsymbol{0}\}$ is an eigenvector. We consider a case where all eigenvalues located in a certain interval are needed, and assume that $A$ is large and sparse.

In 2003, the Sakurai–Sugiura method (SSM) [12] was proposed for computing eigenvalues in a specified region and associated eigenvectors of large sparse non-Hermitian generalized eigenproblems. In the algorithm for the SSM, multiple linear systems must be solved in order to obtain the eigenspace of the target eigenvalues. This step dominates the complexity of the method. Because each linear system can

Y. Futamura (✉) • T. Sakurai
University of Tsukuba, 1-1-1 Tennohdai, Tsukuba, Ibaraki, Japan
e-mail: futamura@cs.tsukuba.ac.jp; sakurai@cs.tsukuba.ac.jp



T. Sakurai et al. (eds.), *Eigenvalue Problems: Algorithms, Software and Applications in Petascale Computing*, Lecture Notes in Computational Science and Engineering 117, https://doi.org/10.1007/978-3-319-62426-6_13

be solved independently, the SSM has the potential to be highly efficient in recent massively parallel computational environments.

While the SSM was originally proposed to be applicable to non-Hermitian problems, several ways have been proposed for taking advantage of the symmetries of matrices. Specifically, in [3, 8, 9], efficient approaches are proposed to solve the linear systems in the SSM that applicable for Hermitian (or real symmetric) standard eigenproblems using the shift invariance of the (block) Krylov subspace. These approaches have been applied to large-scale problems from practical applications.

In some of these applications, the memory requirements for storing vectors are critical for large problems, because the matrices of the problems are sufficiently sparse or computed on-the-fly. In addition, there are cases where eigenvectors are not required, where only eigenvalues must be computed.

In this paper, to accommodate the demands for reducing memory requirements, we propose a technique for computing eigenvalues and the residual norms of the eigenproblem without explicitly calculating eigenvectors. This technique applies in cases where the shifted block conjugate gradient method with residual orthogonalization (SBCGrQ) [3] is applied to solve the linear systems in the algorithm of the SSM.

The remainder of this paper is organized as follows. In Sect. 2, the SSM and SBCGrQ are briefly described. In Sect. 3, we propose a memory-saving technique for computing eigenvalues and residual norms. In Sect. 4, we investigate the effectiveness of the proposed technique through numerical experiments by applying it to three problems derived from practical applications. Finally, concluding remarks are presented in Sect. 5.

## 2 The Sakurai–Sugiura Method with the Shifted Block Conjugate Gradient Method

In this section, we briefly introduce the eigensolver SSM and the linear solver SBCGrQ.

The SSM is a projection method that introduces a contour integration as follows:

$$S_m \equiv \frac{1}{2\pi \mathrm{i}} \oint_{\Gamma} z^m (zI - A)^{-1} V \mathrm{d}z, \quad m = 0, 1, \ldots, M-1 \tag{2}$$

where i is the imaginary unit, $\Gamma$ is a positively oriented Jordan curve on the complex plane, $I$ is the identity matrix of order $n$, and $V \in \mathbb{C}^{n \times L}$ is a matrix formed by linear independent column vectors. The elements of $V$ are randomly chosen when there is no preliminarily given information for the target eigenvectors. Let $n_\Gamma$ be the number of eigenvalues inside $\Gamma$ (counting multiplicity) and $S \equiv [S_0, S_1, \ldots, S_{M-1}]$. According to the Cauchy's integral formula, the range space of $S$ is spanned by the eigenvectors corresponding to the eigenvalues inside $\Gamma$, provided that $LM \geq n_\Gamma$ and that $L$ is greater than or equal to the maximum multiplicity of the eigenvalues inside $\Gamma$. The SSM extracts the eigenpairs from the range space of $S$ with some projection technique. There are several approaches for extracting the eigenpairs, including the Hankel-type method [5, 12], the Rayleigh–Ritz-type method [4, 13] and the Arnoldi-type method [6]. The corresponding solvers are referred to as SS–Hankel, SS–RR, and SS–Arnoldi, respectively. The SSM was originally proposed with $L = 1$ in [12]. Block versions ($L \geq 2$) were proposed to capture degenerate eigenvalues and improve numerical stability [4, 5].

In this study, we use SS–Hankel to solve (1). With SS–Hankel, a sequence of small $L \times L$ matrices $\mathscr{M}_m \equiv V^{\mathrm{H}} S_m$ ($m = 0, 1, \ldots, 2M-1$) are computed. Then, block Hankel matrices $H_{LM} \in \mathbb{C}^{LM \times LM}$ and the shifted Hankel matrix $H_{LM}^{<} \in \mathbb{C}^{LM \times LM}$ are respectively formed as follows:

$$H_{LM} \equiv \begin{pmatrix} \mathscr{M}_0 & \mathscr{M}_1 & \cdots & \mathscr{M}_{M-1} \\ \mathscr{M}_1 & \mathscr{M}_2 & \cdots & \mathscr{M}_M \\ \vdots & \vdots & & \vdots \\ \mathscr{M}_{M-1} & \mathscr{M}_M & \cdots & \mathscr{M}_{2M-2} \end{pmatrix} \tag{3}$$

and

$$H_{LM}^{<} \equiv \begin{pmatrix} \mathscr{M}_1 & \mathscr{M}_2 & \cdots & \mathscr{M}_M \\ \mathscr{M}_2 & \mathscr{M}_3 & \cdots & \mathscr{M}_{M+1} \\ \vdots & \vdots & & \vdots \\ \mathscr{M}_M & \mathscr{M}_{M+1} & \cdots & \mathscr{M}_{2M-1} \end{pmatrix}. \tag{4}$$

The target eigenvalues can be computed with a reduced generalized eigenvalue problem

$$H_{LM}^{<} \boldsymbol{y} = \nu H_{LM} \boldsymbol{y} \tag{5}$$

provided that $\operatorname{rank}(S) = n_\Gamma$, because the eigenvalues of the non-singular part of (5) are equivalent to the target eigenvalues of (1)—i.e. $\lambda = \nu$ [5]. The corresponding eigenvectors are given by $\boldsymbol{u} = S\boldsymbol{y}$.

In order to compute (2) numerically, some quadrature rule is applied. We approximate $S_m$ by

$$\hat{S}_m \equiv \sum_{j=1}^{N} w_j \zeta_j^m (z_j I - A)^{-1} V \tag{6}$$

where $N$ is the number of quadrature points, $z_j$ and $w_j$ are a quadrature point and a weight, respectively, and $\zeta_j \equiv (z_j - \gamma)/\rho$ is a normalized quadrature point defined with a shift parameter $\gamma \in \mathbb{C}$ and a scale parameter $\rho > 0$ such that the absolute values of $\zeta_j$ ($j = 1, 2, \ldots, N$) are close to unity for numerical stability. Let $X_j$ be the solution of a linear system (with multiple right-hand sides):

$$(z_jI - A)X_j = V. \tag{7}$$

For this numerical quadrature case, $\hat{\mathcal{M}}_k$, $\hat{S}$, $\hat{H}_{LM}$, $\hat{H}^<_{LM}$, $\hat{\nu}$, and $\hat{\boldsymbol{y}}$ are defined analogously with $\hat{S}_m$. When the number of the target eigenvalues is much smaller than $n$, the parameters $L$, $M$, and $N$ are typically set so as to be much smaller than $n$ in practice. Strategies for efficient parameter setting are described in [14]. In the reminder of this paper, we assume that $L$, $M$, and $N$ are much smaller than $n$.

In many publications [3, 4, 6, 8, 12, 13], the trapezoidal rule with a ellipsoidal contour path is preferred for the numerical contour integration. In such case, we set $\zeta_j = \cos\theta_j + \mathrm{i}\alpha_{\mathrm{R}}\sin\theta_j$, $w_j = \rho(\alpha_{\mathrm{R}}\cos\theta_j + \mathrm{i}\sin\theta_j)/N$ and $z_j = \gamma + \rho\zeta_j$, where $\theta_j = (2\pi(j-1/2))/N$ $(j = 1, 2, \ldots, N)$. Here, $\rho$, $\gamma$, and $\alpha_{\mathrm{R}} > 0$ denote the center, the horizontal radius and the aspect ratio of the ellipse, respectively.

To solve small generalized eigenvalue problems of $(\hat{H}^<_{LM}, \hat{H}_{LM})$, we convert the problem to a standard eigenproblem with a low-rank approximation of $\hat{H}_{LM}$, using singular value decomposition (SVD). Then, we approximate the eigenvalues and eigenvectors as $\lambda \approx \hat{\lambda} \equiv \gamma + \rho\hat{\nu}$ and $\boldsymbol{u} \approx \hat{\boldsymbol{u}} \equiv \hat{S}\hat{\boldsymbol{y}}$, respectively. The procedure for the low-rank approximation and for solving the small eigenvalue problem is described in Algorithm 1.

In order to compute (6), linear systems with multiple right-hand sides (7) should be solved for each quadrature point $z_j$. The SSM is embarrassingly parallel, because (7) can be solved independently. Owing to this feature, eigensolvers using numerical quadrature (such as the SSM and the FEAST algorithm [11]) have been actively studied.

SBCGrQ was proposed in [3] as a method for solving linear systems with multiple shifts and multiple right-hand sides, and it has been applied to the linear systems in the SSM. SBCGrQ is based on the block CGrQ method (BCGrQ) [2] and the block bi-conjugate gradient (BiCG) method [10]. It is derived from the fact that $n \times L$ residual matrix of the $k$-th iteration of block BiCG, which applied to (7) (denoted by $R_j^{(k)}$), can be represented as

$$R_j^{(k)} = Q^{(k)}\xi_j^{(k)}, \tag{8}$$

---

**Algorithm 1** Hankel_eig

---

**Input:** $\hat{H}^<_{LM}, \hat{H}_{LM}, \delta$ $(0 < \delta \leq 1)$
**Output:** $\hat{\nu}_i, \hat{\boldsymbol{y}}_i, n_{\mathrm{rank}}$ $(i = 1, 2, \ldots, n_{\mathrm{rank}})$
1: Compute SVD: $U\Sigma W^{\mathrm{H}} = \hat{H}_{LM}$ (The diagonals of $\Sigma$ are in descending order)
2: Let $n_{\mathrm{rank}}$ be the number of singular values greater than $\delta\Sigma_{11}$
3: Let $\Sigma_0$ be the leading $n_{\mathrm{rank}} \times n_{\mathrm{rank}}$ submatrix of $\Sigma$
4: Let $U_0$ and $W_0$ be the submatrices of leading $n_{\mathrm{rank}}$ columns of $U$ and $W$, respectively
5: Compute eigenpairs $(\hat{\nu}_i, \boldsymbol{t}_i)$ of $\Sigma_0^{-1/2}U_0^{\mathrm{H}}\hat{H}^<_{LM}W_0\Sigma_0^{-1/2}$
6: $\hat{\boldsymbol{y}}_i = W_0\Sigma_0^{-1/2}\boldsymbol{t}_i$

---

**Algorithm 2** Shifted block conjugate gradient method with residual orthogonalization (SBCGrQ). Here, $O_{n\times L}$ is the $n \times L$-dimensional zero matrix, $I_L$ is the $L$-dimensional unit matrix, qr$(\cdot)$ indicates the QR decomposition, and $\tau > 0$ is the tolerance for the residual norms

**Input:** $A, V, z_j \quad (j = 1, 2, \ldots, N)$
**Output:** $\tilde{X}_j \quad (j = 1, 2, \ldots, N)$

1: $X_j^{(0)} = O_{n\times L}, \xi_j^{(-1)} = \alpha^{(-1)} = I_L,$
2: $Q^{(0)}\rho^{(0)} = \mathrm{qr}(V)$
3: $\xi_j^{(0)} = \Delta^{(0)} = \rho^{(0)}, P_j^{(0)} = P^{(0)} = Q^{(0)}$
4: **for** $k = 0, 1, \ldots$ until $\max_j \|\xi_j^{(k)}\|_{\mathrm{F}} < \tau\|V\|_{\mathrm{F}}$ **do:**
5: $\quad \alpha^{(k)} = \left(P^{(k)\mathrm{H}}(-A)P^{(k)}\right)^{-1}$
6: $\quad Q^{(k+1)}\rho^{(k+1)} = \mathrm{qr}(Q^{(k)} - AP^{(k)}\alpha^{(k)})$
7: $\quad \Delta^{(k+1)} = \rho^{(k+1)}\Delta^{(k)}$
8: $\quad P^{(k+1)} = Q^{(k+1)} + P^{(k)}\rho^{(k+1)\mathrm{H}}$
9: $\quad$ **for** $j = 1, 2, \ldots, N$ **do:**
10: $\quad\quad \xi_j^{(k+1)} = \rho^{(k+1)}\left[I_L + z_j\alpha^{(k)} + \left\{\rho^{(k)} - \xi_j^{(k)}\left(\xi_j^{(k-1)}\right)^{-1}\right\}\left(\alpha^{(k-1)}\right)^{-1}\rho^{(k)\mathrm{H}}\alpha^{(k)}\right]^{-1}\xi_j^{(k)}$
11: $\quad\quad \alpha_j^{(k)} = \alpha^{(k)}\left(\rho^{(k+1)}\right)^{-1}\xi_j^{(k+1)}$
12: $\quad\quad \beta_j^{(k)} = \alpha^{(k)}\left(\rho^{(k+1)}\right)^{-1}\xi_j^{(k+1)}\left(\xi_j^{(k)}\right)^{-1}\left(\alpha^{(k)}\right)^{-1}\rho^{(k+1)\mathrm{H}}$
13: $\quad\quad X_j^{(k+1)} = X_j^{(k)} + P_j^{(k)}\alpha_j^{(k)}$
14: $\quad\quad P_j^{(k+1)} = Q^{(k+1)} + P_j^{(k)}\beta_j^{(k)}$
15: $\quad$ **end for**
16: **end for**
17: $K = k$
18: $\tilde{X}_j = X_j^{(K)}$ for $j = 1, 2, \ldots, N$

provided that $R_j^{(0)} = Q^{(0)}\xi_j^{(0)}$ holds initially. Here $\xi_j^{(k)} \in \mathbb{C}^{L\times L}$ is a non-singular matrix and $Q^{(k)}$ is the orthonormalized residual matrix of the $k$-th iteration of BCGrQ, which is applied to the linear system $-AX = V$.

The pseudo-code for SBCGrQ is shown in Algorithm 2. The outputs $\tilde{X}_j$ $(j = 1, 2, \ldots, N)$ are the approximate solutions to (7). As we can see in the pseudo-code, we can update the solutions to (7) (for $j = 1, 2, \ldots, N$) with only $L$ sparse matrix-vector products per iteration. This is the most attractive property of SBCGrQ. We hereinafter refer to the $k$-loop in Algorithm 2 as the outer iteration. A shifted block CG algorithm is also proposed by Birk et al. [1]. They apply a deflation technique to their algorithm.

## 3 Memory-Saving Technique

In this section, we propose a memory-saving technique to compute the eigenvalues and the residual norms of the eigenproblem in the eigenvalue computation using the SSM and SBCGrQ. This technique is applicable to eigenproblems where eigenvectors are unnecessary.

### 3.1 Memory-Saving Computation of Eigenvalues

We first consider constructing $\hat{H}_M^<$ and $\hat{H}_M$ without explicitly computing and storing $X_j$ $(j = 1, 2, \ldots, N)$ and $\hat{S}_m$ $(m = 0, 1, \ldots, M-1)$. In order to compute

$$\mathscr{M}_m = V^{\mathrm{H}}\hat{S}_m = \sum_{j=1}^{N} w_j \zeta_j^m V^{\mathrm{H}} X_j,$$

we need only $V^{\mathrm{H}}X_j$, rather than $X_j$ itself. According to the 13th and 14th lines in Algorithm 2, $V^{\mathrm{H}}X_j^{(k+1)}$ and $V^{\mathrm{H}}P_j^{(k+1)}$ are given by the recurrences $V^{\mathrm{H}}X_j^{(k+1)} = V^{\mathrm{H}}X_j^{(k)} + V^{\mathrm{H}}P_j^{(k)}\alpha_j^{(k)}$ and $V^{\mathrm{H}}P_j^{(k+1)} = V^{\mathrm{H}}Q^{(k+1)} + V^{\mathrm{H}}P_j^{(k)}\beta_j^{(k)}$, respectively, by multiplying $V^{\mathrm{H}}$ from the left. Owing to the orthogonal property of the sequence of the residual matrices of block CG [10], we have

$$Q^{(k)^{\mathrm{H}}} Q^{(\ell)} = O_{L\times L} \quad (\ell = 0, 1, \ldots, k-1), \tag{9}$$

where $O_{L\times L}$ is the $L \times L$ zero matrix. This implies that $V^{\mathrm{H}}Q^{(k+1)} = O_{L\times L}$. Hence $V^{\mathrm{H}}P_j^{(k+1)} = V^{\mathrm{H}}P_j^{(k)}\beta_j^{(k)}$. Therefore, we can compute $V^{\mathrm{H}}X_j^{(k+1)}$, without the explicit matrix-matrix product involving the dimension $n$, yet with recurrences of the $L \times L$ matrices. Thus, we do not need to store $X_j^{(k)}$ and $P_j^{(k)}$ in order to compute the eigenvalues. Note that, because we cannot compute $\hat{S}$ without explicit $X_j$, we cannot obtain eigenvectors using the technique above.

### 3.2 Memory-Saving Computation of Residual Norms of Eigenproblem

When computing the eigenvalue, residual norms are typically used to predict the accuracy of the approximate eigenvalues. Here, we propose a technique for computing the (absolute) residual norm of the eigenproblem $||A\hat{\boldsymbol{u}} - \hat{\lambda}\hat{\boldsymbol{u}}||_2$ without explicit eigenvectors. Because the square of the residual norm is given by

$$\begin{aligned}
||A\hat{\boldsymbol{u}} - \hat{\lambda}\hat{\boldsymbol{u}}||_2^2 &= (A\hat{\boldsymbol{u}} - \hat{\lambda}\hat{\boldsymbol{u}})^{\mathrm{H}}(A\hat{\boldsymbol{u}} - \hat{\lambda}\hat{\boldsymbol{u}}) \\
&= \hat{\boldsymbol{u}}^{\mathrm{H}}A^2\hat{\boldsymbol{u}} - 2\hat{\lambda}\hat{\boldsymbol{u}}^{\mathrm{H}}A\hat{\boldsymbol{u}} + \hat{\lambda}^2\hat{\boldsymbol{u}}^{\mathrm{H}}\hat{\boldsymbol{u}} \\
&= (\hat{\boldsymbol{y}}^{\mathrm{H}}\hat{S}^{\mathrm{H}}A^2\hat{S}\hat{\boldsymbol{y}} - 2\hat{\lambda}\hat{\boldsymbol{y}}^{\mathrm{H}}\hat{S}^{\mathrm{H}}A\hat{S}\hat{\boldsymbol{y}} + \hat{\lambda}^2\hat{\boldsymbol{y}}^{\mathrm{H}}\hat{S}^{\mathrm{H}}\hat{S}\hat{\boldsymbol{y}}),
\end{aligned}$$

it can be calculated by an order-$LM$ vector $\hat{\boldsymbol{y}}$ and by the order-$LM$ matrices $\Theta_I \equiv \hat{S}^{\mathrm{H}}\hat{S}$, $\Theta_A \equiv \hat{S}^{\mathrm{H}}A\hat{S}$, and $\Theta_{A^2} \equiv \hat{S}^{\mathrm{H}}A^2\hat{S}$.

Now, we consider computing $\Theta_I$, $\Theta_A$, and $\Theta_{A^2}$ without explicit matrix multiplications involving the large matrix $A$. The $(m,\ell)$-th $(L \times L)$ submatrix of $\Theta_F$ $(F = I, A, A^2)$ can be expressed as follows:

$$(\Theta_F)_{m,\ell} = \sum_{i=1}^{N}\sum_{j=1}^{N} \overline{w_i \zeta_i^{m-1}} w_j \zeta_j^{\ell-1} X_i^{\mathrm{H}} F X_j$$

(here, $(\Theta_F)_{m,\ell}$ is represented by $\Theta_F((m-1)L+1 : mL, (\ell-1)L+1 : \ell L)$ in MATLAB notation). Therefore, their calculation requires $X_i^{\mathrm{H}} X_j$, $X_i^{\mathrm{H}} A X_j$, and $X_i^{\mathrm{H}} A^2 X_j$.

Next, let us focus on the SBCGrQ algorithm shown in Algorithm 2. First, let

$$R_j^{(k)} \equiv V - (z_j I - A) X_j^{(k)}$$

be the residual matrix corresponding to $X_j^{(k)}$. Based on (8) and (9), and because $X_j^{(K)}$ can be expressed as $X_j^{(K)} = \sum_{k=0}^{K-1} Q^{(k)} C_j^{(k)}$ with matrices $C_j^{(k)} \in \mathbb{C}^{L \times L}$, we have $X_i^{(K)\mathrm{H}} R_j^{(K)} = O_{L\times L}$. Because we also have

$$X_i^{(K)\mathrm{H}} R_j^{(K)} = X_i^{(K)\mathrm{H}} V - z_j X_i^{(K)\mathrm{H}} X_j^{(K)} + X_i^{(K)\mathrm{H}} A X_j^{(K)},$$

the following holds:

$$X_i^{(K)\mathrm{H}} A X_j^{(K)} = z_j X_i^{(K)\mathrm{H}} X_j^{(K)} - X_i^{(K)\mathrm{H}} V. \tag{10}$$

In addition, because $V^{\mathrm{H}} R_j^{(K)} = \Delta^{(0)\mathrm{H}} Q^{(0)\mathrm{H}} Q^{(K)} \xi_j^{(K)} = O_{L\times L}$ and $V^{\mathrm{H}} R_j^{(K)} = V^{\mathrm{H}} V - z_j V^{\mathrm{H}} X_j^{(K)} + V^{\mathrm{H}} A X_j^{(K)}$, we have

$$V^{\mathrm{H}} A X_j^{(K)} = z_j V^{\mathrm{H}} X_j^{(K)} - V^{\mathrm{H}} V. \tag{11}$$

Based on the equations

$$R_i^{(K)\mathrm{H}} R_j^{(K)} = V^{\mathrm{H}} A X_j^{(K)} - \overline{z_i} X_i^{(K)\mathrm{H}} A X_j^{(K)} + X_i^{(K)\mathrm{H}} A^2 X_j^{(K)}$$

and $R_i^{(K)\mathrm{H}} R_j^{(K)} = \xi_i^{(K)\mathrm{H}} Q^{(K)\mathrm{H}} Q^{(K)} \xi_j^{(K)} = \xi_i^{(K)\mathrm{H}} \xi_j^{(K)}$, and based on (10) and (11), the following holds:

$$X_i^{(K)\mathrm{H}} A^2 X_j^{(K)} = \xi_i^{(K)\mathrm{H}} \xi_j^{(K)} - z_j V^{\mathrm{H}} X_j^{(K)} V^{\mathrm{H}} V + \overline{z_i} X_i^{(K)\mathrm{H}} A X_j^{(K)}.$$

Thus, we can compute $X_i^{(K)\mathrm{H}} A X_j^{(K)}$ and $X_i^{(K)\mathrm{H}} A^2 X_j^{(K)}$ if we have $\xi_j^{(K)\mathrm{H}} \xi_j^{(K)}$, $V^{\mathrm{H}} X_j^{(K)}$, $V^{\mathrm{H}} V$, and $X_i^{(K)\mathrm{H}} X_j^{(K)}$. Note that, we have already derived the recurrences for computing $V^{\mathrm{H}} X_j^{(K)}$ in Sect. 3.1. Therefore, we can compute $X_i^{(K)\mathrm{H}} A X_j^{(K)}$ and $X_i^{(K)\mathrm{H}} A^2 X_j^{(K)}$

without explicitly computing and storing $X_j^{(K)}$ if we have ${X_i^{(K)}}^{\mathrm{H}} X_j^{(K)}$ $(i, j = 1, 2, \ldots, N)$.

We now describe the derivation of the recurrences for computing ${X_i^{(K)}}^{\mathrm{H}} X_j^{(K)}$. Based on (9), and because $X_j^{(0)} = O_{n \times L}$, $P_j^{(0)} = Q^{(0)}$, and $P_j^{(k)}$ can be expressed as $P_j^{(k)} = \sum_{\ell=0}^{k} Q^{(\ell)} G_j^{(\ell)}$ with matrices $G_j^{(\ell)} \in \mathbb{C}^{L \times L}$, both ${Q^{(k+1)}}^{\mathrm{H}} P_j^{(k)} = O_{L \times L}$ and ${X_j^{(k)}}^{\mathrm{H}} Q^{(k+1)} = O_{L \times L}$ hold. Based on these relations and the 13th and 14th lines in Algorithm 2, the following $L \times L$ matrix recurrences are derived:

$$\begin{aligned} {X_i^{(k+1)}}^{\mathrm{H}} X_j^{(k+1)} &= {X_i^{(k)}}^{\mathrm{H}} X_j^{(k)} + {X_i^{(k)}}^{\mathrm{H}} P_j^{(k)} \alpha_j^{(k)} \\ &\quad + {\alpha_i^{(k)}}^{\mathrm{H}} ({X_j^{(k)}}^{\mathrm{H}} P_i^{(k)})^{\mathrm{H}} + {\alpha_i^{(k)}}^{\mathrm{H}} {P_i^{(k)}}^{\mathrm{H}} P_j^{(k)} \alpha_j^{(k)}, \end{aligned} \tag{12}$$

$${X_i^{(k+1)}}^{\mathrm{H}} P_j^{(k+1)} = {X_i^{(k)}}^{\mathrm{H}} P_j^{(k)} \beta_j^{(k)} + {\alpha_i^{(k)}}^{\mathrm{H}} {P_i^{(k)}}^{\mathrm{H}} P_j^{(k)} \beta_j^{(k)}, \tag{13}$$

and

$${P_i^{(k+1)}}^{\mathrm{H}} P_j^{(k+1)} = I_L + {\beta_i^{(k)}}^{\mathrm{H}} {P_i^{(k)}}^{\mathrm{H}} P_j^{(k)} \beta_j^{(k)}, \tag{14}$$

where $I_L$ is the identity matrix of order $L$. Here, the initial values are set such that ${X_i^{(0)}}^{\mathrm{H}} X_j^{(0)} = {X_i^{(0)}}^{\mathrm{H}} P_j^{(0)} = O_{L \times L}$ and ${P_i^{(0)}}^{\mathrm{H}} P_j^{(0)} = I_L$. Because the terms at the right-hand side in (12), (13), and (14) are all known at the end of the $k$-th iteration, we can compute ${X_i^{(K)}}^{\mathrm{H}} X_j^{(K)}$, without the explicitly computing the product of ${X_i^{(K)}}^{\mathrm{H}}$ and $X_j^{(K)}$, yet with recurrences of the $L \times L$ matrices. This indicates that eventually we can compute the residual norms of the eigenproblem without explicit operations and storage involving $X_j^{(k)}$ and $P_j^{(k)}$.

The pseudo-code for the resulting algorithm is shown in Algorithm 3. Both SS–Hankel and SBCGrQ are included in a single pseudo-code. In order to easily recognize that there is no explicit matrix product involving dimension $n$ in the recurrences for ${X_i^{(k)}}^{\mathrm{H}} X_j^{(k)}$, we introduce following symbols: $\eta_j \equiv V^{\mathrm{H}} X_j^{(k)}$, $\omega_j \equiv V^{\mathrm{H}} P_j^{(k)}$, $\chi_{ij}^{(k)} \equiv {X_i^{(k)}}^{\mathrm{H}} X_j^{(k)}$, $\phi_{ij}^{(k)} \equiv {X_i^{(k)}}^{\mathrm{H}} P_j^{(k)}$, $\pi_{ij}^{(k)} \equiv {P_i^{(k)}}^{\mathrm{H}} P_j^{(k)}$, $\kappa_{ij}^{(K)} \equiv {X_i^{(K)}}^{\mathrm{H}} A X_j^{(k)}$ and $\psi_{ij}^{(K)} \equiv {X_i^{(K)}}^{\mathrm{H}} A^2 X_j^{(K)}$. The outputs $\tilde{r}_i$ are equal to $||A\hat{\boldsymbol{u}}_i - \hat{\lambda}_i \hat{\boldsymbol{u}}_i|| / ||\hat{\lambda}_i \hat{\boldsymbol{u}}_i||$ in exact arithmetic. We refer to them as the relative residual norms in this study. In case $\tilde{\lambda}_i = 0$, only the numerator (the absolute residual norm) of $\tilde{r}_i$ may be computed and used to assess the accuracy of the eigenvalues.

We now discuss the memory requirement for our technique. If the matrix $A$ is sufficiently sparse, the storage of $n \times L$ matrices is dominant, and the storage of $L \times L$ matrices is negligible, because $L \ll n$. Thus, we can compare memory requirement by counting the number of $n \times L$ matrices that should be stored in the algorithms of the SSM and SBCGrQ. Using our technique, the SSM does not require $S_m$ $(m = 0, 1, \ldots, M-1)$ and SBCGrQ does not require $X_j$ and $P_j$ $(j = 1, 2, \ldots, N)$. Note

**Algorithm 3** Algorithm of SS-Hankel with SBCGrQ using our proposed memory-saving technique. Here, $O_{n\times L}$ is the $n \times L$-dimensional zero matrix, $I_L$ is the $L$-dimensional unit matrix, $\mathrm{qr}(\cdot)$ indicates the QR decomposition, and $\tau > 0$ is the tolerance for the residual norms

**Input:** $A, V, z_j, w_j, \zeta_j, \gamma, \rho, \delta \quad (j = 1, 2, \ldots, N)$
**Output:** $\tilde{\lambda}_i, \tilde{r}_i$
1: (Beginning of SBCGrQ)
2: $X_j^{(0)} = O_{n\times L}, \xi_j^{(-1)} = \alpha^{(-1)} = I_L,$
3: $Q^{(0)}\rho^{(0)} = \mathrm{qr}(V)$
4: $\xi_j^{(0)} = \Delta^{(0)} = \rho^{(0)}, P_j^{(0)} = P^{(0)} = Q^{(0)}$
5: **for** $k = 0, 1, \ldots$ until $\max_j ||\xi_j^{(k)}||_{\mathrm{F}} < \tau ||V||_{\mathrm{F}}$ **do:**
6: $\quad \alpha^{(k)} = \left(P^{(k)\mathrm{H}}(-A)P^{(k)}\right)^{-1}$
7: $\quad Q^{(k+1)}\rho^{(k+1)} = \mathrm{qr}(Q^{(k)} - AP^{(k)}\alpha^{(k)})$
8: $\quad \Delta^{(k+1)} = \rho^{(k+1)}\Delta^{(k)}$
9: $\quad P^{(k+1)} = Q^{(k+1)} + P^{(k)}\rho^{(k+1)\mathrm{H}}$
10: $\quad$ **for** $j = 1, 2, \ldots, N$ **do:**
11: $\quad\quad \xi_j^{(k+1)} = \rho^{(k+1)}\left[I_L + z_j\alpha^{(k)} + \left\{\rho^{(k)} - \xi_j^{(k)}\left(\xi_j^{(k-1)}\right)^{-1}\right\}\left(\alpha^{(k-1)}\right)^{-1}\rho^{(k)\mathrm{H}}\alpha^{(k)}\right]^{-1}\xi_j^{(k)}$
12: $\quad\quad \alpha_j^{(k)} = \alpha^{(k)}\left(\rho^{(k+1)}\right)^{-1}\xi_j^{(k+1)}$
13: $\quad\quad \beta_j^{(k)} = \alpha^{(k)}\left(\rho^{(k+1)}\right)^{-1}\xi_j^{(k+1)}\left(\xi_j^{(k)}\right)^{-1}\left(\alpha^{(k)}\right)^{-1}\rho^{(k+1)\mathrm{H}}$
14: $\quad\quad \eta_j^{(k+1)} = \eta_j^{(k)} + \omega_j^{(k)}\alpha_j^{(k)}$
15: $\quad\quad \omega_j^{(k+1)} = \omega_j^{(k)}\beta_j^{(k)}$
16: $\quad$ **end for**
17: $\quad$ **for** $i = 1, 2, \ldots, N$ **do:**
18: $\quad\quad$ **for** $j = 1, 2, \ldots, N$ **do:**
19: $\quad\quad\quad \chi_{ij}^{(k+1)} = \chi_{ij}^{(k)} + \phi_{ij}^{(k)}\alpha_j^{(k)} + \alpha_i^{(k)\mathrm{H}}(\phi_{ji}^{(k)})^{\mathrm{H}} + \alpha_i^{(k)\mathrm{H}}\pi_{ij}^{(k)}\alpha_j^{(k)}$
20: $\quad\quad\quad \phi_{ij}^{(k+1)} = \phi_{ij}^{(k)}\beta_j^{(k)} + \alpha_i^{(k)\mathrm{H}}\pi_{ij}^{(k)}\beta_j^{(k)}$
21: $\quad\quad\quad \pi_{ij}^{(k+1)} = I_L + \beta_i^{(k)\mathrm{H}}\pi_{ij}^{(k)}\beta_j^{(k)}$
22: $\quad\quad$ **end for**
23: $\quad$ **end for**
24: **end for**
25: (End of SBCGrQ)
26: $K = k$
27: $\hat{\mathscr{M}}_m = \sum_{j=1}^N w_j\zeta_j^m\eta_j \quad$ for $m = 0, 1, \ldots, 2M-1$
28: **for** $i = 1, 2, \ldots, N$ **do:**
29: $\quad$ **for** $j = 1, 2, \ldots, N$ **do:**
30: $\quad\quad \kappa_{ij}^{(K)} = z_j\chi_{ij}^{(K)} - (\eta_i^{(K)})^{\mathrm{H}}$
31: $\quad\quad \psi_{ij}^{(K)} = \xi_i^{(K)\mathrm{H}}\xi_j^{(K)} - z_j\eta_j^{(K)} + \Delta^{(0)\mathrm{H}}\Delta^{(0)} + \overline{z_i}\kappa_{ij}^{(K)}$
32: $\quad$ **end for**
33: **end for**
34: **for** $m = 1, 2, \ldots, M$ **do:**
35: $\quad$ **for** $\ell = 1, 2, \ldots, M$ **do:**
36: $\quad\quad (\Theta_I)_{m,\ell} = \sum_{i=1}^N\sum_{j=1}^N \overline{w_i\zeta_i^{m-1}}w_j\zeta_j^{\ell-1}\chi_{ij}^{(K)}$
37: $\quad\quad (\Theta_A)_{m,\ell} = \sum_{i=1}^N\sum_{j=1}^N \overline{w_i\zeta_i^{m-1}}w_j\zeta_j^{\ell-1}\kappa_{ij}^{(K)}$
38: $\quad\quad (\Theta_{A^2})_{m,\ell} = \sum_{i=1}^N\sum_{j=1}^N \overline{w_i\zeta_i^{m-1}}w_j\zeta_j^{\ell-1}\psi_{ij}^{(K)}$
39: $\quad$ **end for**
40: **end for**

(continued)

41: Form $\hat{H}_{LM}$ and $\hat{H}^{<}_{LM}$ as (3) and (4)
42: $[\tilde{\nu}_i, \tilde{\boldsymbol{y}}_i, n_{\text{rank}}] = \text{Hankel\_eig}(\hat{H}^{<}_{LM}, \hat{H}_{LM}, \delta)$ (see Algorithm 1)
43: **for** $i = 1, 2, \ldots, n_{\text{rank}}$ **do:**
44: $\tilde{\lambda}_i = \gamma + \rho\tilde{\nu}_i$
45: $\tilde{r}_i = \sqrt{(\tilde{\boldsymbol{y}}_i^{\text{H}}(\Theta_{A^2} - 2\tilde{\lambda}_i\Theta_A + \tilde{\lambda}_i^2\Theta_I)\tilde{\boldsymbol{y}}_i)/(\tilde{\lambda}_i^2\tilde{\boldsymbol{y}}_i^{\text{H}}\Theta_I\tilde{\boldsymbol{y}}_i)}$
46: **end for**

**Table 1** Comparison of the memory requirements of our technique with those of the standard approach

| Variable | Standard approach | Proposed technique |
|---|---|---|
| $V$ | $nL$ | 0 |
| $Q^{(k)}$ | $nL$ | $nL$ |
| $P^{(k)}$ | $nL$ | $nL$ |
| $AP^{(k)}$ | $nL$ | $nL$ |
| $X_j^{(k)}$ $(j = 1, 2, \ldots, N)$ | $NnL$ | 0 |
| $P_j^{(k)}$ $(j = 1, 2, \ldots, N)$ | $NnL$ | 0 |
| Total | $(2N + 4)nL$ | $3nL$ |

that $\boldsymbol{u}$ can be computed by $\tilde{X}_j$ without forming $S$. Hence, there is no need to compute and store $S_m$ also with the standard approach. A comparison between our proposed technique and the standard approach is shown in Table 1. The value of the memory requirement in the table refers to the number of scalar elements that must be stored. We assume that the right-hand side $V$ is overwritten by $Q^{(k)}$ with our technique. As seen in Table 1, our technique requires only $3nL$ elements whereas the standard approach requires $(2N + 4)nL$ elements. Consequently, our technique reduces the memory requirement significantly, because we usually set $N \geq 16$.

Using the proposed technique, we can compute eigenvalues and residual norms of the eigenproblem at a negligible cost when $L, M$, and $N$ are much smaller than $n$. It means that it is easy to compute the residual norms of the eigenproblem at every outer SBCGrQ iteration. The residual norms of eigenpairs are desirable, because in some cases, the highly accurate solution to (7) for a certain $z_j$ does not contribute to the accuracy of the target eigenpairs. We do not have to wait for the convergence of such solution to the linear system if we can obtain the residual norms of the eigenproblem cheaply. This is useful not only in cases where eigenvectors are irrelevant, but also in the cases where eigenvectors are necessary, although in the latter case, the advantages of a small memory requirement have not been investigated.

In the remainder of this paper, we refer to a residual norm that is explicitly computed with $\hat{\boldsymbol{u}}$ as a "true" residual norm. Similarly, we refer to an approximate eigenvalue $\hat{\lambda}$ explicitly computed using $X_j^{(j)}$ as a "true" approximate eigenvalue.

### 3.3 Limitations

As described Sect. 3.2, the basis for our technique for computing the residual norm of the eigenproblem is the computation of its square:

$$||A\hat{u} - \hat{\lambda}\hat{u}||_2^2 = (\hat{y}^{\mathrm{H}}\Theta_{A^2}\hat{y} - 2\hat{\lambda}\hat{y}^{\mathrm{H}}\Theta_{A}\hat{y} + \hat{\lambda}^2\hat{y}^{\mathrm{H}}\Theta_{I}\hat{y}), \tag{15}$$

and the computation for the terms at the right-hand side using the recurrences with (block) bilinear forms. Due to this, a large cancellation error occurs when the residual norm is relatively small. For instance, consider now that the relative difference between $\hat{y}^{\mathrm{H}}\Theta_{A^2}\hat{y} + \hat{\lambda}^2\hat{y}^{\mathrm{H}}\Theta_{I}\hat{y}$ and $2\hat{\lambda}\hat{y}^{\mathrm{H}}\Theta_{A}\hat{y}$ is approximately $10^{-20}$. In such case, a cancellation error with approximately 20 digits occurs in (15). Thus, using our technique with the double precision arithmetic, we cannot correctly compute such relatively small residual norm that causes a cancellation error with more than 16 digits. This large cancellation error was confirmed in the numerical experiments.

In the discussion in the previous section, we assumed that the number of right-hand sides (RHSs) involved in SBCGrQ is $L$. If $L$ is large, however, it is better to employ multiple instances of SBCGrQ and apply them to different subsets of RHSs because as the number of RHSs increases, the computational cost per iteration increases in $O(L^2)$. Unfortunately, our technique is only applicable when a single SBCGrQ instance is employed and when all RHSs are treated collectively.

Furthermore, our technique is not applicable to SS-RR and SS-Arnoldi, because an explicit manipulation of $S$ or $X_j$ is necessary for these methods.

## 4 Numerical Experiments

All numerical experiments described in this section were carried out in MATLAB R2014a. The floating-point arithmetic was executed with double precision. We compared the approximate eigenvalues and the residual norms of the eigenproblem computed by our technique, and the true approximate eigenvalue and the true residual norms of the eigenproblem, respectively. The properties of the matrices used in the examples are shown in Table 2. The intervals of the target eigenvalues and the number of target eigenvalues are also provided. The application areas of the problems are described in each example. For the contour path of the SSM, we used

**Table 2** Properties of the matrices used in the numerical examples

| Example | Matrix type | Size | Interval | # of eigenvalues |
|---|---|---|---|---|
| Ex.1 | Real symmetric | 4000 | $[-13, -7]$ | 20 |
| Ex.2 | Real symmetric | 25,000 | $[-0.78, -0.116]$ | 31 |
| Ex.3 | Complex Hermitian | 3072 | $[0.12, 0.33]$ | 31 |

**Table 3** Fixed parameters in the SSM for all experiments

| Parameter | Description | Value |
|---|---|---|
| $L$ | Number of source vectors | 16 |
| $M$ | Maximum moment size | 8 |
| $N$ | Number of quadrature points | 32 |
| $\alpha_R$ | Aspect ratio of the ellipse | 0.1 |
| $\delta$ | Threshold for the low-rank approximation of $\hat{H}_{LM}$ | $10^{-14}$ |

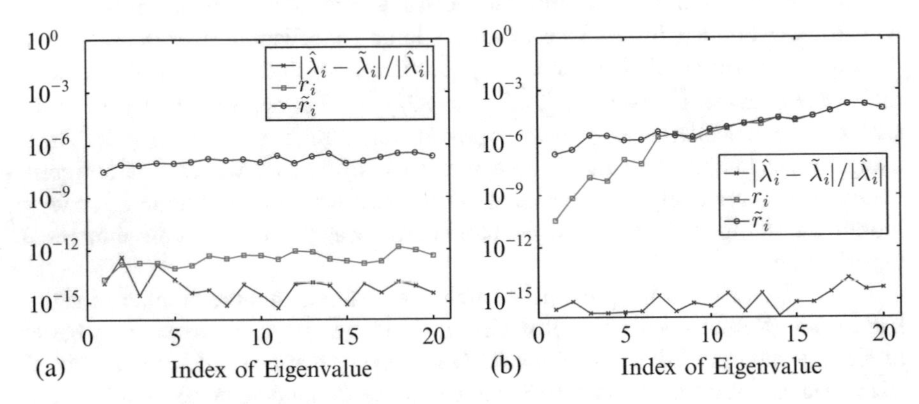


**Fig. 1** Differences in eigenvalues and values of residual norms of Example 1. (**a**) $\tau = 10^{-14}$. (**b**) $\tau = 10^{-3}$

an ellipse that is symmetric with respect to the real axis and intersects at each end of the target interval. We used the trapezoidal rule for the numerical quadrature. The fixed parameters of the SSM for all experiments are shown in Table 3.

## 4.1 *Example 1*

In this example the test problem is derived from a nuclear shell-model calculation [15]. The matrix $A$ is a real symmetric matrix ($n = 4000$). The results of the experiment are shown in Fig. 1. The parameter $\tau$ in the captions of Fig. 1a and b are the parameter of the stopping criterion for SBCGrQ, which appears in Algorithm 2. Thus the difference of the results shown in Fig. 1a and b comes from the difference of the values of $\tau$.

In the figures, $\hat{\lambda}_i$ denotes a true approximate eigenvalue, whereas $\tilde{\lambda}_i$ denotes an approximate eigenvalue computed with our memory-saving technique. The computed eigenvalues are indexed in ascending order. We plot the (relative) difference between $\hat{\lambda}_i$ and $\tilde{\lambda}_i$—i.e, $|\hat{\lambda}_i - \tilde{\lambda}_i|/|\hat{\lambda}_i|$—to show that the approximate eigenvalues are accurately computed with our technique. Note here that $\hat{\lambda}$ and $\tilde{\lambda}$ are equivalent in exact arithmetic. In the figures, we also plot the residual norms of the eigenproblem computed by our technique (denoted by $\tilde{r}_i$) and true residual norms of the eigenproblem (denoted by $r_i$).

For the eigenvalues, as we can see in both Fig. 1a and b, $|\hat{\lambda}_i - \tilde{\lambda}_i|/|\hat{\lambda}_i|$ are smaller than $10^{-12}$. For the case of $\tau = 10^{-3}$, $|\hat{\lambda}_i - \tilde{\lambda}_i|/|\hat{\lambda}_i|$ are sufficiently small with respect to their true residual norms. In Fig. 1a, we can see that the values for $\tilde{r}_i$ are approximately $10^{-7}$, and the values for $r_i$ are smaller than $10^{-11}$. In this case, our technique failed to compute the residual norms correctly. This is due to the large cancellation error that occurs in our technique. We show a result with $\tau = 10^{-3}$ in order to demonstrate that our technique is successful when the relative residual norms of the eigenproblem is not so small as to cause the large cancellation. In contrast to the result in Fig. 1a, we see that our technique can compute the relative residual norms accurately, provided that they are greater than approximately $10^{-6}$. Here, we regard that a residual norm is accurate if its exponents are agree with that of the truc rcsidual norm.

For the application of this example, only rough estimations of eigenvalues are required. Our technique is useful for this kind of application, even though small residual norms cannot be computed correctly.

## 4.2 *Example 2*

In this example, we performed the same experiment as Example 1 with a different test matrix. The test matrix was derived from a real-space density functional calculation [7] and is a real symmetric matrix ($n = 25,000$). The results of the experiment are shown in Fig. 2a and b. In both figures, we can see that $|\hat{\lambda}_i - \tilde{\lambda}_i|/|\hat{\lambda}_i|$ is less than $10^{-12}$. For the case of $\tau = 10^{-4}$, $|\hat{\lambda}_i - \tilde{\lambda}_i|/|\hat{\lambda}_i|$ are sufficiently small with respect to their residual norms. In Fig. 2a, we can see that the values for $\tilde{r}_i$ are

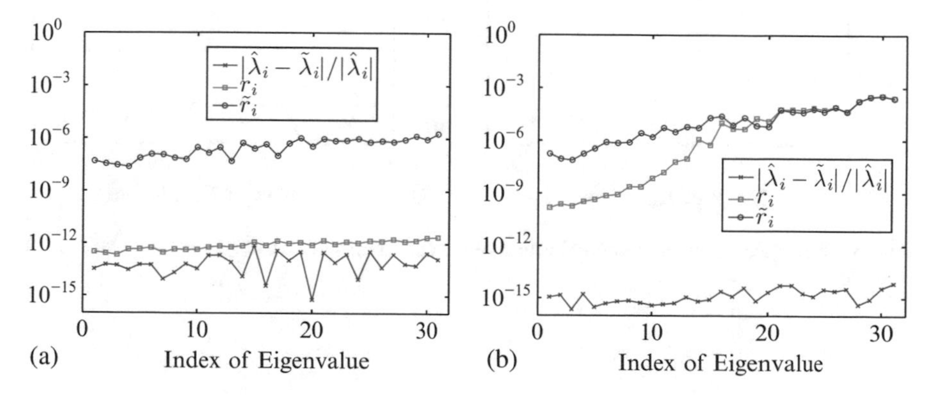


**Fig. 2** Differences in eigenvalues and values of residual norms of Example 2. **(a)** $\tau = 10^{-14}$. **(b)** $\tau = 10^{-4}$

approximately $10^{-7}$, whereas the values for $r_i$ are smaller than $10^{-11}$. Again, this is due to the large cancellation error that occurs with our technique. On the other hand, as seen in Fig. 2b, our technique can compute the relative residual norms accurately when they are greater than approximately $10^{-5}$. In this numerical example, we observe that the differences of the eigenvalues and the values of the residual norms show similar behavior to those observed in Example 1.

## 4.3 Example 3

In this example, we performed the same experiment as the previous two examples with a different test matrix. The test matrix was derived from the computation of quark propagators in quantum chromodynamics. This matrix is a complex Hermitian fermion matrix [9] ($n = 3072$). The results of the experiment are shown in Fig. 3a and b. In both figures, we can see that $|\hat{\lambda}_i - \tilde{\lambda}_i|/|\hat{\lambda}_i|$ are smaller than $10^{-12}$. In Fig. 3a, we can see that the values for $\tilde{r}_i$ are approximately $10^{-7}$, whereas the values for $r_i$ are smaller than $10^{-11}$. Once again, this is due to the large cancellation error that occurs with our technique. On the other hand, as seen in Fig. 3b, our technique can compute the relative residual norms accurately if they are greater than approximately $10^{-4}$. In this numerical example, we also observe that the differences of the eigenvalues and the values of the residual norms show similar behavior to those observed in the previous two examples.

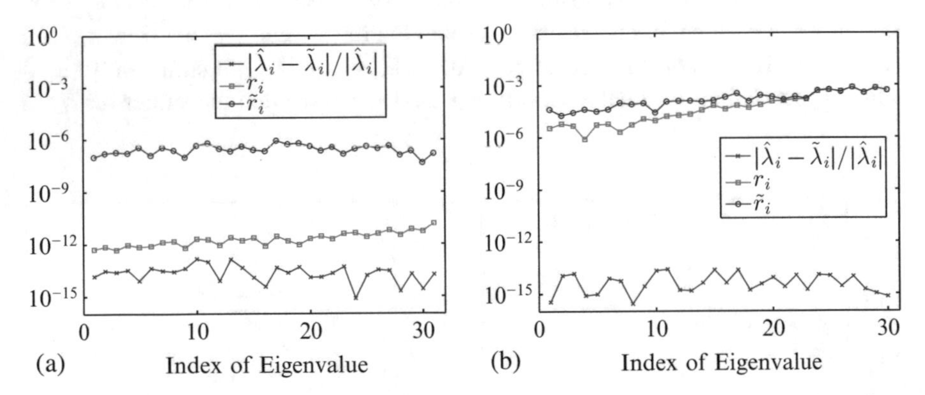


**Fig. 3** Differences in eigenvalues and values of residual norms of Example 3. **(a)** $\tau = 10^{-10}$. **(b)** $\tau = 10^{-2}$

### 4.4 *Monitoring Residual Norms of Eigenproblem While Solving Linear Systems*

As we described in the previous section, we can compute residual norms of the eigenproblem for every outer SBCGrQ iteration with relatively little computational cost. Here, we demonstrate the computation of the relative residual norms of the eigenproblem for every outer SBCGrQ iteration. The test matrix and the parameters were the same as they were in Example 3, and $\tau = 10^{-10}$. The histories of relative residual norms for linear systems and the eigenproblem are shown in Fig. 4. The blue line with markers labeled by **min lin res** indicates the history of the minimum relative residual norm for (7)—i.e., $\min_j ||\xi_j^{(k)}||_F/||V||_F$. The green line with marker labeled by **max lin res** indicates the history of $\max_j ||\xi_j^{(k)}||_F/||V||_F$. The other lines labeled with **eig res *i*** indicate the *i*-th smallest relative residual norms of the eigenproblem of each iteration.

In the figure, we can see that the residual norms of the eigenproblem stagnate from around the 130th iteration because of the cancellation error. The residual norms of the eigenproblem reaches $10^{-6}$ (around the 110th iteration) well before the residual norms of the linear systems reaches $\tau = 10^{-10}$. When $10^{-6}$ is sufficient for these residual norms of eigenproblem and only the corresponding eigenvalues are necessary, the subsequent iterations are wasted. In such case, using our technique, we can skip these wasted iterations.

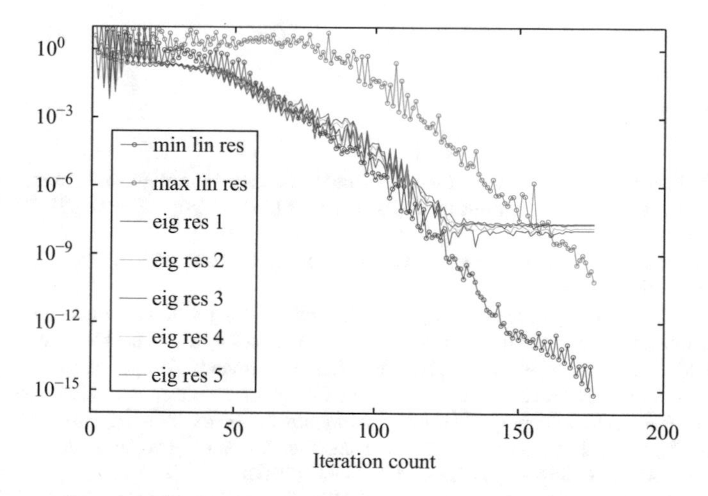


**Fig. 4** Histories of the residual norms of linear systems and the residual norms of the eigenproblem

## 5 Conclusions

We proposed a memory-saving technique for computing eigenvalues and residual norms of the eigenproblem in cases where SBCGrQ is applied in conjunction with the SSM. Our technique drastically reduces the memory consumption needed for applications that eigenvectors are irrelevant, and the memory requirement for the matrix $A$ is negligible. In addition, our technique allows us to compute eigenvalues and residual norms for the eigenproblem cheaply during each outer SBCGrQ iteration. This is beneficial not only in cases where eigenvectors are irrelevant, but also in the cases where eigenvectors are necessary, although in the latter case, the advantages of a small memory requirement have not been investigated. We evaluated the accuracy of eigenvalues and the residual norms of the eigenproblem computed with our approach using three numerical examples from practical applications. However, we observed the occurrences of large cancellation errors when the residual norms of the eigenproblem were small. Thus, we will develop a technique to avoid such cancellation errors in a future work.

**Acknowledgements** The authors would like to thank Dr. Noritaka Shimizu for giving us a matrix derived from a nuclear shell-model. The authors wish to thank Dr. Jun-Ichi Iwata for giving us a matrix derived from a real-space density functional calculation. The authors are grateful to Dr. Hiroshi Ohno for his valuable comments and for providing us a matrix from quantum chromodynamics. The authors would like to thank the anonymous reviewers for their helpful comments and suggestions. This work was supported in part by JST/CREST and MEXT KAKENHI (Grant Nos. 25104701, 25286097).

## References

1. Birk, S., Frommer, A.: A deflated conjugate gradient method for multiple right hand sides and multiple shifts. Numer. Algorithms **67**(3), 507–529 (2013). https://doi.org/10.1007/s11075-013-9805-9
2. Dubrulle, A.A.: Retooling the method of block conjugate gradients. Electron. Trans. Numer. Anal. **12**, 216–233 (2001)
3. Futamura, Y., Sakurai, T., Furuya, S., Iwata, J.I.: Efficient algorithm for linear systems arising in solutions of eigenproblems and its application to electronic-structure calculations. In: Daydé, M., Marques, O., Nakajima, K. (eds.) High Performance Computing for Computational Science - VECPAR 2012. Lecture Notes in Computer Science, vol. 7851, pp. 226–235. Springer, Berlin/Heidelberg (2013). http://dx.doi.org/10.1007/978-3-642-38718-0_23
4. Ikegami, T., Sakurai, T.: Contour integral eigensolver for non-Hermitian systems: a Rayleigh-Ritz-type approach. Taiwan. J. Math. **14**, 825–837 (2010)
5. Ikegami, T., Sakurai, T., Nagashima, U.: A filter diagonalization for generalized eigenvalue problems based on the Sakurai-Sugiura projection method. J. Comput. Appl. Math. **233**(8), 1927–1936 (2010). https://doi.org/10.1016/j.cam.2009.09.029
6. Imakura, A., Du, L., Sakurai, T.: A block Arnoldi-type contour integral spectral projection method for solving generalized eigenvalue problems. Appl. Math. Lett. **32**, 22–27 (2014). https://doi.org/10.1016/j.aml.2014.02.007. http://www.sciencedirect.com/science/article/pii/S0893965914000421

7. Iwata, J.I., Takahashi, D., Oshiyama, A., Boku, T., Shiraishi, K., Okada, S., Yabana, K.: A massively-parallel electronic-structure calculations based on real-space density functional theory. J. Comput. Phys. **229**(6), 2339–2363 (2010)
8. Mizusaki, T., Kaneko, K., Honma, M., Sakurai, T.: Filter diagonalization of shell-model calculations. Phys. Rev. C **82**(2), 10 (2010)
9. Ohno, H., Kuramashi, Y., Sakurai, T., Tadano, H.: A quadrature-based eigensolver with a Krylov subspace method for shifted linear systems for Hermitian eigenproblems in lattice QCD. JSIAM Lett. **2**, 115–118 (2010). http://ci.nii.ac.jp/naid/130000433719/en/
10. O'Leary, D.P.: The block conjugate gradient algorithm and related methods. Linear Algebra Appl. **29**, 293–322 (1980). https://doi.org/10.1016/0024-3795(80)90247-5. Special Volume Dedicated to Alson S. Householder
11. Polizzi, E.: Density-matrix-based algorithm for solving eigenvalue problems. Phys. Rev. B **79**, 115112 (2009). https://doi.org/10.1103/PhysRevB.79.115112
12. Sakurai, T., Sugiura, H.: A projection method for generalized eigenvalue problems using numerical integration. J. Comput. Appl. Math. **159**(1), 119–128 (2003)
13. Sakurai, T., Tadano, H.: CIRR: a Rayleigh-Ritz type method with contour integral for generalized eigenvalue problems. Hokkaido Math. J. **36**, 745–757 (2007)
14. Sakurai, T., Tadano, H., Futamura, Y.: Efficient parameter estimation and implementation of a contour integral-based eigensolver. J. Algorithms Comput. Technol. **7**, 249–269 (2013)
15. Shimizu, N.: Nuclear shell-model code for massive parallel computation, "kshell". arXiv:1310.5431 [nucl-th] (2013)

# Filter Diagonalization Method by Using a Polynomial of a Resolvent as the Filter for a Real Symmetric-Definite Generalized Eigenproblem

**Hiroshi Murakami**


**Abstract** For a real symmetric-definite generalized eigenproblem of size $N$ matrices $A\mathbf{v} = \lambda B\mathbf{v}$ $(B > 0)$, we solve those pairs whose eigenvalues are in a real interval $[a, b]$ by the filter diagonalization method.

In our present study, the filter which we use is a real-part of a polynomial of a resolvent: $F = \text{Re} \sum_{k=1}^{n} \gamma_k \{R(\rho)\}^k$. Here $R(\rho) = (A - \rho B)^{-1}B$ is the resolvent with a non-real complex shift $\rho$, and $\gamma_k$ are coefficients. In our experiments, the (half) degree $n$ is 15 or 20.

By tuning the shift $\rho$ and coefficients $\{\gamma_k\}$ well, the filter passes those eigenvectors well whose eigenvalues are in a neighbor of $[a, b]$, but strongly reduces those ones whose eigenvalues are separated from the interval.

We apply the filter to a set of sufficiently many $B$-orthonormal random vectors $\{\mathbf{x}^{(\ell)}\}$ to obtain another set $\{\mathbf{y}^{(\ell)}\}$. From both sets of vectors and properties of the filter, we construct a basis which approximately spans an invariant subspace whose eigenvalues are in a neighbor of $[a, b]$. An application of the Rayleigh-Ritz procedure to the basis gives approximations of all required eigenpairs.

Experiments for banded problems showed this approach worked in success.


## 1 Introduction

We solve pairs of a real symmetric-definite generalized eigenproblem of size $N$ matrices $A$ and $B$ as:

$$A\,\mathbf{v} = \lambda B\,\mathbf{v} \tag{1}$$

whose eigenvalues are in the specified interval $[a, b]$ by the filter diagonalization method [16]. We define for this kind of eigenproblem, the resolvent with a


H. Murakami (✉)
Tokyo Metropolitan University, 1-1 Minami-Osawa, Hachi-Oji, Tokyo 192-0397, Japan
e-mail: mrkmhrsh@tmu.ac.jp







T. Sakurai et al. (eds.), *Eigenvalue Problems: Algorithms, Software and Applications in Petascale Computing*, Lecture Notes in Computational Science and Engineering 117, https://doi.org/10.1007/978-3-319-62426-6_14

complex-valued shift $\rho$ as:

$$R(\rho) = (A - \rho B)^{-1} B\,. \tag{2}$$

In our previous papers and reports [5–9] and papers of others [1–3], the filter studied or used was a real-part of a complex linear combination of resolvents with complex shifts:

$$F = c_\infty I + \mathrm{Re} \sum_{k=1}^{n} \gamma_k R(\rho_k)\,. \tag{3}$$

Here, $c_\infty$ is a real coefficient, $\gamma_k$, $k{=}1, 2, \ldots, n$ are complex coefficients, and $I$ is the identity matrix. (The reason we take the real-part is to halve the cost of calculation by using the complex-conjugate symmetry). For example, from $n = 6$ to $n = 16$ (or more) resolvents were used.

For a given set of size $N$ column vectors $X$, the action of the resolvent $Y \leftarrow R(\rho)X$ reduces to solve an equation $CY = BX$ for a set of column vectors $Y$. Here, the coefficient matrix is $C = A - \rho B$. When $C$ is banded, the equation may be solved by some direct method using matrix factorization. When $C$ is random sparse, the equation is solved by some iterative method using incomplete matrix factorization. In the application of the filter, the matrix factorization is a large portion of the calculation. The total amount of memory to store the factor is also a very severe constraint in the calculations of large size problems. When many resolvents are used, the total amount of memory requirements is proportional to the number of resolvents applied concurrently.

There are also different but similar approaches and successful studies which are based on the contour integrals and moment calculations [4, 13–15], which also uses many resolvents whose shifts correspond to the integration points.

In this report of study (and in our several previous reports [10–12]), we used only a single resolvent with a complex shift and constructed the filter which is a real-part of a polynomial of the resolvent as:

$$F = c_\infty I + \mathrm{Re} \sum_{k=1}^{n} \gamma_k \, (R(\rho))^k\,. \tag{4}$$

We made some numerical experiments on a test problem to check if this approach is really applicable.

## 2 Present Approach: Filter is Real-Part of Polynomial of Resolvent

For the large eigenproblem, we assume the severest constraint is the amount of memory requirements. Thus, in our present study of the filter diagonalization method, we use a single resolvent in the filter rather than many ones. The filter we use is a real-part of a polynomial of the resolvent. In the filter operation, the same resolvent is applied as many times as the degree of the polynomial. Each time, the application of the resolvent to a set of vectors reduces to the set of solutions of simultaneous linear equations of the same coefficient matrix. To solve the set of simultaneous equations, the coefficient matrix is factored once and the factor is stored. The stored factor is used many times when the resolvent is applied. By the use of a single resolvent rather than many ones, even the transfer function of the filter cannot be made in good shape, but in exchange we obtain advantages of lower memory requirement and reduced number of matrix factorization.

### *2.1 Filter as a Polynomial of a Resolvent and Its Transfer Function*

We consider a real symmetric-definite generalized eigenproblem of size $N$ matrices $A$ and $B$:

$$A\mathbf{v} = \lambda B\mathbf{v}, \text{ where } B > 0\,. \tag{5}$$

The resolvent with a non-real complex shift $\rho$ is:

$$R(\rho) = (A - \rho B)^{-1} B\,. \tag{6}$$

For any pair of the eigenproblem $(\lambda, \mathbf{v})$, we have:

$$R(\rho)\,\mathbf{v} = \frac{1}{\lambda - \rho}\mathbf{v}\,. \tag{7}$$

The filter $F$ is a real-part of a degree $n$ polynomial of the resolvent:

$$F = c_\infty I + \operatorname{Re} \sum_{k=1}^{n} \gamma_k \{R(\rho)\}^k\,. \tag{8}$$

Here, $c_\infty$ is a real number, and $\gamma_k$ are complex numbers. This filter is a real linear operator. For any eigenpair $(\lambda, \mathbf{v})$, we have:

$$F\,\mathbf{v} = f(\lambda)\,\mathbf{v}\,. \tag{9}$$

Here, $f(\lambda)$ is the transfer function of the filter $F$ which is a real rational function of $\lambda$ of degree $2n$ as:

$$f(\lambda) = c_\infty + \mathrm{Re} \sum_{k=1}^{n} \frac{\gamma_k}{(\lambda - \rho)^k} \tag{10}$$

whose only poles are located at a non-real complex number $\rho$ and its complex conjugate (both poles are $n$-th order).

#### 2.1.1 Transfer Function $g(t)$ in Normalized Coordinate $t$

We are to solve those pairs whose eigenvalues are in the specified real interval $[a, b]$. By the linear transformation which maps between $\lambda \in [a, b]$ and $t \in [-1, 1]$, the normalized coordinate $t$ of $\lambda$ is defined as $\lambda = \frac{a+b}{2} + \left(\frac{b-a}{2}\right) t$. We call the interval $t \in [-1, 1]$ as the *passband*, intervals $\mu \leq |t|$ as *stopbands*, and intervals $1 < |t| < \mu$ which are between the passband and stopbands as *transition-bands*.

The transfer function $g(t)$ in the normalized coordinate $t$ is defined by:

$$g(t) = f(\lambda) . \tag{11}$$

To the transfer function $g(t)$, we request the following conditions:

1. $|g(t)| \leq g_{\mathrm{stop}}$ when $t$ is in stopbands. Here, $g_{\mathrm{stop}}$ is a very small positive number.
2. $g_{\mathrm{pass}} \leq g(t)$ when and only when $t$ is in the passband. Here, $g_{\mathrm{pass}}$ is a number much larger than $g_{\mathrm{stop}}$. (The upper-bound of $g(t)$ is about unity. By re-normalization of the filter, which is the multiplication of a constant, we may set the upper-bound to unity later.)

For convenience, we also restrict $g(t)$ to an even function. Then the poles are pure imaginary numbers (Fig. 1).

We just placed the poles of $g(t)$ at pure imaginary numbers $t = \pm\sqrt{-1}$, and the expression of $g(t)$ is written as:

$$g(t) = c'_\infty + \mathrm{Re} \sum_{k=1}^{n} \frac{\alpha_k}{(1 + t\sqrt{-1}\,)^k} . \tag{12}$$

For this expression, to make $g(t)$ an even function, we restrict coefficients $\alpha_k$, $k$=$1, 2, \ldots, n$ as real numbers. The real coefficients are so tuned to make the shape of $g(t)$ satisfies the following two conditions: (1) In the passband $|t| < 1$, the value of $g(t)$ is close to 1, (2) In stopbands $\mu < |t|$, the magnitude of $g(t)$ is very small. (See, Fig. 2). In our present study, coefficients are optimized by a method which is similar to the least-square method.

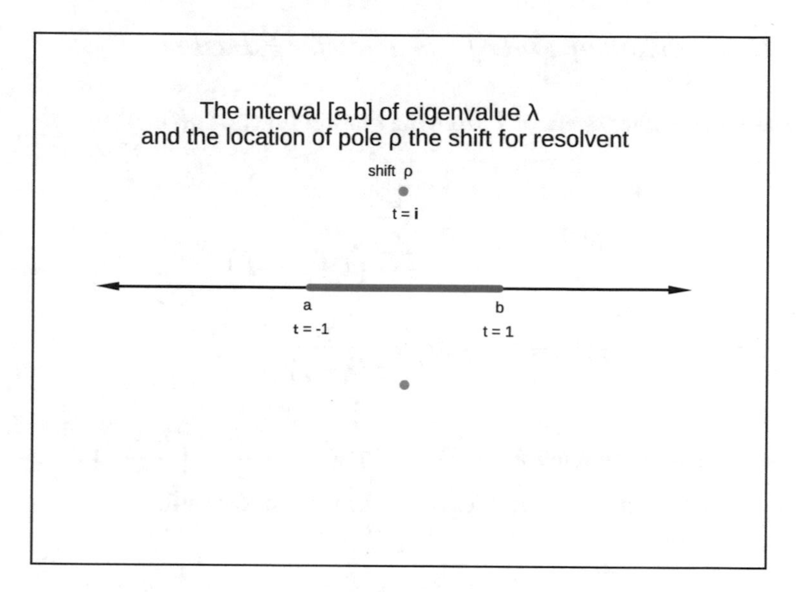


**Fig. 1** Specified interval of eigenvalue and location of poles (shifts)

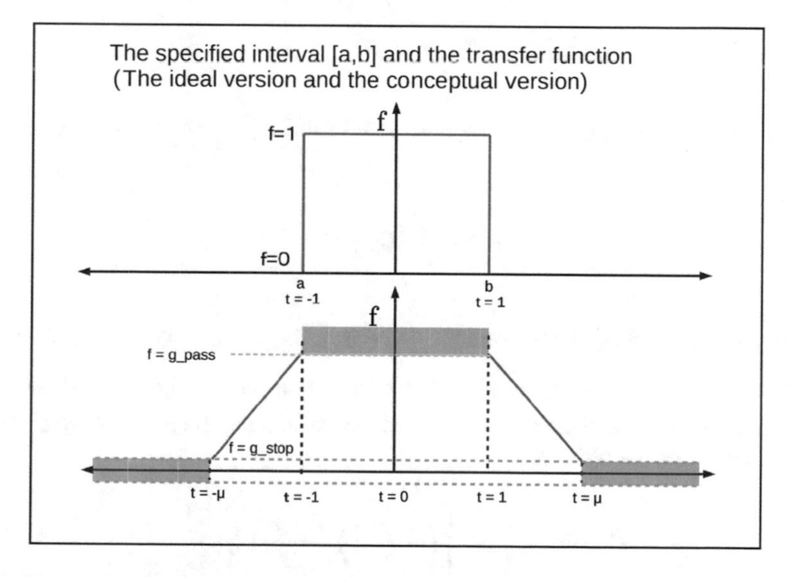


**Fig. 2** Shapes of transfer functions (ideal and conceptual)

## 2.2 Construction of Filter from Transfer Function

We construct the filter operator $F$ from the transfer function $g(t)$.

Since

$$g(t) = c'_{\infty} + \mathrm{Re} \sum_{k=1}^{n} \frac{\alpha_k}{\left(1 + t\sqrt{-1}\right)^k} , \tag{13}$$

$$f(\lambda) = c_{\infty} + \mathrm{Re} \sum_{k=1}^{n} \frac{\gamma_k}{(\lambda - \rho)^k} , \tag{14}$$

and also from both two relations: $f(\lambda) = g(t)$, $\lambda = \dfrac{a+b}{2} + \left(\dfrac{b-a}{2}\right) t$,
we have relations between coefficients and the value of the shift.

$$c'_{\infty} = c_{\infty}, \tag{15}$$

$$\gamma_k = \alpha_k \left(-\sqrt{-1}\right)^k \left(\frac{b-a}{2}\right)^k , k{=}1, 2, \ldots, n , \tag{16}$$

$$\rho = \frac{a+b}{2} + \frac{b-a}{2}\sqrt{-1} . \tag{17}$$

Here after, we simply set the transfer rate at infinity $c_{\infty}$ to zero, thus our filter operator is:

$$F = \mathrm{Re} \sum_{k=1}^{n} \gamma_k \left\{R(\rho)\right\}^k . \tag{18}$$

When the half-width of the interval $\frac{b-a}{2}$ is a large or a small number, then coefficients $\gamma_k = \left(-\frac{b-a}{2}\sqrt{-1}\right)^k \alpha_k$ for higher $k$-th terms might get unnecessary floating point number overflows or underflows, which can be avoided by changing the expression of the filter as:

$$F = \mathrm{Re} \sum_{k=1}^{n} \alpha_k \left\{ \left(-\sqrt{-1}\right) \frac{b-a}{2} R(\rho) \right\}^k . \tag{19}$$

## 2.3 Procedure of Filter Operation

Here, we show the procedure of the filter operation, the action of the filter to a given set of vectors.

**Fig. 3** Procedure of filter operation $Y \leftarrow FX$

$$
\begin{array}{l}
W \leftarrow X\,; \\
Y \leftarrow \mathbf{0}\,; \\
\text{for } k := 1 \text{ to } n \text{ do begin} \\
\quad Z \leftarrow R(\rho)W\,; \\
\quad W \leftarrow \left(-\sqrt{-1}\right) \frac{b-a}{2} Z\,; \\
\quad Y \leftarrow Y + \alpha_k \,\mathrm{Re}\, W \\
\text{end}
\end{array}
$$

Our filter $F$ is specified by the degree $n$, a complex shift $\rho = \frac{a+b}{2} + \frac{b-a}{2}\sqrt{-1}$ and real coefficients $\alpha_k$, $k{=}1, 2, \ldots, n$ as the above expression (19).

Let $X$ and $Y$ are sets of $m$ real column vectors of size $N$ which are represented as real $N{\times}m$ matrices. Then, the filter operation $Y \leftarrow FX$ can be calculated by a procedure shown in Fig. 3. (In the procedure, $W$ and $Z$ are complex $N{\times}m$ matrices for work.)

### 2.4 *Implementation of Resolvent*

To calculate the action of a resolvent $Z \leftarrow R(\rho)\, W$, we first calculate $B\, W$ from $W$, then solve the equation $C Z = B\, W$ for $Z$. Here, the coefficient is $C = A - \rho B$. Since both matrices $A$ and $B$ are real symmetric, $C$ is complex symmetric ($C^T = C$). When both matrices $A$ and $B$ are banded, $C$ is also banded. In our present experiments, the complex modified-Cholesky method without pivoting for banded system is used for the banded complex symmetric matrix $C$, even there might be a potential risk of numerical instability.

In the calculation of Rayleigh quotient inverse-iteration which refines approximated eigenpairs, the shifted matrix is real symmetric but indefinite and very close to singular, therefore the simultaneous linear equation is solved carefully by the banded $LU$ decomposition with partial pivoting without using the symmetry.

## 3 Filter Design

Our filter is a real-part of a polynomials of a resolvent. The coefficients of the polynomial are determined by a kind of least-square method, which is the minimization of the weighted sum of definite integrals in both passband and stopbands. The definite integrals are weighted integrations of square of errors of the transfer function from ideal one. The minimization condition gives a system of linear equation with a symmetric positive definite matrix. However, the equations is numerically highly ill-conditioned. Therefore, to determine accurate coefficients in double precision, we have to make quadruple precision calculation both in the generation of the system of linear equations and in the solution of the generated system by using regularization.

## 3.1 Design by LSQ Method

We show a method to tune $\alpha_k$, the coefficients of $g(t)$, by least square like method. First, we make the change of variable from $t$ to $\theta$ as $t \equiv \tan\theta$, and let $h(\theta) \equiv g(t)$.

$$h(\theta) = \sum_{k=1}^{n} \alpha_k \cos(k\theta)\,(\cos\theta)^k\,. \tag{20}$$

Since $h(\theta)$ is an even function, it is sufficient to consider only in $\theta \in [0, \infty)$. The condition of passband is also considered in $\theta \in [0, \frac{\pi}{4}]$.

### 3.1.1 Method-I

$J_{\text{stop}}$ and $J_{\text{pass}}$ are the integrals in the stopband and in the passband (with weight 1) of the square of difference of the transfer function $h(\theta)$ from the ideal one.

We choose a positive small number $\eta$ and minimize $J \equiv J_{\text{stop}} + \eta\, J_{\text{pass}}$.

For intervals $[0, 1]$ and $[\mu, \infty)$ of $t$ correspond to intervals $[0, \pi/4]$ and $[\tan^{-1}\mu, \pi/2)$ of $\theta$, respectively (these give endpoints of definite integrals for (half of) passband and a stopband). Then we have:

$$J_{\text{stop}} \equiv \int_{\tan^{-1}\mu}^{\pi/2} \{h(\theta)\}^2 d\theta = \sum_{p,q=1}^{n} \alpha_p\, \mathscr{A}_{p,q}\, \alpha_q\,, \tag{21}$$

$$J_{\text{pass}} \equiv \int_{0}^{\pi/4} \{1 - h(\theta)\}^2 d\theta = \sum_{p,q=1}^{n} \alpha_p\, \mathscr{B}_{p,q}\, \alpha_q - 2\sum_{p=1}^{n} \alpha_p\, \mathscr{B}_{p,0} + \text{const}\,. \tag{22}$$

Here,

$$\mathscr{A}_{p,q} \equiv \int_{\tan^{-1}\mu}^{\pi/2} \cos(p\theta)\,\cos(q\theta)\,(\cos\theta)^{p+q}\, d\theta\,, \tag{23}$$

$$\mathscr{B}_{p,q} \equiv \int_{0}^{\pi/4} \cos(p\theta)\,\cos(q\theta)\,(\cos\theta)^{p+q}\, d\theta\,. \tag{24}$$

We calculated numerical values of these definite integrals by analytic closed formulae.

We can easily show that $\cos(p\theta)\cos(q\theta)(\cos\theta)^{p+q}$

$$
\begin{aligned}
&= 2^{-(p+q+2)}(1+e^{-2ip\theta})(1+e^{-2iq\theta})(1+e^{2i\theta})^{p+q} \\
&= 2^{-(p+q+2)}(1+e^{-2ip\theta})(1+e^{-2iq\theta})\sum_{k=0}^{p+q}\binom{p+q}{k}e^{2ik\theta} \\
&= 2^{-(p+q+2)}\sum_{k=0}^{p+q}\binom{p+q}{k}\{e^{2ik\theta}+e^{2i(k-p)\theta}+e^{2i(k-q)\theta}+e^{2i(k-p-q)\theta}\},
\end{aligned}
$$

where $i$ denotes the imaginary unit $\sqrt{-1}$.

We define for an integer $\ell$ and real number $a$ and $b$:

$$
\begin{aligned}
T_\ell &\equiv \int_a^b \cos 2\ell\theta\, d\theta \\
&= \begin{cases} b-a & (\ell=0) \\ (\sin 2\ell b - \sin 2\ell a)/(2\ell) = \{\sin\ell(b-a)\cdot\cos\ell(b+a)\}/\ell & (otherwise) \end{cases}
\end{aligned}
$$

Then for integers $p$ and $q$, we have:

$$
\int_a^b \cos(p\theta)\cos(q\theta)(\cos\theta)^{p+q}\,d\theta = \frac{1}{2^{p+q+2}}\sum_{k=0}^{p+q}\binom{p+q}{k}\{T_k+T_{k-p}+T_{q-k}+T_{p+q-k}\},
$$

which has a symmetry for $p \leftrightarrow q$. We can use another symmetry that $T_k + T_{k-p}$ and $T_{q-k} + T_{p+q-k}$ is interchanged when $k \to p+q-k$. Since $|T_k| \le 1/k$, we calculate the sum so that terms are added in ascending order of magnitudes of binomial coefficients so to reduce rounding errors. Let $w_k \equiv T_k + T_{k-p}$ and $m \equiv p+q$, $c_k \equiv \binom{m}{k}$, the value of integral $v$ is calculated as in Fig. 4.

The minimization condition of $J \equiv J_{\rm stop} + \eta J_{\rm pass}$ is, if we set $b_p \equiv \mathscr{B}_{p,0}$, reduces to a simultaneous linear equations whose coefficient matrix is real symmetric positive definite:

$$
(\mathscr{A} + \eta\mathscr{B})\,\boldsymbol{\alpha} = \eta\,\mathbf{b}\,. \tag{25}
$$

For this linear equation, $\mathscr{A}$ and $\mathscr{B}$ are size $n$ matrices whose elements are $\mathscr{A}_{p,q}$ and $\mathscr{B}_{p,q}$, $p,q=1,2,\ldots,n$, respectively, and also $\boldsymbol{\alpha}$ and $\mathbf{b}$ are column vectors $\boldsymbol{\alpha} \equiv [\alpha_1,\alpha_2,\ldots,\alpha_n]^T$ and $\mathbf{b} \equiv [b_1,b_2,\ldots,b_n]^T$, respectively. We solve this linear equation to obtain the coefficients $\alpha_k$, $k=1,2,\ldots,n$.

**Fig. 4** Procedure to calculate definite integral

```
integral (p, q, a, b) :=
begin
m ← p+q ;
c_0 ← 1 ;
for j := 1 to m do c_j ← c_{j-1} * (m-j+1)/j ;
for j := 0 to m do w_j ← T_j(a,b) + T_{j-p}(a,b) ;
s := 0.0 ;
for j := 0 to m do begin
  if (j < m - j) then
    s ← s + c_j * (w_j + w_{m-j})
  else if (j == m - j) then
    s ← s + c_j * w_j
  else
    exit for
  end if
end;
return v ← s/2^{m+1} ;
end
```

### 3.1.2 Method-II

We assume $\boldsymbol{\alpha}$ as a vector whose 2-norm is a constant, and we first minimize the definite integral in the stopband:

$$J_{\text{stop}} = \sum_{p,q=1}^{n} \alpha_p \mathscr{A}_{p,q} \alpha_q = \boldsymbol{\alpha}^T \mathscr{A} \boldsymbol{\alpha} . \tag{26}$$

If we choose $\boldsymbol{\alpha}$ to the eigenvector of the smallest eigenvalue of the matrix $\mathscr{A}$, then $J_{\text{stop}}$ is the minimum. But if we did so, there is no more freedom left to tune the approximation in the passband. Thus, we introduce the following modification. We choose a suitable small positive number $\epsilon$. If there are $\ell$ eigenvectors whose eigenvalues are under the threshold $\epsilon$, let $S^{(\ell)}$ be the subspace which is spanned by those $\ell$ eigenvectors. Then, it holds $J_{\text{stop}} \leq \epsilon \, ||\boldsymbol{\alpha}||_2^2$ whenever $\boldsymbol{\alpha} \in S^{(\ell)}$.

The minimization condition of $J_{\text{pass}}$ under the constraint $\boldsymbol{\alpha} \in S^{(\ell)}$ reduces to a simultaneous linear equations whose coefficient matrix is of size $\ell$ and symmetric positive definite.

When we extend the subspace (by increasing $\ell$), then $J_{\text{pass}}$ decreases and the approximation in passband become better, however $J_{\text{stop}}$ increases and the approximation in stopband become worse. On the other hand, when we shrink the subspace (by decreasing $\ell$), then $J_{\text{stop}}$ decreases and the approximation in the stopband become better, however $J_{\text{pass}}$ increases and the approximation in passband goes worse.

We have to find a good choice of threshold $\epsilon$ (or $\ell$) considering the balance of both contradicting conditions of approximations in the passband and the stopband.

### 3.2 Examples of Designed Filters

We show in Table 1 (See Tables 2, 3, 4 and Figs. 5, 6, 7), three filters (No.1), (No.2) and (No.3) which are determined by a least-square type method (Method-II). The good thresholds in the method are determined by trials. For the filter (No.1), $\epsilon = 10^{-30}$ is used, which gives $\ell = 2$. For the filter (No.2), $\epsilon = 10^{-25}$ is used, which gives $\ell = 2$. For the filter (No.3), $\epsilon = 10^{-30}$ is used, which gives $\ell = 5$. When $n$ and $\mu$ are given, the result depends only on $\ell$ the rank of subspace. The value of threshold $\epsilon$ is used to obtain the appropriate $\ell$. If about 15-digits reduction ratio in stopbands ($g_{\mathrm{stop}} \approx 10^{-15}$) is desired, we need 30-digits accuracy to calculate the least-square type method. Therefore, we used quadruple precision calculation only in this step to obtain coefficients $\alpha_k$, $k=1, 2, \ldots, n$ in double precision.) It seems the coefficients $\alpha_k$ themselves are numerically very sensitive even the calculation is made in quadruple precision, however it does not matter as long as the shape of the obtained transfer function is good. For the filter (No.2), the value of $\mu$ is set smaller

**Table 1** Filters used in experiments

| Filter | $n$ | $\mu$ | $g_{\mathrm{pass}}$ | $G_{\mathrm{stop}}$ | Coefficients | Graph |
|---|---|---|---|---|---|---|
| (No.1) | 15 | 2.0 | $2.37975\times10^{-4}$ | $1.1\times10^{-15}$ | Table 2 | Figure 5 |
| (No.2) | 15 | 1.5 | $5.46471\times10^{-5}$ | $5.8\times10^{-13}$ | Table 3 | Figure 6 |
| (No.3) | 20 | 2.0 | $1.27268\times10^{-2}$ | $2.6\times10^{-15}$ | Table 4 | Figure 7 |

**Table 2** Filter (No.1): coefficients $\alpha_k$

| $k$ | $\alpha_k$ |
|---|---|
| 1 | 3.10422 91727 23495 E−1 |
| 2 | 3.10422 91727 25609 E−1 |
| 3 | 2.85453 67519 83506 E−1 |
| 4 | 2.35515 19113 67395 E−1 |
| 5 | 1.64913 99494 59607 E−1 |
| 6 | 8.22631 58940 55446 E−2 |
| 7 | −6.57520 79352 44120 E−4 |
| 8 | −7.11802 27019 60262 E−2 |
| 9 | −1.18756 19212 14338 E−1 |
| 10 | −1.37828 28527 33139 E−1 |
| 11 | −1.29654 88587 73316 E−1 |
| 12 | −1.01680 66293 50991 E−1 |
| 13 | −6.60360 83956 00963 E−2 |
| 14 | −3.26587 11429 62141 E−2 |
| 15 | −1.19174 53737 97113 E−2 |

**Table 3** Filter (No.2): coefficients $\alpha_k$

| $k$ | $\alpha_k$ |
|---|---|
| 1 | 2.96820 21545 20158 E−1 |
| 2 | 2.96820 21559 16071 E−1 |
| 3 | 2.75088 15974 67332 E−1 |
| 4 | 2.31624 08572 14527 E−1 |
| 5 | 1.69794 56003 35121 E−1 |
| 6 | 9.63363 20742 38457 E−2 |
| 7 | 2.05451 20416 48405 E−2 |
| 8 | −4.71689 11183 01840 E−2 |
| 9 | −9.79849 96401 27541 E−2 |
| 10 | −1.24548 37945 26314 E−1 |
| 11 | −1.29956 87350 27408 E−1 |
| 12 | −1.07402 42743 15133 E−1 |
| 13 | −8.79229 80353 17280 E−2 |
| 14 | −4.04631 99059 03723 E−2 |
| 15 | −3.14108 29390 18306 E−2 |

**Table 4** Filter (No.3): coefficients $\alpha_k$

| $k$ | $\alpha_k$ |
|---|---|
| 1 | 4.83711 51618 67720 E−1 |
| 2 | 4.83711 51618 86980 E−1 |
| 3 | 3.89953 63967 72419 E−1 |
| 4 | 2.02437 88771 47818 E−1 |
| 5 | −4.47810 12123 49263 E−2 |
| 6 | −2.83593 39733 50968 E−1 |
| 7 | −4.27656 83262 24258 E−1 |
| 8 | −4.04019 57859 22469 E−1 |
| 9 | −1.91913 38100 55309 E−1 |
| 10 | 1.49822 57109 15564 E−1 |
| 11 | 4.82023 35016 46190 E−1 |
| 12 | 6.49169 20356 99877 E−1 |
| 13 | 4.90263 91392 85137 E−1 |
| 14 | 1.69552 59243 54134 E−1 |
| 15 | −5.78530 56855 20654 E−1 |
| 16 | −3.72065 97434 12249 E−1 |
| 17 | −1.44479 33647 97014 E−0 |
| 18 | 7.85556 89830 56973 E−1 |
| 19 | −1.07607 55888 09609 E−0 |
| 20 | 1.70217 32211 25582 E−0 |

than that of filter (No.1), but in exchange the value of $g_{\text{pass}}$ becomes smaller and the value of $g_{\text{stop}}$ becomes larger. For the filter (No.3), the value of $g_{\text{pass}}$ is closer to 1 than that of filter (No.1), which is attained with larger degree $n = 20$.

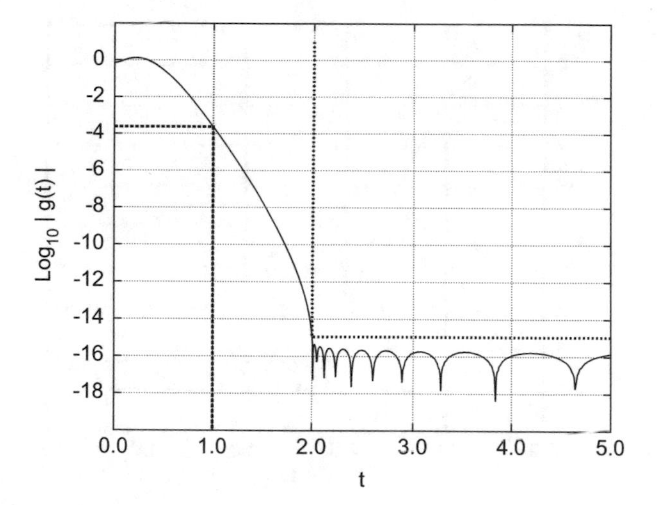


**Fig. 5** Filter (No.1): magnitude of transfer function $|g(t)|$

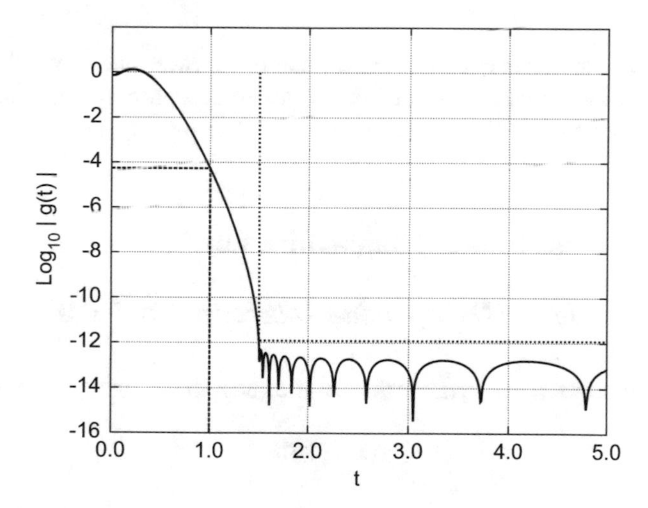


**Fig. 6** Filter (No.2): magnitude of transfer function $|g(t)|$

About the shape parameters of the transfer function $\mu$, $g_{\text{pass}}$ and $g_{\text{stop}}$:

- If $\mu$ is increased, the transition-bands become wider, then it is likely that the number of eigenvectors whose eigenvalues are in the transition-bands increases. The more eigenvectors exist whose eigenvalues are in transition-bands, then the more vectors are required to be filtered to construct an approximation of the basis of the invariant subspace.
- When the max-min ratio of the transfer rate in the passband (related to the reciprocal of $g_{\text{pass}}$) is a large number, the ununiformity of accuracies of approximated pairs tends to be low.

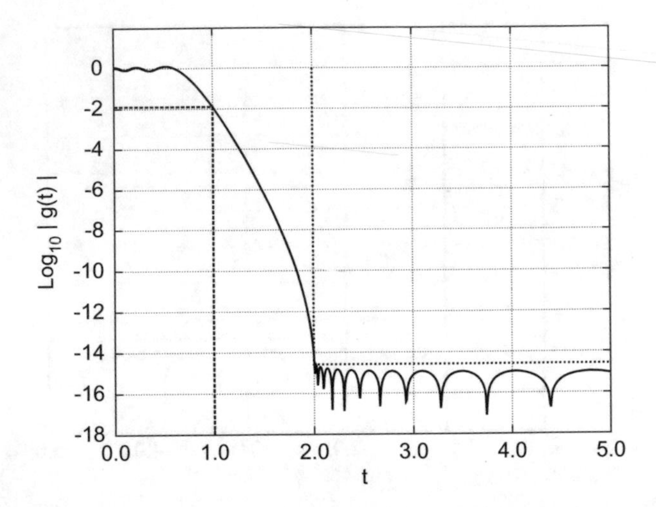


**Fig. 7** Filter (No.3): magnitude of transfer function $|g(t)|$

- When $g_{\mathrm{stop}}$, the upper-bound of magnitude of transfer rate in stopbands, is not very small, the approximation of the invariant subspace will not be good and approximated pairs will be poor.

# 4 Experiments of Filter Diagonalization

## 4.1 Test Problem: 3D-Laplacian Discretized by FEM

Our test problem for a real symmetric-definite generalized eigenproblem:

$$A\,\mathbf{v} = \lambda B\,\mathbf{v} \tag{27}$$

is originated from a discretization by the finite element method approximation of an eigenproblem of the Laplace operator in three dimensions:

$$(-\nabla^2)\,\Psi(x,y,z) = \lambda\,\Psi(x,y,z)\,. \tag{28}$$

The region of the partial differential operator is a cube $[0,\pi]^3$, and the boundary condition is zero-Dirichlet.

For the discretization by the finite element method approximation, each direction of the edge of the cube is equi-divided into $N_1+1$, $N_2+1$, $N_3+1$ sub-intervals. In each finite element, basis functions are tri-linear functions which are products of piece-wise linear function in each direction. The discretization by the finite element method gives a real symmetric-definite generalized eigenproblem of matrices (In this case, both matrices $A$ and $B$ are positive definite, and all eigenvalues are positive real numbers).

The size of both matrices $A$ and $B$ is $N = N_1N_2N_3$. The lower bandwidth of matrices is $1+N_1+N_1N_2$ by a good numbering of basis functions. (Although $A$ and $B$ are quite sparse inside their bands, in our calculation they are treated as if dense inside their bands).

We solve only those eigenpairs $(\lambda, \mathbf{v})$ whose eigenvalues are in a specified interval $[a, b]$. Exact eigenvalues can be calculated by a simple formula. When the numbers of sub-intervals in directions are all different, all eigenvalues are distinct.

**Computer System Environment**

Our calculation is made on a high end class PC system. The CPU is intel Core i7-5960X (3.0 GHz, 8cores, 20 MB L3 cache). Both the turbo mode and the hyper-threading mode of the CPU are disabled from the BIOS menu. The theoretical peak performance of the CPU is 384 GFLOPS in double precision. The memory bus is quad-channel and the total main memory size is 128 GB (8 pieces of DDR4-2133 MHz (PC4-17000) 16 GB memory module). The operating system is CentOS 7 (64bit address). We used intel Fortran compiler ver.15.0.0. for Linux x86_64 with compile options: `-fast,-openmp`.

## 4.2 *Experiment Results*

We solve an eigenproblem of large size whose discretization manner is $(N_1, N_2, N_3) = (50, 60, 70)$. In this case, the size of matrices is $N = 50 \times 60 \times 70 = 210,000$, and the lower bandwidth of matrices is $w_L = 1 + 50 + 50 \times 60 = 3051$.

We solved those pairs whose eigenvalues are in the interval [200, 210] (The true count of such pairs is 91). We chose $m = 200$ for the number of vectors to be filtered. In the calculation of the action of the resolvent, the modified Cholesky factorization for the complex symmetric banded matrix is used. In experiments, three filters (No.1), (No.2) and (No.3) are tested and elapse times are measured in seconds (Table 5). For an approximated eigenpair $(\lambda, \mathbf{v})$ of the generalized eigenproblem,

**Table 5** Elapse times (in s) (matrix size $N$=210, 000)

| Kind of filter | (No.1) | (No.2) | (No.3) |
|---|---|---|---|
| Total filter diagonalization procedure | 2659.68 | 2658.44 | 3318.02 |
| – Generation of random vectors | 0.16 | 0.16 | 0.16 |
| – $B$-Orthonormalization of inputs | 90.83 | 90.82 | 90.87 |
| – Application of the filter | 2273.86 | 2272.92 | 2931.39 |
| – Construction of invariant-subspace | 213.38 | 213.11 | 214.20 |
| – Rayleigh-Ritz procedure | 81.45 | 81.42 | 81.40 |
| Calculation of norms of residuals | 220.22 | 220.45 | 219.93 |
| Memory usage (in GB)(virtual, real) | 21.5(20) | 21.5(20) | 21.5(20) |

the residual of the pair is a vector $\mathbf{r} = (A - \lambda B)\mathbf{v}$. We assume the vector $\mathbf{v}$ of every approximated pair is already normalized in $B$-norm such that $\mathbf{v}^T B\mathbf{v} = 1$. We use $B^{-1}$-norm for the norm to the residual of an approximated pair. Therefore, the norm of residual is $\Delta = \sqrt{\mathbf{r}^T B^{-1}\mathbf{r}}$, where $\mathbf{r} = (A - \lambda B)\mathbf{v}$ and $\mathbf{v}$ is $B$-normalized is assumed. The errors of eigenvalues are calculated by comparisons from exact values by using the formula for this special test problem made by the FEM discretization of Laplace operator in a cube with zero-Dirichlet boundary condition.

- Case of filter (No.1) : The graph in Fig. 8 plots the norm of the residual of each approximated pair. In the middle of the interval of eigenvalue the norm of the residual is about $10^{-10}$, and near the both ends of the interval it is about $10^{-6}$. Their ratio is about $10^4$, which corresponds to the ununiformity of transfer rate of the filter (No.1) in the passband.

  The graph in Fig. 9 plots the absolute error of eigenvalue of each approximated pair. The errors of approximated eigenvalues are less than $10^{-12}$, and approximated eigenvalues are accurate to about 14 digits or more.
- Case of filter (No.2) : The graph in Fig. 10 plots the norm of the residual of each approximated pair. In the middle of the interval of eigenvalue the norm of residual is about $10^{-10}$, and near the both ends of the interval and it is about $10^{-6}$. Their ratio is about $10^4$, which corresponds to the ununiformity of transfer rate of the filter (No.2) in the passband.

  The graph in Fig. 11 plots the absolute error of eigenvalue of each approximated pair. The absolute errors of approximated eigenvalues are less than $10^{-12}$, and approximated eigenvalues are accurate to about 14 digits or more.
- Case of filter (No.3) : The graph in Fig. 12 plots the norm of the residual of each approximated pair. In the middle of the interval of eigenvalue the norm of the residual is about $10^{-10}$, and near the both ends of the interval it is about $10^{-8}$.

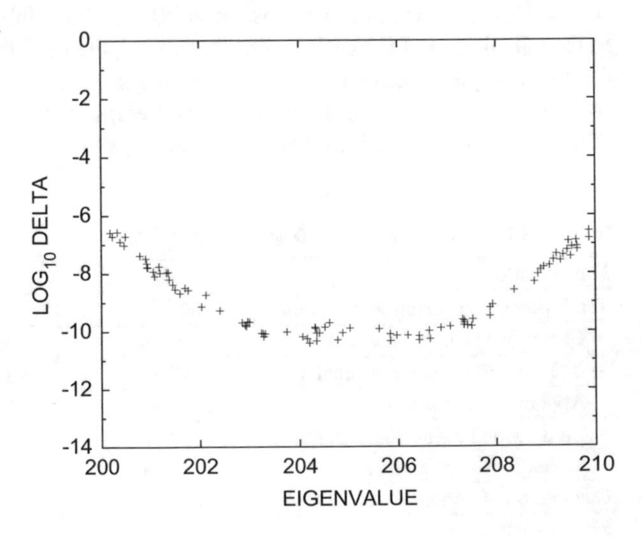


**Fig. 8** Filter (No.1): norm of residual (matrix size $N$=210, 000)

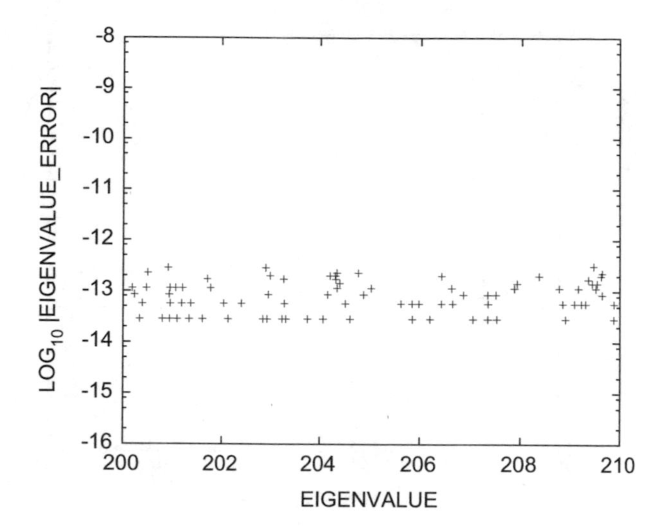


**Fig. 9** Filter (No.1): error of eigenvalue (matrix size $N$=210, 000)

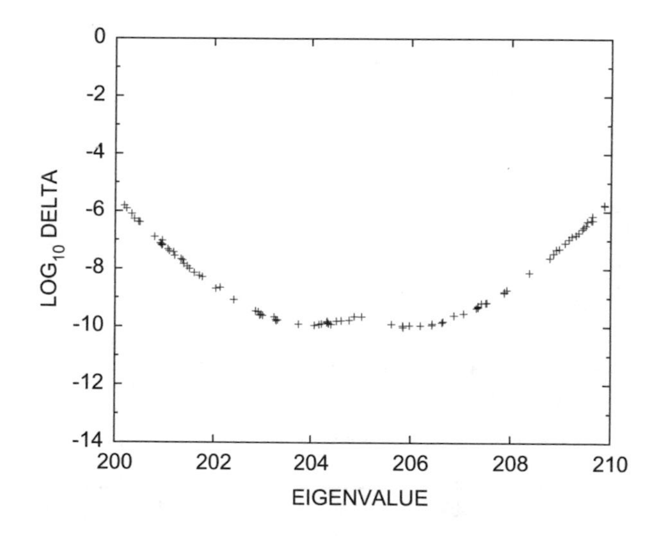


**Fig. 10** Filter (No.2): norm of residual (matrix size $N$=210, 000)

Their ratio is about $10^2$, which corresponds to the ununiformity of transfer rate of the filter (No.3) in the passband.

The graph in Fig. 13 plots the absolute error of eigenvalue of each approximated pair. The absolute errors of approximated eigenvalues are less than $10^{-12}$, and approximated eigenvalues are accurate to about 14 digits or more.

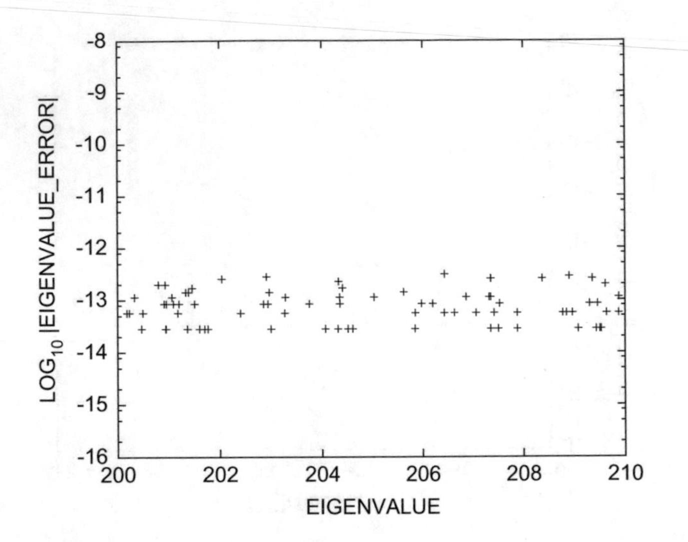


**Fig. 11** Filter (No.2): error of eigenvalue (matrix size $N$=210, 000)

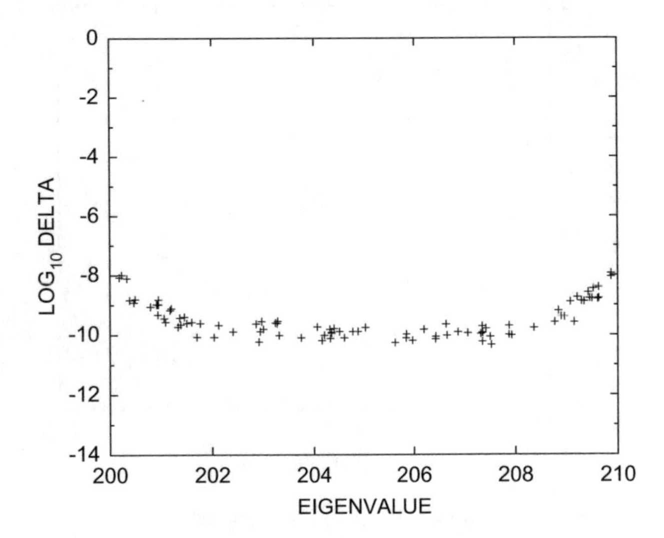


**Fig. 12** Filter (No.3): norm of residual (matrix size $N$=210, 000)

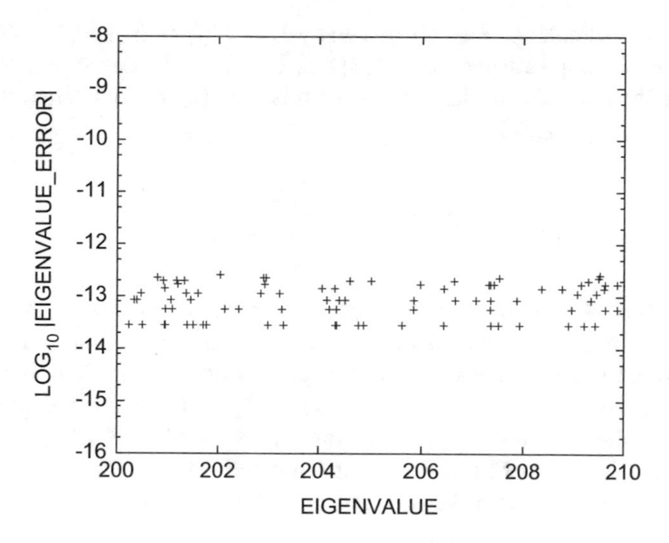


**Fig. 13** Filter (No.3): error of eigenvalue (matrix size $N$=210, 000)

# 5 Timing Comparisons with Elliptic Filters

We compared our present filter to the elliptic filter. Our present filter is a real-part of a polynomial of a resolvent. The elliptic filter is a typical filter which is a real-part of a linear combination of resolvents.

## *5.1 Filter Which is a Real-Part of a Linear Combination of Resolvents*

The filter $\mathscr{F}$ which is a real-part of a linear combination of resolvents is written as:

$$\mathscr{F} = c_\infty I + \mathrm{Re} \sum_{i=1}^{k} \gamma_i R(\rho_i)\,. \tag{29}$$

The coefficient $c_\infty$ is real, coefficients $\gamma_i$ and shifts $\rho_i$ $i$=1, 2, ..., $k$ are complex numbers. We assume shifts are not real numbers and their imaginary parts are positive. An application of $\mathscr{F}$ to a real vector gives a real vector. For any eigenpair $(\lambda, \mathbf{v})$ of the original eigenproblem, we have $\mathscr{F}\mathbf{v} = f(\lambda)\mathbf{v}$. Here $f(\lambda)$ is the transfer function of $\mathscr{F}$, which is the following real rational function of $\lambda$ of degree $2k$:

$$f(\lambda) = c_\infty + \mathrm{Re} \sum_{i=1}^{k} \frac{\gamma_i}{\lambda - \rho_i}\,. \tag{30}$$

The normal coordinate $t$ of $\lambda$ is introduced by $\lambda = \mathscr{L}(t) \equiv (a+b)/2+(b-a)/2\cdot t$, which is a linear map between $t \in [-1, 1]$ and $\lambda \in [a, b]$. We define $g(t)$, the transfer function in the normal coordinate $t$, by the relation $f(\lambda) \equiv g(t)$. Then the form of the transfer function $g(t)$ is:

$$g(t) = c_\infty + \mathrm{Re} \sum_{i=1}^{k} \frac{c_i}{t - \tau_i}, \tag{31}$$

which is a real rational function of degree $2k$. Here, $\rho_i = \mathscr{L}(\tau_i)$, $\gamma_i = \mathscr{L}' \cdot c_i$, where $\mathscr{L}' \equiv (b-a)/2$ is a constant. In reverse, when a real rational function $g(t)$ is given which can be represented in the form of expression (31), then from the real coefficient $c_\infty$, complex coefficients $c_i$, $i$=1, 2, . . ., $k$ and also complex poles $\tau_i$, $i$=1, 2, . . ., $k$, the function $f(\lambda)$ is determined. Thus the filter $\mathscr{F}$ which is a real-part of a linear combination of resolvents is also determined.

We impose conditions for the shape of $g(t)$ on the real axis:

- $|g(t)| \le g_{\mathrm{stop}}$ when $\mu \le |t|$,
- $g_{\mathrm{pass}} \le g(t)$ when $|t| \le 1$, and $\max g(t) = 1$,
- $g_{\mathrm{stop}} < g(t) < g_{\mathrm{pass}}$ when $1 < |t| < \mu$.

We give $\mu$, $g_{\mathrm{pass}}$ and $g_{\mathrm{stop}}$ ($\mu > 1$ and $0 < g_{\mathrm{stop}} \ll g_{\mathrm{pass}} < 1$), then the (half) degree $k$ and the coefficient $c_\infty$ and coefficients and poles $c_i$, $\tau_i$, $i$=1, 2, . . ., $k$ are so determined to satisfy the shape conditions.

## 5.2 Elliptic Filters for Comparisons

For comparisons, we choose the *elliptic filter* (also called *Cauer filter* or *Zolotarev filter*) as the filter which is a real-part of a linear combination of resolvents, which comes from the theory of best approximation by rational function. The elliptic filter is so powerful that it can choose the value of $\mu$ any close to unity, $g_{\mathrm{p}}$ any close to unity, and also the value of $g_{\mathrm{s}}$ any small if (half) degree $k$ is raised. But in our experiments to make comparisons, we choose three elliptic filters (No.E1), (No.E2) and (No.E3), which have similar shape parameters ($\mu$, $g_{\mathrm{pass}}$, $g_{\mathrm{stop}}$) to our present filters (No.1), (No.2) and (No.3) respectively (Table 6). We set the same values of $\mu$ and $g_{\mathrm{pass}}$ between the present filters and the corresponding elliptic filters, but since the (half) degree $n$ of the elliptic filter must be an integer, the true values $g_{\mathrm{stop}}$ for the elliptic filters are not the same but chosen smaller (better in shape). For each elliptic filters we used, complex poles in the upper half complex plane and their complex coefficients of $g(t)$ are tabulated in Tables 7, 8 and 9, respectively. Figures 14, 15 and 16 plot graphs of transfer functions $g(t)$ of elliptic filters for only $t \ge 0$ since they are even functions.

**Table 6** Elliptic filters used in comparisons

| Filter | $k$ | $\mu$ | $g_{\text{pass}}$ | $g_{\text{stop}}$ | Coefficients | Graph |
|---|---|---|---|---|---|---|
| (No.E1) | 8 | 2.0 | $2.37975\times10^{-4}$ | $4.15\times10^{-17}$ | Table 7 | Figure 14 |
| (No.E2) | 7 | 1.5 | $5.46471\times10^{-5}$ | $7.80\times10^{-14}$ | Table 8 | Figure 15 |
| (No.E3) | 8 | 2.0 | $1.27268\times10^{-2}$ | $2.25\times10^{-15}$ | Table 9 | Figure 16 |

## 5.3 *Elapse Times of Filtering*

For an elliptic filter which is a real-part of a linear combination of resolvents, the applications of resolvents to a set of vectors can be made in parallel, however when the applications are made in parallel, the larger memory space is required especially when the applications of resolvents are made by solving linear equations by direct method (matrix factorization method). Therefore, in this experiment, the applications of resolvents are made *one by one* to keep the memory requirement low.

In both present filters and elliptic filters, in an application of a resolvent $R(\rho_i) \equiv (A-\rho_i B)^{-1}B$ to a set of vectors, the linear equation with complex symmetric matrix $C = A - \rho_i B$ is solved by the complex version of modified Cholesky method $C = LDL^T$ for banded matrix $C$ and it is calculated by the same program code. For the elliptic filters, in the calculation of $R(\rho_i)X$ for the set of $m$ vectors $X$, we make a matrix multiplication $X' = BX$ once, since $X'$ is the common right-hand-sides of the set of linear equations for every $\rho_i$.

We are to solve the same eigenproblem from FEM discretization of Laplacian problem as before whose discretization manner is $(N_1, N_2, N_3) = (50, 60, 70)$. The size of matrices of the eigenproblem is $N = 50\times60\times70 = 210,000$, and the lower bandwidth of matrices is $w_L = 1 + 50 + 50\times60 = 3051$.

The only difference from the previous experiment is the kind of filters used, therefore elapse times are compared for the filtering procedure.

For present filter (No.1), (No.2) and (No.3), the count of matrix decomposition is only once, however $n$ the number of repeats of a matrix multiplication by $B$ followed by solution of a set of simultaneous linear equations by using the matrix factors is 15, 15 and 20, respectively. For elliptic filter (No.E1), (No.E2) and (No.E3), the count of matrix decompositions $k$ is 8, 7 and 8, respectively.

We measured elapse times to filter a set of $m$ vectors for the cases $m = 30$ and $m = 200$, by using present filters (No.1), (No.2) and (No.3) and elliptic filters (No.E1), (No.E2) and (No.E3), which are shown in Table 10. In the case of $m = 30$, the elapse times are about 3 times less for present filters compared with elliptic ones, therefore, when $m = 30$ the use of present filter reduces the elapse time for filtering than elliptic filter. But in the case of $m = 200$, the elapse times were not so much different between elliptic filters and present filters.

When the size of matrices $A$ and $B$ of the eigenproblem is $N$ and its bandwidth is $w$, the amount of computation to factor a symmetric banded matrix $C$ is $O(Nw^2)$, and it is $O(Nwm)$ to solve the set of simultaneous linear equations with $m$ right-

**Table 7** Elliptic filter (No.E1) for $\mu = 2.0$, $g_{\text{pass}} = 2.3797\times10^{-4}$, $g_{\text{stop}} = 1.1\times10^{-15}$ (The half degree $k = 8$ and the true $g_{\text{stop}} = 4.1457\times10^{-17}$)

| $c_\infty$=0.41457 27735 04193 0E−16 | | |
|---|---|---|
| $i$ | $\tau_i$ | $c_i$ |
| 1 | (−0.98333 59161 66002 6E+0, 0.32764 47565 36014 5E−3) | (0.32282 04928 71600 5E−5, −0.32761 86707 18423 1E−3) |
| 2 | (−0.84933 21816 52866 9E+0, 0.98905 43009 62692 8E−3) | (0.32349 46453 11294 0E−5, −0.98897 60764 00351 2E−3) |
| 3 | (−0.58304 77825 30795 7E+0, 0.16083 78800 99375 6E−2) | (0.26969 42336 27235 4E−5, −0.16082 52907 59052 8E−2) |
| 4 | (−0.20878 57279 01429 9E+0, 0.20129 18707 11943 9E−2) | (0.10982 56974 97660 9E−5, −0.20127 62426 40295 4E−2) |
| 5 | (0.20878 57279 01427 2E+0, 0.20129 18707 11944 1E−2) | (−0.10982 56974 97659 5E−5, −0.20127 62426 40295 5E−2) |
| 6 | (0.58304 77825 30793 6E+0, 0.16083 78800 99375 9E−2) | (−0.26969 42336 27234 6E−5, −0.16082 52907 59053 1E−2) |
| 7 | (0.84933 21816 52865 9E+0, 0.98905 43009 62696 0E−3) | (−0.32349 46453 11294 0E−5, −0.98897 60764 00354 3E−3) |
| 8 | (0.98333 59161 66002 2E+0, 0.32764 47565 36016 7E−3) | (−0.32282 04928 71598 6E−5, −0.32761 86707 18425 4E−3) |

**Table 8** Elliptic filter (No.E2) for $\mu = 1.5$, $g_{\text{pass}} = 5.4647{\times}10^{-5}$, $g_{\text{stop}} = 5.8{\times}10^{-13}$ (The half degree $k = 7$ and the true $g_{\text{stop}} = 7.7975{\times}10^{-14}$)

$c_\infty = 0.0$

| $i$ | $\tau_i$ | $c_i$ |
|---|---|---|
| 1 | (−0.98131 05913 82855 9E+0, 0.17707 38116 11888 8E−3) | (0.85480 75525 93772 8E−6, −0.17707 04878 81648 8E−3) |
| 2 | (−0.82397 85781 93720 0E+0, 0.57606 38840 41885 7E−3) | (0.10256 96227 55537 0E−5, −0.57605 32864 32075 0E−3) |
| 3 | (−0.48589 61141 89152 9E+0, 0.10060 52142 56393 2E−2) | (0.88797 41935 99554 4E−6, −0.10060 34220 85259 1E−2) |
| 4 | (0.00000 00000 00000 0E+0, 0.12166 69145 31958 6E−2) | (0.00000 00000 00000 0E+0, −0.12166 47849 65604 5E−2) |
| 5 | (0.48589 61141 89152 2E+0, 0.10060 52142 56393 3E−2) | (−0.88797 41935 99553 8E−6, −0.10060 34220 85259 1E−2) |
| 6 | (0.82397 85781 93719 6E+0, 0.57606 38840 41886 9E−3) | (−0.10256 96227 55537 2E−5, −0.57605 32864 32076 1E−3) |
| 7 | (0.98131 05913 82855 9E+0, 0.17707 38116 11888 6E−3) | (−0.85480 75525 93770 0E−6, −0.17707 04878 81648 6E−3) |

**Table 9** Elliptic filter (No.E3) for $\mu = 2.0$, $g_{\text{pass}} = 1.2727\times10^{-2}$, $g_{\text{stop}} = 2.6\times10^{-15}$ (The half degree $k = 8$ and the true $g_{\text{stop}} = 2.2452\times10^{-15}$)

$c_\infty$= 0.22451 63375 99703 6E−14

| $i$ | $\tau_i$ | $c_i$ |
|---|---|---|
| 1 | (−0.98342 13559 84046 3E+0, 0.24060 91027 92044 1E−2) | (0.17336 66188 99784 1E−3, −0.23958 11873 77108 1E−2) |
| 2 | (−0.84941 78015 80027 0E+0, 0.72633 17228 26009 7E−2) | (0.17373 54957 94905 3E−3, −0.72324 92487 29854 6E−2) |
| 3 | (−0.58311 91649 98342 8E+0, 0.11811 70538 57957 5E−1) | (0.14484 96328 39890 6E−3, −0.11762 09595 25342 0E−1) |
| 4 | (−0.20881 47970 76505 9E+0, 0.14782 83729 13307 9E−1) | (0.58988 50525 82562 5E−4, −0.14721 25294 03648 0E−1) |
| 5 | (0.20881 47970 76503 1E+0, 0.14782 83729 13308 0E−1) | (−0.58988 50525 82555 1E−4, −0.14721 25294 03648 1E−1) |
| 6 | (0.58311 91649 98340 7E+0, 0.11811 70538 57957 8E−1) | (−0.14484 96328 39890 2E−3, −0.11762 09595 25342 3E−1) |
| 7 | (0.84941 78015 80026 0E+0, 0.72633 17228 26012 2E−2) | (−0.17373 54957 94905 3E−3, −0.72324 92487 29857 0E−2) |
| 8 | (0.98342 13559 84046 0E+0, 0.24060 91027 92045 8E−2) | (−0.17336 66188 99783 1E−3, −0.23958 11873 77109 8E−2) |

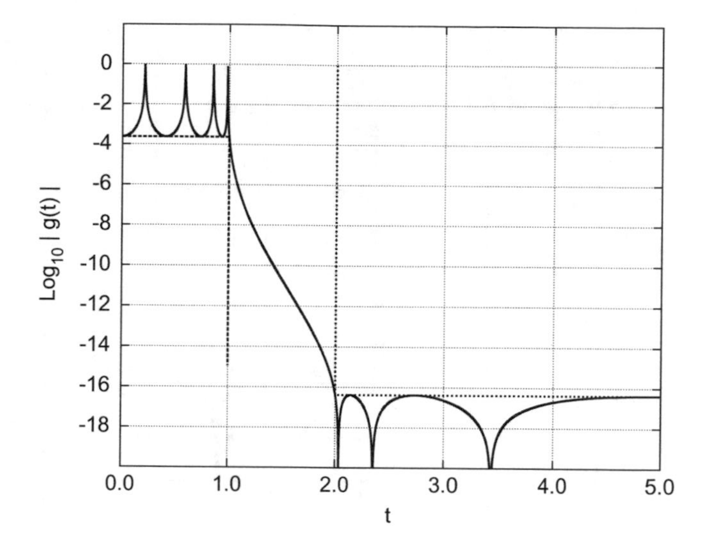


**Fig. 14** Filter (No.E1): magnitude of transfer function $|g(t)|$

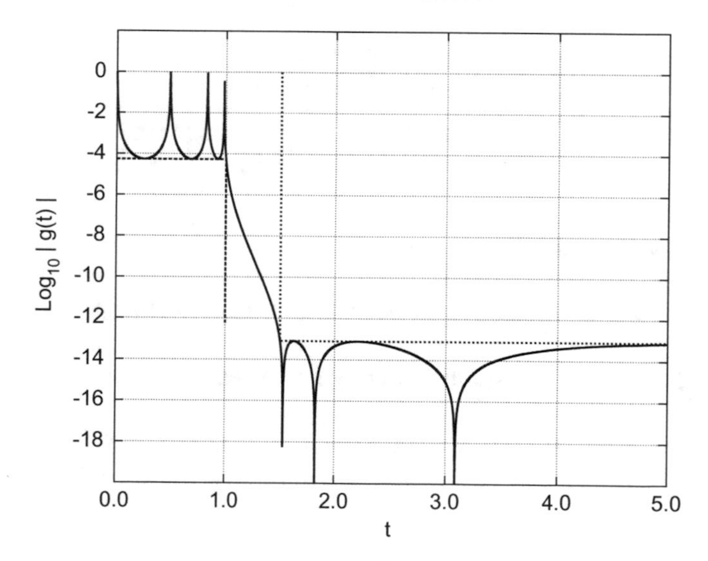


**Fig. 15** Filter (No.E2): magnitude of transfer function $|g(t)|$

hand-sides after the matrix is factored. The amount of computation to multiply a symmetric banded matrix $B$ to a set of $m$ vectors is also $O(Nwm)$.

Thus, the elapse time to factor a symmetric banded matrix $C$ of size $N$ with bandwidth $w$ is $T_{\text{decompose}} \approx t_1 N w^2$, the elapse time to solve a set of $m$ of simultaneous linear equations using the matrix factor is $T_{\text{solve}} \approx t_2 N w m$ and the elapse time to multiply a set of $m$ vectors to a symmetric banded matrix $B$ of size $N$ with bandwidth $w$ is $T_{\text{mulB}} \approx t_3 N w m$. Here, $t_1$, $t_2$ and $t_3$ are some coefficients. Then,

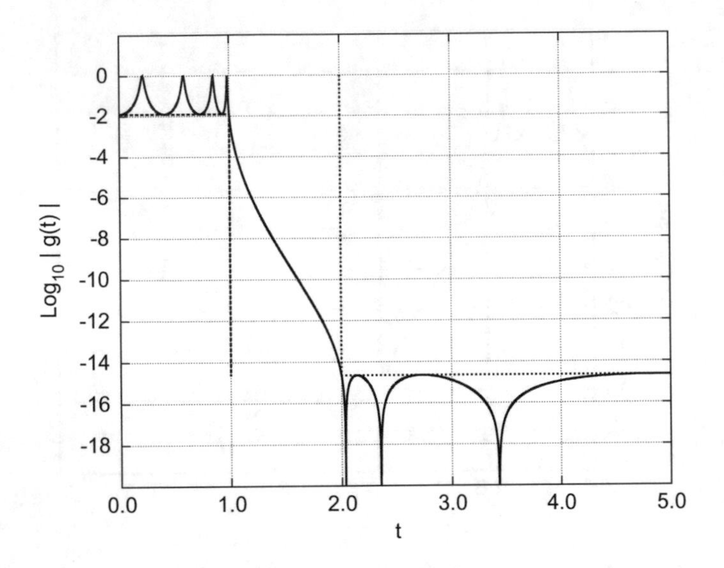


**Fig. 16** Filter (No.E3): magnitude of transfer function $|g(t)|$

**Table 10** Elapse times (in s) for filtering $m$ vectors

| $m$ | (No.1) | (No.2) | (No.3) | (No.E1) | (No.E2) | (No.E3) |
|---|---|---|---|---|---|---|
| 30 | 621.8 | 622.3 | 728.3 | 2516.7 | 2210.2 | 2517.9 |
| 200 | 2273.9 | 2272.9 | 2931.4 | 3352.1 | 3023.7 | 3352.0 |

the elapse time of present filter can be written as:

$$T_{\text{present}} = T_{\text{decompose}} + n \times (T_{\text{mulB}} + T_{\text{solve}} + O(Nm)) \,, \tag{32}$$

and the elapse time of elliptic filter by using $k$ resolvents is written as:

$$T_{\text{elliptic}} = T_{\text{mulB}} + k \times (T_{\text{decompose}} + T_{\text{solve}} + O(Nm)) \,. \tag{33}$$

We have

$$\begin{aligned} T_{\text{present}}/N &\approx t_1 w^2 + (t_2 + t_3) nwm, \\ T_{\text{elliptic}}/N &\approx t_1 k w^2 + (t_2 k + t_3) wm \,. \end{aligned} \tag{34}$$

The coefficients $t_1$, $t_2$ and $t_3$ depends on the system and the manner of calculation. Since the calculation of matrix decomposition has much higher locality of data references than the matrix-vector multiplication or the solution of linear equation using the matrix factors, the coefficient $t_1$ must be smaller than $t_2$ or $t_3$. From the above expression (34), when $m$ is small and ignorable, the elapse time of present filter could be nearly $k$ times faster than the elliptic filter (with no parallel resolvent calculation) since it makes just one matrix decomposition. However, as $m$ increases, the advantage of the present approach seems reduced.

## 6 Conclusion

For a real symmetric-definite generalized eigenproblem, the filter diagonalization method solves those eigenpairs whose eigenvalues are in the specified interval. In this present study, the filter we used is a real-part of a polynomial of a single resolvent rather than a real-part of a linear combination of many resolvents to take advantages of reductions of the amount of memory requirement and also computation. In numerical experiments, we obtained good results.

When the filter is a real-part of a linear combination of many (8 to 16) resolvents, for each resolvents the applications to a set of vectors can be made in parallel. For our present method, when the filter is a real-part of a polynomial of a resolvent, applications of the resolvent as many times as the degree of the polynomial can be made only in sequential. However, even the potential parallelism is reduced, the present method has an advantage that it requires only single resolvent, therefore the amount of storage requirement is low and also the total amount of computation can be reduced. Once a single matrix which corresponds to the resolvent is decomposed and factors are stored, each application of the resolvent can be calculated easily and fast. Another difficulty of the present type of filter is that the shape of the transfer function is not so good as the shape of the filter which is a real-part of a linear combination of many resolvents such as elliptic filter.

In this paper, three filters are constructed and they are shown with their polynomial cocfficients and shape parameters. By using these three filters, we made some experiments of the filter diagonalization. For a generalized eigenproblem which is derived from FEM discretization of the Laplace operator over a cubic region with zero Dirichlet boundary condition, we solved some internal eigenpairs whose eigenvalues are in the specified interval. We compared eigenvalues of approximated pairs with exact ones, and found their agreements were good, which showed our approach worked as expected.

## References


1. Austin, A.P., Trefethen, L.N.: Computing eigenvalues of real symmetric matrices with rational filters in real arithmetic. SIAM J. Sci. Comput. **37**(3), A1365–A1387 (2015)
2. Galgon, M., Krämer, L., Lang, B.: The FEAST algorithm for large eigenvalue problems. Proc. Appl. Math. Mech. **11**, 747–748 (2011)
3. Güttel, S., Polizzi, E., Tang, P.T.P., Viaud, G.: Zolotarev quadrature rules and load balancing for the FEAST eigensolver. SIAM J. Sci. Comput. **37**(4), A2100–A2122 (2015)
4. Ikegami, T., Sakurai, T., Nagashima, U.: A filter diagonalization for generalized eigenvalue problems based on the Sakurai–Sugiura projection method. J. Comput. Appl. Math. **233**(8), 1927–1936 (2010)
5. Murakami, H.: A filter diagonalization method by the linear combination of resolvents. IPSJ Trans. ACS-21 **49**(SIG2), 66–87 (2008, written in Japanese)
6. Murakami, H.: Filter designs for the symmetric eigenproblems to solve eigenpairs whose eigenvalues are in the specified interval. IPSJ Trans. ACS-31 **3**(3), 1–21 (2010, written in Japanese)

7. Murakami, H.: Construction of the approximate invariant subspace of a symmetric generalized eigenproblem by the filter operator. IPSJ Proc. SACSIS 2011, 332–339 (2011, written in Japanese).
8. Murakami, H.: Construction of the approximate invariant subspace of a symmetric generalized eigenproblem by the filter operator. IPSJ Trans. ACS-35 **4**(4), 51–64 (2011, written in Japanese).
9. Murakami, H.: Filter diagonalization method for a Hermitian definite generalized eigenproblem by using a linear combination of resolvents as the filter. IPSJ Tans. ACS-45 **7**(1), 57–72 (2014, written in Japanese)
10. Murakami, H.: On filter diagonalization method. In: Abstracts of JSIAM 2014 annual meeting, pp. 329–330 (August 2014, written in Japanese)
11. Murakami, H.: An experiment of filter diagonalization method for symmetric definite generalized eigenproblem which uses a filter constructed from a resolvent. IPSJ SIG Technical Reports, vol. 2015-HPC-149(7), pp. 1–16 (June, 2015, written in Japanese)
12. Murakami, H.: Filter diagonalization method for real symmetric definite generalized eigenproblem whose filter is a polynomial of a resolvent. In: Abstracts of EPASA 2015 at Tsukuba, p. 28 (single page poster abstract) (September 2015)
13. Polizzi, E.: A density matrix-based algorithm for solving eigenvalue problems. Phys. Rev. B **79**(1), 115112(6pages) (2009)
14. Sakurai, T., Sugiura, H.: A projection method for generalized eigenvalue problems using numerical integration. J. Comput. Appl. Math. **159**, 119–128 (2003)
15. Sakurai, T., Tadano, H.: CIRR: a Rayleigh-Ritz type method with contour integral for generalized eigenvalue problems. Hokkaido Math. J. **36**(4), 745–757 (2007)
16. Toledo, S., Rabani, E.: Very large electronic structure calculations using an out-of-core filter-diagonalization method. J. Comput. Phys. **180**(1), 256–269 (2002)

# Off-Diagonal Perturbation, First-Order Approximation and Quadratic Residual Bounds for Matrix Eigenvalue Problems

**Yuji Nakatsukasa**

**Abstract** When a symmetric block diagonal matrix $\begin{bmatrix} A_1 & \\ & A_2 \end{bmatrix}$ undergoes an off-diagonal perturbation $\begin{bmatrix} A_1 & E_{12} \\ E_{12}^T & A_2 \end{bmatrix}$, the eigenvalues of these matrices are known to differ only by $O(\frac{\|E_{12}\|^2}{\text{gap}})$, which scales *quadratically* with the norm of the perturbation. Here gap measures the distance between eigenvalues, and plays a key role in the constant. Closely related is the first-order perturbation expansion for simple eigenvalues of a matrix. It turns out that the accuracy of the first-order approximation is also $O(\frac{\|E\|^2}{\text{gap}})$, where $E$ is the perturbation matrix. Also connected is the residual bounds of approximate eigenvalues obtained by the Rayleigh-Ritz process, whose accuracy again scales quadratically in the residual, and inverse-proportionally with the gap between eigenvalues. All these are tightly linked, but the connection appears to be rarely discussed. This work elucidates this connection by showing that all these results can be understood in a unifying manner via the quadratic perturbation bounds of block diagonal matrices undergoing off-diagonal perturbation. These results are essentially known for a wide range of eigenvalue problems: symmetric eigenproblems (for which the explicit constant can be derived), nonsymmetric and generalized eigenvalue problems. We also extend such results to matrix polynomials, and show that the accuracy of a first-order expansion also scales as $O(\frac{\|E\|^2}{\text{gap}})$, and argue that two-sided projection methods are to be preferred to one-sided projection for nonsymmetric eigenproblems, to obtain higher accuracy in the computed eigenvalues.

## 1 Introduction

Classical eigenvalue perturbation theory studies bounds or approximations to the eigenvalues and eigenvectors of $A + E$ for some "small" $E$ (such as small norm or low rank), given the knowledge of some information on $A$, such as an eigenpair such

---

Y. Nakatsukasa (✉)
Department of Mathematical Informatics, The University of Tokyo, Tokyo 113-8656, Japan
e-mail: nakatsukasa@mist.i.u-tokyo.ac.jp



T. Sakurai et al. (eds.), *Eigenvalue Problems: Algorithms, Software and Applications in Petascale Computing*, Lecture Notes in Computational Science and Engineering 117, https://doi.org/10.1007/978-3-319-62426-6_15

that $Ax_i = \lambda_i(A)x_i$. Such results are of interest in large-scale scientific computing because, for example, (i) given the information of $A$, they give estimates for the eigenvalues and eigenvectors of $A+E$ that can be obtained cheaply (for example the first-order expansion (1)), and (ii) they can be used to give quantitative analysis for the accuracy of the computed eigenpairs. See [7] and [15, Ch. IV] for an overview of eigenvalue perturbation theory.

The eigenvalues of two unstructured matrices $A$ and $A + E$ generally differ by $O(\|E\|_2)$, or sometimes more (as large as $O(\|E\|_2^{1/n})$ in the worst case, when defective eigenvalues are present). However, there are important situations when eigenvalues behave more nicely than such general bounds suggest, and this work focuses on such cases.

Among the most well-known results for the (simplest and most well-understood) symmetric case $A = A^T \in \mathbb{R}^{n\times n}$ are Weyl's theorem $|\lambda_i(A) - \lambda_i(A+E)| \le \|E\|$ (throughout, we employ the spectral norm $\|A\| = \sigma_{\max}(A)$ for matrices, and 2-norm for vectors), and the first-order perturbation expansion for simple eigenvalues (e.g. [4, § 7.2.2])

$$\lambda_i(A+E) = \lambda_i(A) + \frac{x^T E x}{x^T x} + O(\|E\|^2). \tag{1}$$

Here $x \neq 0$ is an eigenvector such that $Ax = \lambda_i(A)x$. Note that this gives an approximation $\lambda_i(A) + \frac{x^T A x}{x^T x}$ to $\lambda_i(A+E)$, rather than a bound as in Weyl's theorem.

This work revolves around the less well-known (but, we argue, equally important) quadratic perturbation bounds for eigenvalues of block-diagonal matrices that undergo off-diagonal perturbation.

**Theorem 1 ([9, 10])** *Let $A, E \in \mathbb{R}^{n\times n}$ are symmetric matrices partitioned as*

$$A = \begin{bmatrix} A_1 & 0 \\ 0 & A_2 \end{bmatrix}, \quad E = \begin{bmatrix} 0 & E_1^T \\ E_1 & 0 \end{bmatrix}, \tag{2}$$

*then [10]*

$$|\lambda_i(A+E) - \lambda_i(A)| \le \frac{\|E\|^2}{\mathrm{gap}_i}, \tag{3}$$

*and a slightly tighter bound [9] holds:*

$$|\lambda_i(A+E) - \lambda_i(A)| \le \frac{2\|E\|^2}{\mathrm{gap}_i + \sqrt{\mathrm{gap}_i^2 + 4\|E\|^2}} \le \frac{\|E\|^2}{\mathrm{gap}_i}. \tag{4}$$

*Here* $\text{gap}_i$ *is defined by*

$$\text{gap}_i := \begin{cases} \min_{\lambda_j \in \lambda(A_2)} |\lambda_i - \lambda_j| & \textit{if} \quad \lambda_i \in \lambda(A_1) \\ \min_{\lambda_j \in \lambda(A_1)} |\lambda_i - \lambda_j| & \textit{if} \quad \lambda_i \in \lambda(A_2). \end{cases} \tag{5}$$

Note that the bound in (3) scales quadratically with the perturbation $\|E\|$, which is significantly smaller than Weyl's bound $\|E\|$ when $\|E\|$ is small. We shall look at (3) from many different viewpoints, and one central goal of this work is to reveal the implications of such quadratic bounds. For simplicity, we refer to situations as in (2), where a block-diagonal matrix undergoes off-diagonal perturbation, as *off-diagonal perturbation*.

We note that the bounds in (4) are sharp in the sense that without further information, there are examples where equality is attained. This sharpness can be confirmed by verifying that the first bound in (4) is exact for $2 \times 2$ matrices, and it reduces to the second bound in the limit $\|E\| \to 0$.

Also of interest in this work is quadratic residual bounds [6], which claims for symmetric eigenvalue problems that defining the residual by $r = A\hat{x} - \hat{\lambda}\hat{x}$ where $\hat{x}$ is an approximate eigenvector with $\|\hat{x}\| = 1$ and $\hat{\lambda} = \hat{x}^T A \hat{x}$ is the Rayleigh quotient, we have

$$|\lambda - \hat{\lambda}| \leq \frac{\|r\|^2}{\text{gap}_i}, \tag{6}$$

where $\text{gap}_i$ is as in (5) with $A_1$ taken to be $1 \times 1$. The notion of gap is subtly different between (6) and those in the literature, e.g. [1, § 4.8], [13, Thm. 11.7.1]: we explain this more in Sect. 2.1. More generally, with the Ritz values $\{\hat{\lambda}_i\}_{i=1}^k$ for a symmetric matrix obtained as the eigenvalues of $X^T A X$ where $X \in \mathbb{R}^{n \times k}$ has orthonormal columns, we have

$$|\lambda - \hat{\lambda}| \leq \frac{\|R\|^2}{\text{gap}_i}, \tag{7}$$

where $R = AX - X\Lambda$ is the matrix of residuals, $\text{gap}_i$ is again as in (5), which is "widened", resulting in an improved bound. We derive this below in Sect. 2.2.

The first-order expansion (1) and the off-diagonal quadratic perturbation bounds (3) are closely connected: specifically, the first-order perturbation expansion (1) explains why off-diagonal perturbation results in quadratic eigenvalue perturbation bounds (without information on the constant $\frac{1}{\text{gap}_i}$). Conversely, using (3) one can obtain (1), and moreover obtain the constant $\frac{1}{\text{gap}_i}$ hidden in the trailing term $O(\|E\|^2)$. We shall see that the residual bound can also be regarded as a consequence of (3). In other words, (3) can be regarded as a fundamental fact that implies many results in eigenvalue perturbation theory.

These connections are known to experts in eigenvalue perturbation theory, but to the author's knowledge there is no literature that states them explicitly. One goal of this work is to clarify this connection, which holds not only for symmetric eigenvalue problems but also for nonsymmetric and generalized eigenvalue problems. All this is not exactly new, in that they are simply observations that connect results in the literature.

The second goal of this work is to extend such results to polynomial eigenvalue problems. For polynomial eigenvalue problems, the first-order perturbation expansion (1) is known [16], but no result seems to be available on off-diagonal perturbation analogous to (3). We shall obtain such result, and show that if

$$P(\lambda) = \begin{bmatrix} P_1(\lambda) & 0 \\ 0 & P_2(\lambda) \end{bmatrix}, \quad E = \begin{bmatrix} 0 & E_{12}(\lambda) \\ E_{21}(\lambda) & 0 \end{bmatrix}, \tag{8}$$

then

$$\lambda_i(P) - \lambda_i(P+E) \leq c \frac{\|E(\lambda_i(P))\|^2}{\mathrm{gap}_i}, \tag{9}$$

for some $c$, which depends on the conditioning of the eigenvectors. The point we wish to convey here is that the eigenvalue gap plays the same role even in polynomial eigenvalue problems. Note that $E(\lambda_i(P))$ is the value of the matrix polynomial $E(\lambda)$ (representing the perturbation) evaluated at $\lambda = \lambda_i(P)$.

All in all, in this note we investigate the quadratic eigenvalue perturbation bounds under off-diagonal perturbation such as (3) and (9) from different viewpoints, and reveal some of their practical ramifications.

The only reason we stated the above results for symmetric matrices is for simplicity; extensions to nonsymmetric and generalized eigenproblems are available [8, 10, 15]. We structure this note similarly: we first discuss the symmetric case in Sect. 2, then deal with nonsymmetric and generalized eigenvalue problems, and treat polynomial eigenvalue problems at the end.

In what follows, for simplicity we normalize any right eigenvector to have unit norm, and we scale the left eigenvector via the orthogonality relation such as $y^T x = 1$ or $y^T P(\lambda_i)' x = 1$. $\lambda_i(A)$ denotes the $i$th eigenvalue of $A$, arranged in ascending order if $A$ is symmetric, and otherwise its ordering is insignificant: $\lambda_i(A)$ denotes a specific eigenvalue of $A$.

## 2 Symmetric Case

We start by treating the simplest case of symmetric eigenvalue problems; entirely analogous results hold for the complex Hermitian case.

As advertised, let us first explain how (1) implies off-diagonal perturbation should result in quadratic eigenvalue perturbation bounds. Let the matrices $A, E$ be as in (2), and $\lambda_i(A) \in \lambda(A_1)$ with $\mathrm{gap}_i > 0$. Then $Ax = \lambda_i(A)x$ with eigenvector structure $x = \begin{bmatrix} x_1 \\ 0 \end{bmatrix}$, and so by substituting into (1) we obtain

$$\begin{aligned}\lambda_i(A+E) &= \lambda_i(A) + x^T Ex + O(\|E\|^2)\\ &= \lambda_i(A) + \begin{bmatrix} x_1 \\ 0 \end{bmatrix}^T \begin{bmatrix} 0 & E_1^T \\ E_1 & 0 \end{bmatrix} \begin{bmatrix} x_1 \\ 0 \end{bmatrix} + O(\|E\|^2)\\ &= \lambda_i(A) + O(\|E\|^2),\end{aligned}$$

in which we note that $\begin{bmatrix} x_1 \\ 0 \end{bmatrix}^T \begin{bmatrix} 0 & E_1^T \\ E_1 & 0 \end{bmatrix} \begin{bmatrix} x_1 \\ 0 \end{bmatrix} = 0$ due to the block structure. That is, the first-order term in (1) disappears because of the structure. Thus the first term in the perturbation expansion scales quadratically with the perturbation $\|E\|$.

## 2.1 *First-Order Expansion and Its Constant via Off-Diagonal Bounds*

We now turn to the connection in the opposite direction and derive the first-order expansion (1) using the quadratic bounds (3). We are not claiming the derivation here is simpler than the standard method of differentiating the equation $Ax = \lambda x$ and left-multiplying the left eigenvector (see [4, § 7.2.2] or [16]). However, as we shall see, the derivation here reveals the constant in front of the quadratic term $\|E\|^2$. In Sect. 3.4 we also give an explanation based on Gerschgorin's theorem.

Using (3) we shall derive the following result, which can be seen as a variant of (1) that reveals the constant hidden in $O(\|E\|^2)$. Note below that $\widetilde{\text{gap}}_i$ can be regard as a modified gap.

**Proposition 1** *Let* $A, E \in \mathbb{R}^{n \times n}$ *be symmetric matrices. Then*

$$|\lambda_i(A+E) - (\lambda_i(A) + \frac{x^T Ex}{x^T x})| \le \frac{\|E\|^2}{\widetilde{\text{gap}}_i}, \tag{10}$$

*where* $\widetilde{\text{gap}}_i = \max(0, \min_{j \neq i} |\lambda_i + \frac{x^T Ex}{x^T x} - \lambda_j| - \|E\|)$, *and* $\widetilde{\text{gap}}_i \to \text{gap}_i$ *as* $E \to 0$.

*Proof* Consider the eigenvalue decomposition

$$X^T AX = \begin{bmatrix} \lambda_1 & & \\ & \ddots & \\ & & \lambda_n \end{bmatrix},$$

where $X = [x_1, \dots, x_n]$ is an orthogonal eigenvector matrix. Then

$$X^T(A+E)X = \begin{bmatrix} \lambda_1 & & \\ & \ddots & \\ & & \lambda_n \end{bmatrix} + X^T E X,$$

whose $(i,i)$ element is $\lambda_i + x_i^T E x_i$. We can then apply a permutation matrix $P$ that moves the $i$th position to the first (the specific choice of $P$ does not matter), which gives

$$P^T X^T (A+E) X P = \begin{bmatrix} \lambda_i + x_i^T E x_i & & x_i^T E x_j \\ & \ddots & \\ x_i^T E x_j & & \lambda_n + x_n^T E x_n \end{bmatrix}.$$

we now partition this matrix and write $P^T X^T (A+E) X P = \begin{bmatrix} A_1 & \\ & A_2 \end{bmatrix} + \begin{bmatrix} 0 & E_1^T \\ E_1 & 0 \end{bmatrix}$, where $A_1 = \lambda_i + x_i^T E x_i$ is $1 \times 1$ (highlighted in red), a scalar, hence $A_1 = \lambda_i + x_i^T E x_i$. Noting that $\lambda_i(P^T X^T (A+E) X P) = \lambda_i(A+E)$, we now use Theorem 1 to obtain

$$|\lambda_i(A+E) - (\lambda_i(A) + x_i^T E x_i)| \le \frac{\|E\|^2}{\widetilde{\text{gap}}_i},$$

where $\widetilde{\text{gap}}_i := \max(0, \text{gap}_i - \|E\|)$. This updated gap is obtained by using Weyl's bound for the lower-right $(n-1) \times (n-1)$ part of $P^T X^T (A+E) X P$, which is altered from $A_2$ by the lower-right part of $P^T X^T E X P$. This establishes (10) (and hence also the first-order expansion (1)). □

Note that $\widetilde{\text{gap}}_i$ is different from $\text{gap}_i$: as alluded to after (6), this difference is reflected in the formal statements of the residual bounds and quadratic off-diagonal perturbation bounds in the following sense: in (6) the gap is between an approximate and exact eigenvalue. In (5) the gap is between two approximate eigenvalues, namely the eigenvalues of $A$. While this subtlety is certainly present, we shall not expound on this further as this difference diminishes as $E \to 0$. Furthermore, they both convey the same message that the accuracy scaled inverse-proportionally to the gap.

## 2.2 Connection to Residual Bounds

Now we explain how the residual bounds (6), (7) can be obtained from the off-diagonal quadratic perturbation bound (3).

Recall that the Rayleigh-Ritz process employs a subspace spanned by a matrix $Q \in \mathbb{R}^{n\times k}$ with orthonormal columns with $k < n$, and computes the $k \times k$ symmetric eigenproblem $Q^TAQ = V_Q\Lambda_QV_Q^T$, from which one extracts the Ritz values $\mathrm{diag}(\Lambda_Q)$, and Ritz vectors $QV_Q$. Here we examine the accuracy of the Ritz values, and derive (6) using (3).

Consider $Q^\perp \in \mathbb{R}^{n\times(n-k)}$, which spans the orthogonal complement of $Q$ so that $[Q\, Q^\perp]$ is a square orthogonal matrix. Then we have

$$[Q\, Q^\perp]^TA[Q\, Q^\perp] = \begin{bmatrix} A_{11} & A_{12} \\ A_{21} & A_{22} \end{bmatrix}.$$

Then $Q^TAQ = A_{11}$, and the problem essentially reduces to quantifying the accuracy of the eigenvalues of $A_{11}$ as approximants to $k$ of the eigenvalues of $A$. This is exactly the problem treated in Theorem 1, and gives (6): again, the gap in the literature differs subtly from what we get here. The above argument more generally gives (7). Note how the definition of gap differs between (6) and (7); the gap is usually much wider in (7), giving better bounds if the residual norms are of comparable magnitudes.

## 3 Non-symmetric Eigenvalue Problems

The main message of the last section was the significance of the off-diagonal quadratic perturbation bound (3) in symmetric eigenvalue problems. We now turn to the analogous results for more general eigenvalue problems, focusing on nonsymmetric standard eigenproblems.

### 3.1 *Statements*

Here we display the extensions of results in the previous section to nonsymmetric matrices. When $A$ is nonsymmetric, the first-order perturbation expansion (1) becomes

$$\lambda_i(A+E) = \lambda_i(A) + \frac{y^TEx}{y^Tx} + O(\|E\|^2), \tag{11}$$

where $Ax = \lambda_i(A)x$ as before and $y$ is a left eigenvector, that is, $y^TA = y^T\lambda_i(A)$. Here and below, $\lambda_i(A)$ denotes an eigenvalue of $A$ (not necessarily ordered, as they can be nonreal) and $\lambda_i(A+E)$ denotes an eigenvalue that spawns from $\lambda_i(A)$ in that $\lambda_i(A+tE)$ is continuous in $t \in \mathbb{R}$ and $\lambda_i(A+tE) \to \lambda_i(A)$ as $t \to 0$. The expansion (11) holds for any scaling of $x, y$; we scale them so that $\|x\| = 1$ and $y^Tx = 1$.

The analogue of the key result (3) becomes [4, Ch. 7] the following (ignoring the nontrivial issue of "ordering" the eigenvalues, which can be nonreal). Note below that neither $A$ nor $E$ is assumed to be symmetric, but the block structure is preserved.

**Theorem 2 ([10])** *Let $A, E \in \mathbb{R}^{n\times n}$ be matrices partitioned as*

$$A = \begin{bmatrix} A_1 & 0 \\ 0 & A_2 \end{bmatrix}, \quad E = \begin{bmatrix} 0 & E_{12} \\ E_{21} & 0 \end{bmatrix}. \tag{12}$$

*Then [10, Thm. 5]*

$$\lambda_i(A+E) - \lambda_i(A) \le c\frac{\|E_{12}\|\|E_{21}\|}{\mathrm{gap}_i}, \tag{13}$$

*where $c$ is the product of condition numbers of eigenvector matrices of $A+E$ and $A$.* Theorem 2 is less sharp than Theorem 1: for example, it is not exact for $2 \times 2$ case. It nonetheless suffices for the argument here.

Bounds like (13) are often stated in terms of the quantity sep [14, § 4.2], which here is $\mathrm{sep}_i = 1/\|(A_2 - \lambda_i(A))^{-1}\|$. Note that $\mathrm{sep}_i = \mathrm{gap}_i$ when the matrices are symmetric (or normal), and $\mathrm{sep}_i$ takes into account the conditioning of the eigenvector matrix. In this note, for simplicity we absorb this effect in the constant $c$, in order to highlight the role played by the gap throughout eigenvalue problems.

Finally, the residual bound in the nonsymmetric case becomes [14, Thm. 4.2.12]

$$|\lambda - \hat{\lambda}| \le c\frac{\|r_r\|\|r_l\|}{\mathrm{gap}_i}, \tag{14}$$

where

$$\hat{\lambda} = \frac{y^T A x}{y^T x}$$

is the Ritz value (via two-sided projection) and $r_r, r_l$ are the right and left residual vectors defined by

$$r_r = Ax - \hat{\lambda}x, \qquad r_l = y^T A - \hat{\lambda} y^T. \tag{15}$$

More generally, for block matrices (or projection onto a $k > 1$-dimensional subspace is employed), we have

$$|\lambda_i - \hat{\lambda}_i| \le c\frac{\|R_r\|\|R_l\|}{\mathrm{gap}_i}, \tag{16}$$

where

$$\hat{\lambda}_i = \lambda_i(Y^T AX - \lambda Y^T X)$$

are the eigenvalues of the matrix pencil $Y^TAX - \lambda Y^TX$ (sometimes called Ritz values via two-sided projection), and denoting $\hat{\Lambda} = \mathrm{diag}(\hat{\lambda}_1, \ldots, \hat{\lambda}_k)$, $R_r, R_l$ are the right and left residual matrices defined by

$$R_r = AX - X\hat{\Lambda}, \qquad R_l = Y^TA - \hat{\Lambda}Y^T. \tag{17}$$

Below we follow the same line of argument as in Sect. 2 and derive the first-order expansion (11) and residual bound (14) using the off-diagonal quadratic perturbation bound (13).

### 3.2 First-Order Expansion and Its Constant via Off-Diagonal Bounds

Let us establish an analogue of Proposition 1 for the nonsymmetric case.

**Proposition 2** *Let $A, E \in \mathbb{R}^{n\times n}$ be nonsymmetric matrices, and let $(\lambda_i(A), x, y)$ be an eigentriplet of $A$ such that $Ax = \lambda_i(A)x$ and $y^TA = \lambda_i(A)y^T$ where $\lambda_i(A)$ is a simple eigenvalue and $x, y \in \mathbb{C}^n$ are nonzero right and left eigenvectors. Then*

$$|\lambda_i(A+E) - (\lambda_i(A) + \frac{y^TEx}{y^Tx})| \leq \frac{\|E\|^2}{\widetilde{\mathrm{gap}}_i}, \tag{18}$$

*where $\widetilde{\mathrm{gap}}_i := \max(0, \mathrm{gap}_i - \tilde{c}\|E\|)$, where $\mathrm{gap}_i$ is the gap between $\lambda_i(A)$ and the rest of the eigenvalues of $A$.*

To establish (18) and therefore (11), assume that $A$ is diagonalizable (this assumption is mainly for simplicity: it can be relaxed to just $\lambda_i(A)$ being simple, or even to multiple eigenvalues as long as they are not defective) and consider the eigenvalue decomposition

$$X^{-1}AX = \begin{bmatrix} \lambda_1 & & \\ & \ddots & \\ & & \lambda_n \end{bmatrix},$$

where $X = [x_1, \ldots, x_n]$ is a nonsingular eigenvector matrix. Then

$$X^{-1}(A+E)X = \begin{bmatrix} \lambda_1 & & \\ & \ddots & \\ & & \lambda_n \end{bmatrix} + X^{-1}EX,$$

where recalling that $[y_1, \ldots, y_n]^T = X^{-1}$, the $(i, i)$ element is $\lambda_i + y_i^TEx_i$. We can then apply a permutation matrix $P$ that moves the $i$th position to the first, which gives

$$P^T X^{-1}(A+E)XP = \begin{bmatrix} \lambda_i + y_i^T E x_i & & y_i^T E x_j \\ & \ddots & \\ y_i^T E x_j & & \lambda_n + y_n^T E x_n \end{bmatrix}. \tag{19}$$

We now partition this matrix and write $P^T X^T(A+E)XP = \begin{bmatrix} A_1 & \\ & A_2 \end{bmatrix} + \begin{bmatrix} 0 & E_1^T \\ E_1 & 0 \end{bmatrix}$, where $A_1$ is $1 \times 1$, a scalar, hence $A_1 = \lambda_i + y_i^T E x_i$. Noting that $\lambda_i(P^T X^{-1}(A+E)XP) = \lambda_i(A+E)$, we now use Theorem 2 to obtain

$$|\lambda_i(A+E) - (\lambda_i(A) + y_i^T E x_i)| \le c\frac{\|E\|^2}{\widetilde{\text{gap}}_i},$$

where $\widetilde{\text{gap}}_i := \max(0, \text{gap}_i - \tilde{c}\|E\|)$; this is a lower bound for the gap between $\lambda_i(A) + y_i^T E x_i$ and the eigenvalues of the $(n-1)\times(n-1)$ bottom-right part of (19), and $\tilde{c}$ depends on its eigenvector matrix. This establishes (18), and hence also the first-order expansion (11).

Let us discuss the case where $\lambda_i$ is a multiple eigenvalue. When $\lambda_i$ is defective, belonging to a Jordan block of size $k > 1$, then it is known that an $O(\epsilon)$ perturbation can perturb the $k$ eigenvalues by $O(\epsilon^{1/k})$. Thus instead of a expansion with a linear leading term, one needs to deal with an expansion where the lowest-order term is $\epsilon^{1/k}$; see [2] for details and more. On the other hand, if $\lambda_i$ is multiple but nondefective (i.e., semisimple), then one can essentially follow the above argument, taking the colored block to be $k \times k$ (one needs to treat the $k$ eigenvalues together as otherwise the gap would be 0) and diagonalizing it by a similarity transformation, to obtain an analogous first-order expansion for the $k$ eigenvalues spawned from $\lambda_i$. Generally, semisimple eigenvalues much in the same way as simple eigenvalues in perturbation analysis.

## *3.3 Connection to Residual Bounds*

Now we explain how the residual bound (14) can be obtained from (13).

For the nonsymmetric case, we analyze the two-sided projection method, which spanned by two matrices: $X \in \mathbb{R}^{n\times k}$, (hoped to approximate some *right* eigenvectors), usually but not necessarily with orthonormal columns, and $Y \in \mathbb{R}^{n\times k}$ (hoped to approximate the same *left* eigenvectors); however, the simple choice $Y = X$ is quite common and natural in view of the Schur form.

We then compute the $k\times k$ generalized eigendecomposition $V_Y^T(Y^TAX, Y^TX)V_X = (\Lambda_{XY}, I)$, which reduces to a standard eigenproblem if we choose $Y$ so that $Y^TX = I$. One then extracts the approximate eigenvalues (sometimes also called Ritz values) as $\text{diag}(\Lambda_{XY})$, and approximate right and left eigenvectors (Ritz vectors) $XV_X$ and $YV_Y$. Here we examine the accuracy of the Ritz values, and derive (16) using Theorem 2.

For simplicity we discuss the case $Y^TX = I_k$. Let $X_2, Y_2$ be such that $[X\, X_2]$, $[Y\, Y_2]$ are nonsingular matrices and $[Y\, Y_2]^T[X\, X_2] = I_n$. Then we write

$$[Y\, Y_2]^T A [X\, X_2] = \begin{bmatrix} A_{11} & A_{12} \\ A_{21} & A_{22} \end{bmatrix}.$$

Then $Y^TAX = A_{11}$, and the problem essentially reduces to quantifying the accuracy of the eigenvalues of $A_{11}$ as approximants to $k$ of the eigenvalues of $A$. This is exactly the problem treated in Theorem 2, in which $\|A_{21}\|$ corresponds to the right residual $\|R_r\|$ and $\|A_{21}\|$ to $\|R_l\|$, leading to (16).

We note that the residual bounds become linear in $\|R_r\|$ if we use a one-sided projection method with $Y = X$, as then $\|R_l\|$ will be $O(1)$ rather than $O(\|R_r\|)$. This indicates that it is worth using two-sided projection when an approximation to the left eigenvectors is available.

### 3.4 Gerschgorin's Viewpoint

Here we explain the same quadratic scaling $|\lambda_i(A+E) - \widehat{\lambda}_i| \leq \frac{\|E\|^2}{\text{gap}_i}$ from the viewpoint of Gerschgorin's theorem. We could have included such treatment in the symmetric case, but we have deferred its treatment until now since no simplification accrues in the symmetric case. Gerschgorin's theorem states that

$$\lambda(A) \in \bigcup_i \Gamma_i, \qquad \Gamma_i = \{z \in \mathbb{C} \mid |z - a_{ii}| \leq \sum_{j \neq i} |a_{ij}|\},$$

that is, the eigenvalues of $A$ lie in the union of Gerschgorin disks $\Gamma_i$ of radius $\sum_{j\neq i} |a_{ij}|$ centered at $a_{ii}$. Now we focus on $\lambda_i$, and denoting by $\epsilon$ an entry bounded by $|\epsilon| \leq \|E\|$, we see that

$$P^TX^{-1}(A+E)XP = \begin{bmatrix} \lambda_i & & & \\ & \lambda_1 & & \\ & & \ddots & \\ & & & \lambda_n \end{bmatrix} + \begin{bmatrix} \epsilon_i & \epsilon & \dots & \epsilon \\ \epsilon & \epsilon & \cdots & \epsilon \\ \vdots & \vdots & \ddots & \vdots \\ \epsilon & \epsilon & \cdots & \epsilon \end{bmatrix}.$$

If $\lambda_i$ is a simple eigenvalue and $E$ is sufficiently small, we will have $\Gamma_j \cap \Gamma_i = \phi$ for $j \neq i$, which means there is exactly one eigenvalue lying in $\Gamma_i$. Let $\delta$ be a quantity smaller than $\text{gap}_i = \min_{j\neq i} |\lambda_i - \lambda_j|$. Then using the diagonal matrix

$D = \mathrm{diag}(\frac{\epsilon_i}{\delta}, 1, \dots, 1)$ we have

$$DP^T X^{-1}(A+E)XPD^{-1} = \begin{bmatrix} \lambda_i & & & \\ & \lambda_1 & & \\ & & \ddots & \\ & & & \lambda_n \end{bmatrix} + \begin{bmatrix} \epsilon_i & \frac{\epsilon^2}{\delta} & \cdots & \frac{\epsilon^2}{\delta} \\ \delta & \epsilon & \cdots & \epsilon \\ \vdots & \vdots & \ddots & \vdots \\ \delta & \epsilon & \cdots & \epsilon \end{bmatrix}.$$

Now with Gerschgorin's theorem applied to this matrix, $\Gamma_i \cap \Gamma_j = \phi$ still holds, and $\Gamma_i : |z - (\lambda_i + \epsilon_i)| \le O(\frac{\epsilon^2}{\delta})$. Note the radius of $\Gamma_i$ is now $O(\frac{\epsilon^2}{\mathrm{gap}_i})$. Noting that $\epsilon_i = y^T Ex/(y^T x)$ where $x, y$ are the left/right eigenvectors of $A$ corresponding to $\lambda_i$, it follows that $\lambda_i + \epsilon_i = \lambda_i + y^T Ex/(y^T x)$ approximates an eigenvalue of $A + E$ to $O(\frac{\epsilon^2}{\mathrm{gap}_i})$ accuracy.

The above diagonal scaling technique combined with Gerschgorin's theorem is again commonly used, for example in [15, Ch. IV].

### 3.5 *Extensions to Generalized Eigenproblem*

Analogous results for generalized eigenvalue problems can be established, using quadratic off-diagonal perturbation bounds presented in [8]. In particular, the Gerschgorin argument can be used for establishing quadratic perturbation bounds for generalized nonsymmetric eigenvalue problems; see the last section of [11]. We omit the details here.

## 4 Polynomial Eigenvalue Problems

We now turn to polynomial eigenvalue problems. In a polynomial eigenvalue problem, one is to find $x \neq 0$ such that $P(\lambda)x = 0$ where $P(\lambda) = \sum_{i=0}^k \lambda^i A_i \in \mathbb{C}[\lambda]^{n\times n}$ is a matrix polynomial. Let $E(\lambda) \in \mathbb{C}[\lambda]^{n\times n}$ be another matrix polynomial, representing a perturbation to $P(\lambda)$. The first-order perturbation expansion of an eigenvalue $\lambda_i(P(\lambda))$, with $\lambda_i((P+tE)(\lambda))$ depending continuously on $t$ (as in (11)), is known [16] to be

$$\lambda_i((P+E)(\lambda)) = \lambda_i(P(\lambda)) - \frac{y^T E(\lambda_i P(\lambda))x}{y^T P'(\lambda)x} + O(\|E(\lambda_i P(\lambda))\|_2^2). \tag{20}$$

The denominator $y^T P'(\lambda)x$ in the first-order term is known to be nonzero when $\lambda_i(P(\lambda))$ is simple. The expansion (20) is in fact valid without the restriction that $E(\lambda)$ is a matrix polynomial of the same or less degree as $P(\lambda)$, but here we focus on such cases (as otherwise the number of eigenvalues is not controlled). We can verify that (20) reduces to the expansions (1) and (11) in the special case where $P(\lambda)$ represents a linear standard eigenvalue problem $P(\lambda) = \lambda I - A$.

### 4.1 Analysis via Linearization

The most common approach to studying polynomial eigenvalue problems, both in theory and practice is linearization [3]. Here we follow this standard approach to examine the accuracy of first-order expansion (20) and to derive quadratic perturbation bounds for matrix polynomials. The most well known and widely used linearization is the companion linearization. For a *monic* matrix polynomial $P(\lambda) = \sum_{i=0}^{k} A_i\lambda^i$ with $A_k = I$, the companion linearization is defined by

$$C = \begin{bmatrix} -A_{k-1} & \cdots & -A_1 & -A_0 \\ I & & & \\ & \ddots & & \\ & & I & \end{bmatrix}. \tag{21}$$

This $kn \times kn$ matrix clearly has $kn$ eigenvalues, which match those of $P(\lambda)$, so we can write $\lambda_i(C) = \lambda_i(P(\lambda))$. The right eigenvectors of $P(\lambda)$ and $C$ are related by the Vandermonde structure as follows: if $P(\lambda_i)x_i = 0$, then

$$C \begin{bmatrix} \lambda_i^{k-1}x \\ \vdots \\ \lambda_i x \\ x \end{bmatrix} = \lambda_i \begin{bmatrix} \lambda_i^{k-1}x \\ \vdots \\ \lambda_i x \\ x \end{bmatrix}. \tag{22}$$

In view of the first-order expansion, we also need the left eigenvector of $C$. Let $y$ be a left eigenvector of $P$ such that $y^T P(\lambda) = 0$. Then the left eigenvector of $C$ has the structure of the Horner shift [5, eq. (3.12)]

$$\begin{bmatrix} y \\ (\lambda_i A_k + A_{k-1})y \\ \vdots \\ (\lambda_i^k A_k + \lambda^{k-1}A_{k-1} + \cdots + A_1)y \end{bmatrix}^T C = \lambda_i \begin{bmatrix} y \\ (\lambda_i A_k + A_{k-1})y \\ \vdots \\ (\lambda_i^k A_k + \lambda^{k-1}A_{k-1} + \cdots + A_1)y \end{bmatrix}^T. \tag{23}$$

We denote the right and left eigenvectors of $C$ by $\underline{x}$ and $\underline{y}$ respectively, and use (22) and (23) for (20) to obtain the first-order expansion of the eigenvalue $\lambda_i$ of $P$ as

$$\lambda_i((P+E)(\lambda)) = \lambda_i + \frac{y^T(\sum_{j=0}^{k} \lambda_i^i E_i)x}{y^T(\sum_{j=1}^{k} \lambda_i^{i-1} A_i)x} + O(\|E(\lambda_i)\|^2). \tag{24}$$

On the other hand, denoting by $C + \Delta C$ the companion linearization associated with $P + E$, the expansion with respect to $C$ becomes (using (11) with $A \leftarrow C$)

$$\lambda_i(C + \Delta C) = \lambda_i + \frac{\underline{y}^T(\Delta C)\underline{x}}{\underline{y}^T\underline{x}} + O(\|\Delta C\|^2), \tag{25}$$

which, in view of (22) and (23), is equivalent to (24); this is to be expected because $P + E$ and $C + \Delta C$ have the same eigenvalues.

The value in the equivalence between (24) and (25) is that with (25), we can invoke the analysis for linear eigenvalue problems to examine the eigenvalues of $P$ and its perturbed variant. Indeed, assuming $\lambda_i$ is a simple eigenvalue, the exact same arguments as in Sect. 3 shows that the second-order term in the expansion (25) can be written as $c\frac{\|\Delta C\|^2}{\text{gap}_i}$. Note that this allows for general perturbation in the matrix $C$, whereas the perturbation of interest here is structured, because, as we can see in (23), the only elements in $C$ that depend on $P$ are those in the first block row. In any case, we have proven the following result.

**Theorem 3** *Let $P(\lambda) = \sum_{i=0}^{k} \lambda^i A_i \in \mathbb{C}[\lambda]^{n\times n}$ be a monic matrix polynomials of degree k, and $E(\lambda) = \sum_{i=0}^{k-1} \lambda^i E_i \in \mathbb{C}[\lambda]^{n\times n}$. Let $(\lambda_i, x_i, y_i)$ be a simple eigentriple of $P(\lambda)$. Then*

$$\lambda_i((P+E)(\lambda)) = \lambda_i + \frac{y^T(\sum_{j=0}^{k} \lambda_i^i E_i)x}{y^T(\sum_{j=1}^{k} \lambda_i^{i-1} A_i)x} + c\frac{\|E(\lambda_i)\|^2}{\text{gap}_i}, \tag{26}$$

*where* $\text{gap}_i = \min_{j\neq i} |\lambda_i - \lambda_j(P(\lambda))|$ *and c depends on the conditioning of the eigenvector matrix of C in* (21).

We have not yet examined whether the un-structured perturbation results in a constant that is smaller than the unstructured counterpart by (25) would indicate. To examine whether this happens, we turn to MATLAB experiments in which we construct a random matrix polynomial $P(\lambda)$ companion matrix as in (21), compute an eigentriple $(\lambda_i, x_i, y_i)$, then examine the perturbation in $\lambda_i$ when we introduce perturbation in $C$ in two different forms:

1. Perturb only the first block row by norm $\epsilon$,
2. Perturb the whole matrix $C$ by norm $\epsilon$,

for some small $\epsilon$, which here we set to $10^{-4}$. We then examine the difference in the accuracy of $\lambda_i + y^T(\Delta C)x/(y^T x)$ as an approximation to an eigenvalue of the perturbed matrix; clearly, since the second type includes the first as a special case, the second would lead to a larger perturbation in the worst case. We experimented with various $n$ and $k = 2, 3, \ldots, 10$, with randomly generated perturbation matrices, and observed that there is never a significant difference between the sensitivity of $\lambda_i$ under the two types of perturbations. That said, making this observation precise seems nontrivial, and we leave it as an open problem.

## 4.2 Quadratic Bounds by Off-Diagonal Perturbation

We now turn to off-diagonal perturbation and derive a bound analogous to (9).

**Theorem 4** *Let*

$$P(\lambda) = \begin{bmatrix} P_1(\lambda) & 0 \\ 0 & P_2(\lambda) \end{bmatrix}, \quad E(\lambda) = \begin{bmatrix} 0 & E_{12}(\lambda) \\ E_{21}(\lambda) & 0 \end{bmatrix} \tag{27}$$

*be matrix polynomials, with $P(\lambda)$ being monic and degree k, and $E(\lambda)$ of degree $k-1$ or less. $P_1(\lambda), P_2(\lambda)$ are square. Then*

$$\lambda_i(P(\lambda)) - \lambda_i((P+E)(\lambda)) \le c\frac{\|E(\lambda_i(P))\|^2}{\mathrm{gap}_i}, \tag{28}$$

*where* $\mathrm{gap}_i = \min_{j \neq i} |\lambda_i(P(\lambda)) - \lambda_j(P(\lambda))|$ *and c depends on the conditioning of the eigenvector matrix of C in* (21). *Moreover, the bound with "widened" gap holds:*

$$\lambda_i(P(\lambda)) - \lambda_i((P+E)(\lambda)) \le c\frac{\|E(\lambda_i(P))\|^2}{\min_j |\lambda_i(P_1(\lambda)) - \lambda_j(P_2(\lambda))|}. \tag{29}$$

*Proof* For (28), the argument is simple as we now have all the essential tools. Note that for any eigenvalue of $P_1(\lambda)$ that is not an eigenvalue of $P_2(\lambda)$, the left and right eigenvectors have the block zero structure $\begin{bmatrix} x_1 \\ 0 \end{bmatrix}$ and $\begin{bmatrix} y_1 \\ 0 \end{bmatrix}$. Plugging this into (26), we obtain (28).

To establish (29), first recall the eigenvector structure of $C$ in (22) and (23). Let $n$ and $\ell$ denote the size of $P(\lambda)$ and $P_1(\lambda)$, and $e_i$ the $i$th column of $I_k$. Then defining

$$X_1 = \oplus_{i=1}^{k} \left( e_i \otimes \left[ \begin{smallmatrix} I_\ell \\ 0_{(n-\ell)\times\ell} \end{smallmatrix} \right] \right), \quad X_2 = \oplus_{i=1}^{k} \left( e_i \otimes \left[ \begin{smallmatrix} 0_{\ell\times(n-\ell)} \\ I_{n-\ell} \end{smallmatrix} \right] \right)$$

and $X = [X_1 \ X_2] \in \mathbb{R}^{nk\times nk}$, we have $X = X^{-1}$ and $X^{-1}CX = \mathrm{diag}(C_1, C_2)$ where $\lambda(C_1) = \lambda(P_1(\lambda))$ and $\lambda(C_2) = \lambda(P_2(\lambda))$, and the perturbation $E$ results in off-diagonal perturbation: $X^{-1}(C+E)X = \left[ \begin{smallmatrix} C_1 & E_1 \\ E_2 & C_2 \end{smallmatrix} \right]$. We then invoke Theorem 2 to complete the proof. □

Note that the above argument takes the opposite route from before: now we are using the first-order expansion to obtain the quadratic off-diagonal perturbation bound.

Observe in (28) that what matters for the perturbation in $\lambda_i$ is the magnitude of $E(\lambda)$ evaluated at $\lambda = \lambda_i$; for example, the perturbation is zero if $E(\lambda_i) = 0$, even if $E(\lambda)$ takes large values away from $\lambda_i$.

### 4.2.1 Accuracy of Eigenvalues Obtained by Projection Methods

Another implication of Theorem 4 can be observed on an approximate eigenpair $(\hat{\lambda}_i, \hat{x}_i)$ obtained via a projection method applied to polynomial eigenvalue problems. Consider for simplicity a symmetric matrix polynomial $P(\lambda)$. Suppose $(\hat{\lambda}_i, \hat{x}_i)$ is obtained by solving $V^T P(\hat{\lambda}_i) V y_i = 0$ for some orthonormal matrix $V \in \mathbb{C}^{n\times k}$, $k < n$, with $\hat{x}_i = V y_i$. Then we can write, using an orthogonal matrix $[V\ V^\perp]$,

$$[V\ V^\perp]^T P(\lambda)[V\ V^\perp] = \begin{bmatrix} P_1(\lambda) & E_{12}(\lambda) \\ E_{21}(\lambda) & P_2(\lambda) \end{bmatrix},$$

where $P_1(\hat{\lambda}_i)$ has $\hat{\lambda}_i$ as an exact eigenvalue, and the residual $\|P(\lambda_i)\hat{x}_i\|$ (which is computable) is bounded by $\|E_{12}(\hat{\lambda}_i)\| = \|E_{21}(\hat{\lambda}_i)\|$ (usually not computable). Thus by the above theorem it follows that the computed eigenvalue $\hat{\lambda}_i$ has accuracy $O(\frac{\|E(\lambda_i(P))\|^2}{\mathrm{gap}_i}) = O(\frac{\|P(\hat{\lambda}_i)\hat{x}_i\|^2}{\mathrm{gap}_i})$.

Note that the same type of quadratic bound follows for *nonsymmetric* matrix polynomials, provided that we employ a two-sided projection method in which we work with $Y^T P(\lambda) X$ where $Y$ and $X$ approximate the desired left and right eigenspaces respectively. This is exactly the same situation as in linear eigenvalue problems, for which we need two-sided projection to obtain quadratic eigenvalue convergence in the nonsymmetric case. Put another way, because the left and right eigenvectors are the same for symmetric eigenvalue problems, the Rayleigh-Ritz method automatically approximates both the left and right eigenvectors simultaneously. The apparent difference in convergence speed for symmetric and nonsymmetric eigenvalue problems (which is present e.g. in the QR algorithm and Rayleigh quotient iteration) comes from the fact that the algorithm is implicitly employing a one-sided projection method, not because the convergence is inherently hindered by lack of symmetry.

## 5 Discussion

This work examined the ramifications of the fact that off-diagonal perturbation of a block diagonal matrix (or matrix polynomial) result in perturbation in the eigenvalues that scale quadratically with the norm of the perturbation. The quadratic scaling hinges on the block structure of the matrices as in (3) or (27), which the eigenvectors inherit. In fact, even tighter bounds can be obtained if further block structure is present, such as block tridiagonal [12]. In addition to some indicated in the text, possible future directions include investigating the accuracy in the expansion and residual bounds and in such cases, examine the implications in terms of the eigenvectors, and overcoming the case where the gap is too small for the bounds to be of use. Eigenvalue perturbation theory is a well-established yet active and useful area of research.

## References


1. Bai, Z., Demmel, J., Dongarra, J., Ruhe, A., van der Vorst, H.: Templates for the Solution of Algebraic Eigenvalue Problems: A Practical Guide. SIAM, Philadelphia, PA (2000)
2. De Terán, F., Dopico, F.M., Moro, J.: First order spectral perturbation theory of square singular matrix pencils. Linear Algebra Appl. **429**(2), 548–576 (2008)
3. Gohberg, I., Lancaster, P., Rodman, L.: Matrix Polynomials. SIAM, Philadelphia (2009). Unabridged republication of book first published by Academic Press in 1982
4. Golub, G.H., Van Loan, C.F.: Matrix Computations, 4th edn. The Johns Hopkins University Press, Baltimore (2012)
5. Higham, N.J., Li, R.-C., Tisseur, F.: Backward error of polynomial eigenproblems solved by linearization. SIAM J. Matrix Anal. Appl. **29**(4), 1218–1241 (2007)
6. Kahan, W., Parlett, B.N., Jiang, E.: Residual bounds on approximate eigensystems of nonnormal matrices. SIAM J. Numer. Anal. **19**(3), 470–484 (1982)
7. Li, R.-C.: Matrix perturbation theory. In: Hogben, L., Brualdi, R., Greenbaum, A., Mathias, R. (eds.) Handbook of Linear Algebra, chapter 15. CRC Press, Boca Raton, FL (2006)
8. Li, R.-C., Nakatsukasa, Y., Truhar, N., Xu, S.: Perturbation of partitioned Hermitian generalized eigenvalue problem. SIAM J. Matrix Anal. Appl. **32**(2), 642–663 (2011)
9. Li, C.-K., Li, R.-C.: A note on eigenvalues of perturbed Hermitian matrices. Linear Algebra Appl. **395**, 183–190 (2005)
10. Mathias, R.: Quadratic residual bounds for the Hermitian eigenvalue problem. SIAM J. Matrix Anal. Appl. **19**(2), 541–550 (1998)
11. Nakatsukasa, Y.: Gerschgorin's theorem for generalized eigenvalue problems in the Euclidean metric. Math. Comput. **80**(276), 2127–2142 (2011)
12. Nakatsukasa, Y.: Eigenvalue perturbation bounds for Hermitian block tridiagonal matrices. Appl. Numer. Math. **62**(1), 67–78 (2012)
13. Parlett, B.N.: The Symmetric Eigenvalue Problem. SIAM, Philadelphia (1998)
14. Stewart, G.W.: Matrix Algorithms Volume II: Eigensystems. SIAM, Philadelphia (2001)
15. Stewart, G.W., Sun, J.-G.: Matrix Perturbation Theory (Computer Science and Scientific Computing). Academic Press, Boston (1990)
16. Tisseur, F.: Backward error and condition of polynomial eigenvalue problems. Linear Algebra Appl. **309**(1), 339–361 (2000)

# An Elementary Derivation of the Projection Method for Nonlinear Eigenvalue Problems Based on Complex Contour Integration

**Yusaku Yamamoto**

**Abstract** The Sakurai-Sugiura (SS) projection method for the generalized eigenvalue problem has been extended to the nonlinear eigenvalue problem $A(z)\mathbf{w} = \mathbf{0}$, where $A(z)$ is an analytic matrix valued function, by several authors. To the best of the authors' knowledge, existing derivations of these methods rely on canonical forms of an analytic matrix function such as the Smith form or the theorem of Keldysh. While these theorems are powerful tools, they require advanced knowledge of both analysis and linear algebra and are rarely mentioned even in advanced textbooks of linear algebra. In this paper, we present an elementary derivation of the SS-type algorithm for the nonlinear eigenvalue problem, assuming that the wanted eigenvalues are all simple. Our derivation uses only the analyticity of the eigenvalues and eigenvectors of a parametrized matrix $A(z)$, which is a standard result in matrix perturbation theory. Thus we expect that our approach will provide an easily accessible path to the theory of nonlinear SS-type methods.

## 1 Introduction

Given an $n \times n$ matrix $A(z)$ whose elements are analytic function of a complex parameter $z$, we consider the problem of finding the values of $z$ for which the linear simultaneous equation $A(z)\mathbf{w} = \mathbf{0}$ has a nonzero solution $\mathbf{w}$. Such a problem is known as the *nonlinear eigenvalue problem* and the value of $z$ and $\mathbf{w}$ that satisfy this condition are called the eigenvalue and the eigenvector, respectively. The nonlinear eigenvalue problem arises in many fields of scientific and engineering computing, such as the electronic structure calculation, nonlinear elasticity and theoretical fluid dynamics.

There are several algorithms for solving the nonlinear eigenvalue problem, including the multivariate Newton's method [14] and its variants [13], the nonlinear Arnoldi method [21], the nonlinear Jacobi-Davidson method [3] and methods based

Y. Yamamoto (✉)
The University of Electro-Communications, 1-5-1 Chofugaoka, Chofu, Tokyo 182-8585, Japan
e-mail: yusaku.yamamoto@uec.ac.jp



T. Sakurai et al. (eds.), *Eigenvalue Problems: Algorithms, Software and Applications in Petascale Computing*, Lecture Notes in Computational Science and Engineering 117, https://doi.org/10.1007/978-3-319-62426-6_16

on complex contour integration [1, 2, 4, 22]. Among them, the last class of methods have a unique feature that they can compute all the eigenvalues in a specified region on the complex plane enclosed by a Jordan curve (i.e., simple closed curve) $\Gamma$. In addition, they have large grain parallelism since the function evaluations for numerical integration can be done for each sample point independently. In fact, [1] reports that nearly linear speedup can be achieved even in a Grid environment, where the interconnection network among the computing nodes is relatively weak.

These algorithms can be viewed as nonlinear extensions of the Sakurai-Sugiura (SS) method for the generalized eigenvalue problem $A\mathbf{x} = \lambda B\mathbf{x}$ [16]. To find the eigenvalues within a closed Jordan curve $\Gamma$ in the complex plane, the SS method computes the moments $\mu_p = \int_\Gamma z^p \mathbf{u}^*(A - zB)^{-1}\mathbf{v}\, dz$, where $\mathbf{u}$ and $\mathbf{v}$ are some constant vectors, and extracts the information of the eigenvalues from the moments. To justify the algorithm, Weierstrass's canonical form [5] for (linear) matrix pencils is used. Similarly, existing derivations of the SS-type algorithms for the nonlinear eigenvalue problem rely on canonical forms of the analytic matrix function $A(z)$. Specifically, Asakura et al. uses the Smith form for analytic matrix functions [6], while Beyn and Yokota et al. employ the theorem of Keldysh [11, 12]. These theorems are intricate structure theorems, which give canonical representations of $A(z)$ that are valid on the *whole* domain enclosed by $\Gamma$. On the other hand, they require advanced knowledge of both analysis and linear algebra and are rarely introduced even in advanced textbooks of linear algebra.

In this paper, we present an elementary derivation of the SS-type method for the nonlinear eigenvalue problem, assuming that all the eigenvalues of $A(z)$ in $\Gamma$ are simple. Instead of the whole domain enclosed by $\Gamma$, we consider an *infinitesimally small* circle $\Gamma_i^\epsilon$ around each eigenvalue $z_i$. This allows us to use the analyticity of the eigenvalues and eigenvectors of a parametrized matrix $A(z)$, which is a well-known result in matrix perturbation theory [9, p. 117][10, Chapter 2, Sections 1 & 2], to evaluate the contour integral along $\Gamma_i^\epsilon$. Then we aggregate the contributions from each $\Gamma_i^\epsilon$ to evaluate the contour integral along $\Gamma$. This is sufficient for theoretical justification of the nonlinear SS-type algorithm in the case of simple eigenvalues. We believe that this provides an easily accessible approach to the theory of SS-type methods for the nonlinear eigenvalue problem. We emphasize that our focus here is not to propose a new algorithm for the nonlinear eigenvalue problem, but to provide an elementary derivation of the SS-type nonlinear eigensolver.

This paper is structured as follows: In Sect. 2, we develop a theory for computing the eigenvalues of $A(z)$ based on the complex contour integral. The algorithm based on this theory is presented in Sect. 3. Section 4 gives some numerical results. Finally, we give some concluding remarks in Sect. 5.

Throughout this paper, we use capital letters to denote matrices, bold small letters to denote vectors, roman small letters and Greek letters to denote scalars. $A^T$ and $A^*$ denote the transpose and the Hermitian conjugate of a matrix $A$, respectively. $I_n$ denotes the identity matrix of dimension $n$. For $\mathbf{x} \in \mathbf{C}^n$, $\{\mathbf{x}\}$ denotes a subspace of $\mathbf{C}^n$ spanned by $\mathbf{x}$.

## 2 The Theory

Let $A(z)$ be an $n \times n$ matrix whose elements are analytic functions of a complex parameter $z$ in some region of the complex plane. Let $\Gamma$ be a closed Jordan curve within that region and assume that $A(z)$ has $m$ eigenvalues $z_1, z_2, \ldots, z_m$ within $\Gamma$. We further assume that they are simple eigenvalues, that is, simple zeroes of $\det(A(z))$, and the number $m$ is known. In the following, we refer to $z_1, z_2, \ldots, z_m$ as *nonlinear eigenvalues* of $A(z)$.

For a fixed value of $z$, $A(z)$ is a constant matrix and therefore has $n$ eigenvalues. We refer to them as *linear eigenvalues* of $A(z)$. Also, we call the eigenvectors of a constant matrix $A(z)$ *linear eigenvectors*. If $z_i$ is a nonlinear eigenvalue of $A(z)$, $A(z_i)$ is singular and at least one of the linear eigenvalues of $A(z_i)$ is zero. Moreover, since $z_1, z_2, \ldots, z_m$ are simple eigenvalues, only one of the $n$ linear eigenvalues become zero at each of them. We denote the linear eigenvalue that becomes zero at $z = z_i$ by $\lambda_i(z)$. Note that $z_i$ is a simple zero of $\lambda_i(z)$ because otherwise $z_i$ will not be a simple zero of $\det(A(z))$.

Since $\lambda_i(z)$ is a continuous function of $A(z)$ near $z_i$ [10, p. 93, Theorem 2.3], it remains to be a simple linear eigenvalue of $A(z)$ in the neighborhood of $z = z_i$ and has one-dimensional right and left eigenspaces. Let $\mathbf{x}_i(z)$ and $\mathbf{y}_i(z)$ be the (linear) left and right eigenvectors, respectively, chosen so that $\mathbf{y}_i^*(z)\mathbf{x}_i(z) = 1$. Also, let $X_i(z) \in \mathbf{C}^{n\times(n-1)}$ and $Y_i(z) \in \mathbf{C}^{n\times(n-1)}$ be matrices whose column vectors are the basis of the orthogonal complementary subspaces of $\{\mathbf{y}_i(z)\}$ and $\{\mathbf{x}_i(z)\}$, respectively, and which satisfy $Y_i^*(z)X_i(z) = I_{n-1}$. From these definitions, we have

$$\begin{bmatrix} \mathbf{y}_i^*(z) \\ Y_i^*(z) \end{bmatrix} \begin{bmatrix} \mathbf{x}_i(z) \; X_i(z) \end{bmatrix} = \begin{bmatrix} \mathbf{y}_i^*(z)\mathbf{x}_i(z) & \mathbf{y}_i^*(z)X_i(z) \\ Y_i^*(z)\mathbf{x}_i(z) & Y_i^*(z)X_i(z) \end{bmatrix} = \begin{bmatrix} 1 & \mathbf{0}^T \\ \mathbf{0} & I_{n-1} \end{bmatrix}. \tag{1}$$

Note that $\mathbf{x}_i(z)$, $\mathbf{y}_i(z)$, $X_i(z)$ and $Y_i(z)$ are not yet uniquely determined under these conditions.

Now we show the following basic lemma.

**Lemma 2.1** *Let $\Gamma_i^\epsilon$ be a circle with center $z_i$ and radius $\epsilon$. For sufficiently small $\epsilon$, $\lambda_i(z)$ is an analytic function of $z$ within $\Gamma_i^\epsilon$ and all of $\mathbf{x}_i(z)$, $\mathbf{y}_i(z)$, $X_i(z)$ and $Y_i(z)$ can be chosen to be analytic functions of $z$ within $\Gamma_i^\epsilon$.*

*Proof* For a sufficiently small $\epsilon$, $\lambda_i(z)$ is a simple linear eigenvalue of $A(z)$ everywhere in $\Gamma_i^\epsilon$. In this case, it is well known that $\lambda_i(z)$ is an analytic function of $z$ in $\Gamma_i^\epsilon$. See [9, p. 117] for the proof. Let $P_i(z) \in \mathbf{C}^{n\times n}$ be a projection operator on the right eigenvector of $A(z)$ belonging to $\lambda_i(z)$ along the left eigenvector. It is also shown in [10, p. 93, Theorem 2.3] that $P_i(z)$ is an analytic function of $z$ in $\Gamma_i^\epsilon$ for sufficiently small $\epsilon$.

Now, let $\mathbf{x}_i^{(0)} \neq \mathbf{0}$ be a (linear) right eigenvector of $A(z_i)$ corresponding to $\lambda_i(z_i)$ and set $\mathbf{x}_i(z) = P_i(z)\mathbf{x}_i^{(0)}$. Then $\mathbf{x}_i(z)$ is an analytic function of $z$ and belongs to the right eigenspace of $\lambda_i(z)$. Moreover, since $\mathbf{x}_i(z_i) = P_i(z_i)\mathbf{x}_i^{(0)} = \mathbf{x}_i^{(0)} \neq \mathbf{0}$, $\mathbf{x}_i(z)$

remains nonzero within $\Gamma_i^\epsilon$ if $\epsilon$ is sufficiently small. Thus we can adopt $\mathbf{x}_i(z)$ as a (linear) right eigenvector corresponding to $\lambda_i(z)$.

Next, let $\tilde{\mathbf{y}}_i^{(0)} \neq \mathbf{0}$ be a (linear) left eigenvector of $A(z_i)$ corresponding to $\lambda_i(z_i)$ and $X_i^{(0)} \in \mathbf{C}^{n\times(n-1)}$ be a matrix whose column vectors are the basis of the orthogonal complementary subspace of $\{\tilde{\mathbf{y}}_i^{(0)}\}$. Set $X_i(z) = (I - P_i(z))\, X_i^{(0)}$. Then, $X_i(z)$ is an analytic function of $z$. Also, its column vectors are orthogonal to the (linear) left eigenvector of $A(z)$ corresponding to $\lambda_i(z)$, which we denote by $\tilde{\mathbf{y}}_i(z)$, since

$$\tilde{\mathbf{y}}_i^*(z)X_i(z) = \tilde{\mathbf{y}}_i^*(z)\,(I - P_i(z))\, X_i^{(0)} = \left(\tilde{\mathbf{y}}_i^*(z) - \tilde{\mathbf{y}}_i^*(z)\right) X_i^{(0)} = \mathbf{0}^T, \tag{2}$$

where we used the fact that $P_i^*(z)$ is a projection operator on the left eigenvector along the right eigenvector. Moreover, since $X_i(z_i) = (I - P_i(z_i))\, X_i^{(0)} = X_i^{(0)}$, $X_i(z)$ remains to be rank $n-1$ within $\Gamma_i^\epsilon$ if $\epsilon$ is sufficiently small. In this situation, the column vectors of $X_i(z)$ constitute the basis of the orthogonal complementary subspace.

Finally we note that for sufficiently small $\epsilon$, the matrix $[\mathbf{x}_i(z)\ X_i(z)]$ is of full rank since the column vectors of $X_i(z)$ are orthogonal to the left eigenvector, while the left and right eigenvectors are not orthogonal for a simple eigenvalue [15]. Hence we can define a vector $\mathbf{y}_i(z)$ and a matrix $Y_i(z) \in \mathbf{C}^{n\times(n-1)}$ by

$$\begin{bmatrix} \mathbf{y}_i^*(z) \\ Y_i^*(z) \end{bmatrix} = \left[\mathbf{x}_i(z)\ X_i(z)\right]^{-1}. \tag{3}$$

It is clear that $\mathbf{y}_i(z)$ and $Y_i(z)$ are analytic functions of $z$ and $\mathbf{x}_i(z)$, $\mathbf{y}_i(z)$, $X_i(z)$ and $Y_i(z)$ satisfy Eq. (1). From Eq. (1), it is apparent that $Y_i(z)$ is of rank $n-1$ and its columns are the basis of the orthogonal subspace of $\mathbf{x}_i(z)$. Finally, $\mathbf{y}_i(z)$ is a (linear) left eigenvector corresponding to $\lambda_i(z)$ since it is orthogonal to the columns of $X_i(z)$ and the eigenspace is one-dimensional.

Thus we have constructed $\mathbf{x}_i(z)$, $\mathbf{y}_i(z)$, $X_i(z)$ and $Y_i(z)$ that satisfy all the requirements of the lemma. □

Using the result of Lemma 2.1 and Eq. (1), we can expand $A(z)$ in $\Gamma_i^\epsilon$ as

$$\begin{aligned} A(z) &= \left[\mathbf{x}_i(z)\ X_i(z)\right] \begin{bmatrix} \mathbf{y}_i^*(z) \\ Y_i^*(z) \end{bmatrix} A(z) \left[\mathbf{x}_i(z)\ X_i(z)\right] \begin{bmatrix} \mathbf{y}_i^*(z) \\ Y_i^*(z) \end{bmatrix} \\ &= \left[\mathbf{x}_i(z)\ X_i(z)\right] \begin{bmatrix} \mathbf{y}_i^*(z)A(z)\mathbf{x}_i(z) & \mathbf{y}_i^*(z)A(z)X_i(z) \\ Y_i^*(z)A(z)\mathbf{x}_i(z) & Y_i^*(z)A(z)X_i(z) \end{bmatrix} \begin{bmatrix} \mathbf{y}_i^*(z) \\ Y_i^*(z) \end{bmatrix} \\ &= \left[\mathbf{x}_i(z)\ X_i(z)\right] \begin{bmatrix} \lambda_i(z) & \mathbf{0}^T \\ \mathbf{0} & Y_i^*(z)A(z)X_i(z) \end{bmatrix} \begin{bmatrix} \mathbf{y}_i^*(z) \\ Y_i^*(z) \end{bmatrix} \end{aligned} \tag{4}$$

where all the elements and submatrices appearing in the last line are analytic functions of $z$.

As for the submatrix $Y_i^*(z)A(z)X_i(z)$, we can show the following lemma.

**Lemma 2.2** *For sufficiently small* $\epsilon$, $Y_i^*(z)A(z)X_i(z)$ *is nonsingular within* $\Gamma_i^\epsilon$.

*Proof* Since $\epsilon$ is sufficiently small, we can assume that there is no other nonlinear eigenvalues of $A(z)$ in $\Gamma_i^\epsilon$ than $z_i$.

Now, assume that $Y_i^*(z)A(z)X_i(z)$ is singular at some point $z = \hat{z}$ in $\Gamma_i^\epsilon$. Then there is a nonzero vector $\mathbf{p} \in \mathbf{C}^{n-1}$ such that $Y_i^*(\hat{z})A(\hat{z})X_i(\hat{z})\mathbf{p} = \mathbf{0}$. It then follows from Eqs. (4) and (1) that $X_i(\hat{z})\mathbf{p}$ is a (linear) right eigenvector of $A(\hat{z})$ corresponding to the linear eigenvalue 0. Hence, $\hat{z}$ is a nonlinear eigenvalue of $A(z)$. But because $A(z)$ has no other nonlinear eigenvalues than $z_i$ in $\Gamma_i^\epsilon$, we have $z = z_i$. On the other hand, $\mathbf{x}_i(\hat{z}) = \mathbf{x}_i(z_i)$ is also a (linear) right eigenvector of $A(\hat{z})$ corresponding to the linear eigenvalue 0. Since the matrix $[\mathbf{x}_i(\hat{z})\ X_i(\hat{z})]$ is of full rank (see Eq. (1)), $\mathbf{x}_i(\hat{z})$ and $X_i(\hat{z})\mathbf{p}$ are linearly independent. Thus the null space of $A(\hat{z})$ is at least two-dimensional. But this contradicts the assumption that $\hat{z} = z_i$ is a simple zero of $\det(A(z))$. Hence $Y_i^*(z)A(z)X_i(z)$ must be nonsingular within $\Gamma_i^\epsilon$. □

Combining Lemma 2.2 with Eq. (4), we have the following expansion of $A(z)^{-1}$ valid everywhere in $\Gamma_i^\epsilon$ except at $z = z_i$:

$$A(z)^{-1} = \left[\mathbf{x}_i(z)\ X_i(z)\right] \begin{bmatrix} \lambda_i(z)^{-1} & \mathbf{0}^T \\ \mathbf{0} & \left\{Y_i^*(z)A(z)X_i(z)\right\}^{-1} \end{bmatrix} \begin{bmatrix} \mathbf{y}_i^*(z) \\ Y_i^*(z) \end{bmatrix}. \tag{5}$$

In the right hand side, $\lambda_i(z)$ is analytic except at $z = z_i$. All other elements and submatrices are analytic everywhere in $\Gamma_i^\epsilon$. Note that $\left\{Y_i^*(z)A(z)X_i(z)\right\}^{-1}$ is analytic because $A(z)$, $X_i(z)$ and $Y_i(z)$ are analytic (see Lemma 2.1) and $Y_i^*(z)A(z)X_i(z)$ is nonsingular, as proved in Lemma 2.2.

We now define the complex moments $\mu_1, \mu_2, \ldots, \mu_{2m-1}$ by complex contour integration as

$$\mu_p(\mathbf{u}, \mathbf{v}) = \frac{1}{2\pi i} \oint_\Gamma z^p \mathbf{u}^* A(z)^{-1} A'(z) \mathbf{v}\, dz, \tag{6}$$

where $\mathbf{u}$ and $\mathbf{v}$ are some constant vectors in $\mathbf{C}^n$. The next lemma shows that these complex moments contain information on the nonlinear eigenvalues of $A(z)$ in $\Gamma$.

**Lemma 2.3** *The complex moments can be written as*

$$\mu_p(\mathbf{u}, \mathbf{v}) = \sum_{i=1}^{m} \nu_i(\mathbf{u}, \mathbf{v}) z_i^p, \tag{7}$$

*where* $\{\nu_i(\mathbf{u}, \mathbf{v})\}_{i=1}^m$ *are some complex numbers. Moreover,* $\{\nu_i(\mathbf{u}, \mathbf{v})\}_{i=1}^m$ *are nonzero for generic* $\mathbf{u}$ *and* $\mathbf{v}$.

*Proof* Let $\Gamma_i^\epsilon$ $(i = 1, \ldots, m)$ be a circle with center $z_i$ and with sufficiently small radius $\epsilon$. In $\Gamma$, the integrand is analytic everywhere except inside $\Gamma_1^\epsilon, \ldots, \Gamma_m^\epsilon$, so we only need to consider the integration along $\Gamma_i^\epsilon$.

Since $\epsilon$ is sufficiently small, Lemma 2.1 ensures that we can choose analytic $\mathbf{x}_i(z)$, $\mathbf{y}_i(z)$, $X_i(z)$ and $Y_i(z)$ within $\Gamma_i^\epsilon$. Of course, $\lambda_i(z)$ is also analytic in $\Gamma_i^\epsilon$. In addition, since $\lambda_i(z)$ has a simple zero at $z = z_i$, it can be expressed as $\lambda_i(z) = (z - z_i)p_i(z)$, where $p_i(z)$ is analytic and nonzero in $\Gamma_i^\epsilon$.

Now, from Eq. (5), we have

$$A(z)^{-1} = \lambda_i(z)^{-1}\mathbf{x}_i(z)\mathbf{y}_i^*(z) + X_i(z)\left\{Y_i^*(z)A(z)X_i(z)\right\}^{-1} Y_i^*(z) \tag{8}$$

By differentiating Eq. (4) with respect to $z$, we have

$$\begin{aligned} A'(z) = \lambda_i{}'(z)\mathbf{x}_i(z)\mathbf{y}_i^*(z) + \lambda_i(z)\left\{\mathbf{x}_i(z)\mathbf{y}_i^*(z)\right\}' \\ + \left\{X_i(z)Y_i^*(z)A(z)X_i(z)Y_i^*(z)\right\}' \end{aligned} \tag{9}$$

Combining Eqs. (8) and (9), we have

$$\begin{aligned} A(z)^{-1}A'(z) &= \lambda_i(z)^{-1}\lambda_i{}'(z)\mathbf{x}_i(z)\mathbf{y}_i^*(z) + \mathbf{x}_i(z)\mathbf{y}_i^*(z)\left\{\mathbf{x}_i(z)\mathbf{y}_i^*(z)\right\}' \\ &\quad + \lambda_i(z)\mathbf{x}_i(z)\mathbf{y}_i^*(z)\left\{X_i(z)Y_i^*(z)A(z)X_i(z)Y_i^*(z)\right\}' \\ &\quad + X_i(z)\left\{Y_i^*(z)A(z)X_i(z)\right\}^{-1} Y_i^*(z)A'(z) \\ &= \frac{1}{z - z_i}\cdot\mathbf{x}_i(z)\mathbf{y}_i^*(z) + \frac{p_i{}'(z)}{p_i(z)}\mathbf{x}_i(z)\mathbf{y}_i^*(z) \\ &\quad + \quad \mathbf{x}_i(z)\mathbf{y}_i^*(z)\left\{\mathbf{x}_i(z)\mathbf{y}_i^*(z)\right\}' \\ &\quad + \frac{1}{z - z_i}\cdot\frac{1}{p_i(z)}\cdot\mathbf{x}_i(z)\mathbf{y}_i^*(z)\left\{X_i(z)Y_i^*(z)A(z)X_i(z)Y_i^*(z)\right\}' \\ &\quad + X_i(z)\left\{Y_i^*(z)A(z)X_i(z)\right\}^{-1} Y_i^*(z)A'(z). \end{aligned} \tag{10}$$

Note that in the rightmost hand side of Eq. (10), the second, third and fifth terms are analytic and vanish by contour integration. Hence,

$$\begin{aligned} &\frac{1}{2\pi i}\oint_{\Gamma_i^\epsilon} z^p\mathbf{u}^*A(z)^{-1}A'(z)\mathbf{v}\,dz \\ &= \frac{1}{2\pi i}\oint_{\Gamma_i^\epsilon} z^p\mathbf{u}^*\left[\frac{1}{z - z_i}\cdot\mathbf{x}_i(z)\mathbf{y}_i^*(z)\right. \\ &\quad \left. + \frac{1}{z - z_i}\cdot\frac{1}{p_i(z)}\cdot\mathbf{x}_i(z)\mathbf{y}_i^*(z)\left\{X_i(z)Y_i^*(z)A(z)X_i(z)Y_i^*(z)\right\}'\right]\mathbf{v}\,dz \\ &= \nu_i(\mathbf{u}, \mathbf{v})z_i^p, \end{aligned} \tag{11}$$

where

$$\nu_i(\mathbf{u},\mathbf{v}) = \mathbf{u}^* \mathbf{x}_i(z_i)\mathbf{y}_i^*(z_i) \left[ I_n + \frac{1}{p_i(z_i)} \{X_i(z_i)Y_i^*(z_i)A(z_i)X_i(z_i)Y_i^*(z_i)\}' \right] \mathbf{v}. \tag{12}$$

In deriving the last equality of Eq. (11), we used the fact that all the factors in the integrand except $1/(z - z_i)$ is analytic in $\Gamma_i^\epsilon$. $\nu_i(\mathbf{u},\mathbf{v})$ is nonzero for generic $\mathbf{u}$ and $\mathbf{v}$, since $\nu_i(\mathbf{u},\mathbf{v})$ can be written as $\mathbf{u}^* X \mathbf{v}$, where $X$ is a nonzero constant matrix.

Finally, we have

$$\begin{aligned}\mu_p(\mathbf{u},\mathbf{v}) &= \frac{1}{2\pi i}\oint_\Gamma z^p \mathbf{u}^* A(z)^{-1} A'(z)\mathbf{v}\, dz \\ &= \sum_{i=1}^m \frac{1}{2\pi i}\oint_{\Gamma_i^\epsilon} z^p \mathbf{u}^* A(z)^{-1} A'(z)\mathbf{v}\, dz \\ &= \sum_{i=1}^m \nu_i(\mathbf{u},\mathbf{v}) z_i^p. \end{aligned} \tag{13}$$

This completes the proof. □

Note that we could adopt the definition

$$\mu_p(\mathbf{u},\mathbf{v}) = \frac{1}{2\pi i}\oint_\Gamma z^p \mathbf{u}^* A'(z) A(z)^{-1}\mathbf{v}\, dz, \tag{14}$$

instead of Eq. (6) and get the same result, although the expression for $\nu_i(\mathbf{u},\mathbf{v})$ in Eq. (12) is slightly different. So the order of $A(z)^{-1}$ and $A'(z)$ does not actually matter.

Once $\{\mu_p(\mathbf{u},\mathbf{v})\}_{p=0}^{2m-1}$ have been computed, we can extract the information on the nonlinear eigenvalues $\{z_i\}_{i=1}^m$ from them in the same way as in the algorithm for the linear eigenvalue problem [16]. To this end, we first define two Hankel matrices $H_m$ and $H_m^<$ by

$$H_m = \begin{pmatrix} \mu_0 & \mu_1 & \cdots & \mu_{m-1} \\ \mu_1 & \mu_2 & \cdots & \mu_m \\ \vdots & \vdots & \ddots & \vdots \\ \mu_{m-1} & \mu_m & \cdots & \mu_{2m-2} \end{pmatrix}, \quad H_m^< = \begin{pmatrix} \mu_1 & \mu_2 & \cdots & \mu_m \\ \mu_2 & \mu_3 & \cdots & \mu_{m+1} \\ \vdots & \vdots & \ddots & \vdots \\ \mu_m & \mu_{m+1} & \cdots & \mu_{2m-1} \end{pmatrix}. \tag{15}$$

Here we have suppressed the dependence of $\mu_p$ on $\mathbf{u}$ and $\mathbf{v}$ for brevity. The next theorem shows how to compute the nonlinear eigenvalues from $H_m$ and $H_m^<$. This is exactly the same theorem used for the linear eigenvalue problem in [16], but we include the proof for completeness.

**Theorem 2.4** *Assume that $A(z)$ has $m$ simple nonlinear eigenvalues $z_1, z_2, \ldots, z_m$ within $\Gamma$. Assume further that $\nu_i$'s defined by Eq. (12) are nonzero for $i = 1, \ldots, m$. Then, $z_1, z_2, \ldots, z_m$ are given as the $m$ eigenvalues of the matrix pencil $H_m^< - \lambda H_m$ defined by Eq. (15).*

*Proof* Define a Vandermonde matrix $V_m$ and two diagonal matrices $D_m$ and $\Lambda_m$ by

$$V_m = \begin{pmatrix} 1 & 1 & \cdots & 1 \\ z_1 & z_2 & \cdots & z_m \\ \vdots & \vdots & \vdots & \vdots \\ z_1^{m-1} & z_2^{m-1} & \cdots & z_m^{m-1} \end{pmatrix}, \tag{16}$$

$$D_m = \mathrm{diag}(\nu_1, \nu_2, \cdots, \nu_m), \tag{17}$$

$$\Lambda_m = \mathrm{diag}(z_1, z_2, \cdots, z_m). \tag{18}$$

Then it is easy to see that $H_m = V_m D_m V_m^T$ and $H_m^< = V_m D_m \Lambda_m V_m^T$. Since $\nu_i \neq 0$ $(i = 1, \ldots, m)$, $D_m$ is nonsingular. Also, since the $m$ nonlinear eigenvalues are distinct, $V_m$ is nonsingular. Thus we have

$$\begin{aligned} &\lambda \text{ is an eigenvalue of } H_m^< - \lambda H_m. \\ \Leftrightarrow\ & H_m^< - \lambda H_m \text{ is singular.} \\ \Leftrightarrow\ & \Lambda_m - \lambda I_m \text{ is singular.} \\ \Leftrightarrow\ & \exists k, \lambda = z_k. \end{aligned} \tag{19}$$

This completes the proof. □

We can also compute the (nonlinear) eigenvectors corresponding to $z_1, z_2, \ldots, z_m$ by slightly modifying the lemma and the theorem stated above. Let $n$-dimensional vectors $\mathbf{s}_0, \mathbf{s}_1, \ldots, \mathbf{s}_{m-1}$ be defined by

$$\mathbf{s}_p(\mathbf{v}) = \frac{1}{2\pi i} \oint_\Gamma z^p A(z)^{-1} A'(z) \mathbf{v}\, dz \quad (p = 0, 1, \ldots, m-1). \tag{20}$$

Then we have the following lemma.

**Lemma 2.5** *The vector $\mathbf{s}_p$ can be written as*

$$\mathbf{s}_p(\mathbf{v}) = \sum_{i=1}^{m} z_i^p \sigma_i(\mathbf{v}) \mathbf{x}_i(z_i), \tag{21}$$

*where $\{\sigma_i(\mathbf{v})\}_{i=1}^m$ are some complex numbers. Moreover, $\{\sigma_i(\mathbf{v})\}_{i=1}^m$ are nonzero for generic $\mathbf{v}$.*

*Proof* Let $\mathbf{e}_j$ be the $j$-th column of $I_n$. Then we have from Eqs. (20), (13) and (12),

$$\begin{aligned}
\mathbf{s}_p(\mathbf{v}) &= \sum_{j=1}^{n} \mathbf{e}_j \frac{1}{2\pi i} \oint_{\Gamma} z^p \mathbf{e}_j^* A(z)^{-1} A'(z) \mathbf{v}\, dz \\
&= \sum_{j=1}^{n} \mathbf{e}_j \mu_p(\mathbf{e}_j, \mathbf{v}) \\
&= \sum_{j=1}^{n} \sum_{i=1}^{m} \mathbf{e}_j \nu_i(\mathbf{e}_j, \mathbf{v}) z_i^p \\
&= \sum_{i=1}^{m} z_i^p \mathbf{x}_i(z_i) \mathbf{y}_i^*(z_i) \left[ I_n + \frac{1}{p_i(z_i)} \{X_i(z_i) Y_i^*(z_i) A(z_i) X_i(z_i) Y_i^*(z_i)\}' \right] \mathbf{v} \\
&= \sum_{i=1}^{m} z_i^p \sigma_i(\mathbf{v}) \mathbf{x}_i(z_i),
\end{aligned} \tag{22}$$

where

$$\sigma_i(\mathbf{v}) = \mathbf{y}_i^*(z_i) \left[ I_n + \frac{1}{p_i(z_i)} \{X_i(z_i) Y_i^*(z_i) A(z_i) X_i(z_i) Y_i^*(z_i)\}' \right] \mathbf{v}. \tag{23}$$

Apparently, $\sigma_i(\mathbf{v})$ is nonzero for generic $\mathbf{v}$. □

Denote by $\mathbf{w}_i$ the (nonlinear) eigenvector of $A(z)$ corresponding to the eigenvalue $z_i$, that is, $\mathbf{w}_i = \mathbf{x}_i(z_i)$. Then $\mathbf{w}_1, \mathbf{w}_2, \ldots, \mathbf{w}_m$ can be computed as follows.

**Theorem 2.6** *If $\sigma_i \neq 0$ for $i = 1, \ldots, m$, the eigenvectors are given by*

$$[\mathbf{w}_1, \mathbf{w}_2, \ldots, \mathbf{w}_m] = [\mathbf{s}_0, \mathbf{s}_1, \ldots, \mathbf{s}_{m-1}]\, V_m^{-T}. \tag{24}$$

*Proof* From Lemma 2.5, $[\mathbf{s}_0, \mathbf{s}_1, \ldots, \mathbf{s}_{m-1}]$ can be written as

$$\begin{aligned}
&[\mathbf{s}_0, \mathbf{s}_1, \ldots, \mathbf{s}_{m-1}] \\
&= \left[ \sum_{i=1}^{m} z_i^0 \sigma_i(\mathbf{v}) \mathbf{x}_i(z_i), \sum_{i=1}^{m} z_i^1 \sigma_i(\mathbf{v}) \mathbf{x}_i(z_i), \ldots, \sum_{i=1}^{m} z_i^{m-1} \sigma_i(\mathbf{v}) \mathbf{x}_i(z_i) \right] \\
&= [\sigma_1 \mathbf{x}_1(z_1), \sigma_2 \mathbf{x}_2(z_2), \ldots, \sigma_m \mathbf{x}_m(z_m)]\, V_m^T.
\end{aligned} \tag{25}$$

Hence,

$$[\sigma_1 \mathbf{x}_1(z_1), \sigma_2 \mathbf{x}_2(z_2), \ldots, \sigma_m \mathbf{x}_m(z_m)] = [\mathbf{s}_0, \mathbf{s}_1, \ldots, \mathbf{s}_{m-1}]\, V_m^{-T}. \tag{26}$$

The theorem follows by noting that if $\sigma_i \neq 0$, $\sigma_i \mathbf{x}_i(z_i)$ is a nonzero vector that satisfies $A(z_i)\mathbf{x}_i(z_i) = \lambda_i(z_i)\mathbf{x}_i(z_i) = \mathbf{0}$ and is itself a nonlinear eigenvector corresponding to $z_i$. □

# 3 The Algorithm

In this section, we present an algorithm for computing the nonlinear eigenvalues of $A(z)$ that lie within $\Gamma$ based on the theory developed in the previous section. For simplicity, we restrict ourselves to the case where $\Gamma$ is a circle centered at the origin and with radius $r$.

In the algorithm, we need to approximate the contour integrals in Eqs. (6) and (20) with some quadrature. Since they are integrals of an analytic function over the entire period, we use the trapezoidal rule [19, 20], which converges exponentially and therefore is an excellent method for the task. When the number of sample points is $K$, Eqs. (6) and (20) become

$$\mu_p(\mathbf{u}, \mathbf{v}) = \frac{r^{p+1}}{K} \sum_{j=0}^{K-1} \omega_K^{(p+1)j} \mathbf{u}^* A(r\omega_K^j)^{-1} A'(r\omega_K^j)\mathbf{v}, \tag{27}$$

$$\mathbf{s}_p(\mathbf{v}) = \frac{r^{p+1}}{K} \sum_{j=0}^{K-1} \omega_K^{(p+1)j} A(r\omega_K^j)^{-1} A'(r\omega_K^j)\mathbf{v}, \tag{28}$$

respectively, where $\omega_K = \exp\left(\frac{2\pi i}{K}\right)$.

Using these expressions, the algorithm can be written as in Algorithm 1.

[Algorithm1: Finding the eigenvalues in $\Gamma$ and corresponding eigenvectors]
⟨1⟩ Input $n, m, r, K$, $\mathbf{u}$ and $\mathbf{v}$ĄD
⟨2⟩ $\omega_K = \exp\left(\frac{2\pi i}{K}\right)$
⟨3⟩ **for** $j = 0, 1, \ldots, K-1$
⟨4⟩ $\xi_j = r\omega_K^j$
⟨5⟩ $\mathbf{t}_j = A(\xi_j)^{-1} A'(\xi_j)\mathbf{v}$
⟨6⟩ **end for**
⟨7⟩ **for** $p = 0, 1, \ldots, 2m-1$
⟨8⟩ $\mathbf{s}_p = \frac{r^{p+1}}{K} \sum_{j=0}^{K-1} \omega_K^{(p+1)j} \mathbf{t}_j$
⟨9⟩ $\mu_p = \mathbf{u}^* \mathbf{s}_p$
⟨10⟩ **end for**
⟨11⟩ Construct $H_m$ and $H_m^<$ from $\mu_0, \mu_1, \ldots, \mu_{2m-1}$.
⟨12⟩ Find the eigenvalues $z_1, z_2, \ldots, z_m$ of $H_m^< - \lambda H_m$.
⟨13⟩ Compute $[\mathbf{w}_1, \mathbf{w}_2, \ldots, \mathbf{w}_m] = [\mathbf{s}_0, \mathbf{s}_1, \ldots, \mathbf{s}_{m-1}]\, V_m^{-T}$
using the matrix $V_m$ defined by Eq. (16).
⟨14⟩ Output $z_1, z_2, \ldots, z_m$ and $\mathbf{w}_1, \mathbf{w}, \ldots, \mathbf{w}_m$.

Concerning the use of Algorithm 1, several remarks are in order.

1. In this algorithm, the computationally dominant part is step 5, where the solution of linear equations with coefficient matrix $A(\xi_j)$ for $j = 0, 1, \ldots, K-1$ is needed. This operation is repeated for $K$ different values of $\xi_j$. However, as is clear from the algorithm, these $K$ operations can be done completely in parallel. Thus the algorithm has large-grain parallelism.
2. In step 13, since $V_m$ is a Vandermonde matrix, multiplying $V_m^{-T}$ can be done using a specialized solver for Vandermonde systems [7]. This is faster and more accurate than first constructing $V_m$ explicitly and then using a general-purpose solver such as the Gaussian elimination.
3. Though this algorithm presupposes that $m$, the number of eigenvalues in $\Gamma$, is known in advance, this is often not the case. When $m$ is unknown, we can choose some integer $M$, which hopefully satisfies $M \geq m$, run the algorithm by replacing $m$ with $M$, and compute $\{\nu_i\}_{i=1}^{M}$ by $\nu_i = \mathbf{e}_i^T V_M^{-1} H_M V_M^{-T} \mathbf{e}_i$. In this case, $M - m$ of $\{z_1, z_2, \ldots, z_m\}$ are *spurious eigenvalues* that do not correspond to the nonlinear eigenvalues of $A(z)$ in $\Gamma$. These spurious eigenvalues can be distinguished from the true ones since the corresponding $|\nu_i|$'s are very small. This technique was proposed in [17] for the (linear) generalized eigenvalue problem and its detailed analysis is given in [1]. There is also a technique to determine $m$ using the singular value decomposition of $H_M$. See [8] for details.

## 4 Numerical Examples

In this section, we give numerical examples of our Algorithm 1. The experiments were performed on a PC with a Xeon processor and Red Hat Linux using the Gnu C++ compiler. We used LAPACK routines to solve the linear simultaneous equation with coefficient matrix $A(\xi_j)$ and to find the eigenvalues of the matrix pencil $H_m^{<} - \lambda H_m$.

*Example 1* Our first example is a small symmetric quadratic eigenvalue problem taken from [18]:

$$A(z) = \begin{bmatrix} -10\lambda^2+\lambda+10 & & & & \\ 2\lambda^2+2\lambda+2 & -11\lambda^2+\lambda+9 & & \text{sym.} & \\ -\lambda^2+\lambda-1 & 2\lambda^2+2\lambda+3 & -12\lambda^2+10 & & \\ \lambda^2+2\lambda+2 & -2\lambda^2+\lambda-1 & -\lambda^2-2\lambda+2 & -10\lambda^2+2\lambda+12 & \\ 3\lambda^2+\lambda-2 & -\lambda^2+3\lambda-2 & \lambda^2-2\lambda-1 & 2\lambda^2+3\lambda+1 & -11\lambda^2+3\lambda+10 \end{bmatrix}. \tag{29}$$

This problem has ten distinct eigenvalues and their values are (to three decimals) [13]:

$$\begin{array}{ccccc} -1.27 & -1.08 & -1.0048 & -0.779 & -0.512 \\ 0.502 & 0.880 & 0.937 & 1.47 & 1.96. \end{array} \tag{30}$$

**Table 1** Computed eigenvalues, their residuals and the values of $\nu_i$ for Example 1

| $i$ | Eigenvalue $z_i$ | Residual | $\nu_i$ |
|---|---|---|---|
| 1 | $+8.14126840 + 3.48672318i$ | $6.95 \times 10^{-1}$ | $+0.00000000 + 0.00000000i$ |
| 2 | $\mathbf{-0.51176193 - 0.00000000i}$ | $4.62 \times 10^{-11}$ | $+0.10296094 - 0.15679519i$ |
| 3 | $-1.07716760 + 0.00000072i$ | $2.23 \times 10^{-6}$ | $-0.00000929 - 0.00000217i$ |
| 4 | $-1.00483822 + 0.00000000i$ | $7.37 \times 10^{-11}$ | $-1.22943061 + 0.25648078i$ |
| 5 | $\mathbf{-0.77909458 - 0.00000000i}$ | $5.33 \times 10^{-11}$ | $+1.65030210 + 0.71552727i$ |
| 6 | $\mathbf{+0.50241527 + 0.00000000i}$ | $4.54 \times 10^{-12}$ | $-0.03125232 + 0.34335409i$ |
| 7 | $\mathbf{+0.87992728 + 0.00000000i}$ | $3.23 \times 10^{-11}$ | $-0.06281529 - 0.23562418i$ |
| 8 | $\mathbf{+0.93655066 - 0.00000000i}$ | $5.81 \times 10^{-11}$ | $+0.00985445 + 0.10564969i$ |

Five eigenvalues of $A(z)$ are given in bold

We applied our method with $r = 1.0$ to find the eigenvalues in the unit disk with center at the origin. There are five eigenvalues of $A(z)$ in this circle. We set $M = 8$ (see item (iii) of the previous subsection) and $K = 128$. The computed eigenvalues $z_i$ of $H_M^{<} - \lambda H_M$ is shown in Table 1, along with the residual of the computed eigenvectors $\mathbf{w}_i$ and the values of $\nu_i$. Here the residual is defined by $\| A(z_i)\mathbf{w}_i \| / (\| A(z_i) \|_\infty \| \mathbf{w}_i \|)$.

Among the eight computed eigenvalues, $z_2$ and $z_5$ through $z_8$ are inside the circle and have relatively large value of $|\nu_i|$. Thus we know that they are wanted eigenvalues. In fact, they have small residuals of order $10^{-11}$. Hence we can say that we have succeeded in finding all the five eigenvalues in the circle and the corresponding eigenvectors with high accuracy.

On the other hand, $z_4$ is located outside the circle and $z_1$ and $z_3$ have small value of $|\nu_i|$. This shows that they are either unwanted or spurious eigenvalues. Among these three eigenvalues, $z_4$ has a large value of $|\nu_i|$ and its residual is as small as that for the inner eigenvalues. Thus it seems that this is a true outer eigenvalue that has been computed accurately. This occurs because the effect of the poles of $\mathbf{u}^* A(z)^{-1} A'(z)\mathbf{v}$ just outside the circle remains due to numerical integration. This phenomenon occurs also in the algorithm using $\mathrm{Tr}(A(z)^{-1}A'(z))$ and is analyzed in [1] in detail.

*Example 2* Our next example is a medium size problem whose elements have both linear and exponential dependence on $z$. Specifically,

$$A(z) = A - zI_n + \epsilon B(z), \tag{31}$$

where $A$ is a real nonsymmetric matrix whose elements follow uniform random distribution in $[0, 1]$, $B(z)$ is an anti-diagonal matrix with antidiagonal elements $e^z$ and $\epsilon$ is a parameter that determines the degree of nonlinearity. This test matrix is used in [1]. In the present example, $n = 500$ and we applied our method with $r = 0.7$. It is known that there are ten eigenvalues in the circle. We set $M = 12$ and $K = 128$. The result are shown in Table 2.

**Table 2** Computed eigenvalues, their residuals and the values of $\nu_i$ for Example 2

| $i$ | Eigenvalue $z_i$ | Residual | $\nu_i$ |
|---|---|---|---|
| 1 | **+0.18966905 + 0.63706191i** | $2.13 \times 10^{-14}$ | $+1.17984929 - 0.22981043i$ |
| 2 | $-0.43247175 - 0.71100593i$ | $3.68 \times 10^{-6}$ | $+0.00000000 - 0.00000000i$ |
| 3 | **−0.51507157 − 0.45900079i** | $4.72 \times 10^{-15}$ | $-0.98601173 - 6.36155661i$ |
| 4 | **+0.18966905 − 0.63706191i** | $8.81 \times 10^{-15}$ | $-1.41428395 - 1.01002310i$ |
| 5 | **+0.59154350 − 0.25937027i** | $8.23 \times 10^{-15}$ | $+5.09208563 - 0.07378654i$ |
| 6 | **+0.59154350 + 0.25937027i** | $1.49 \times 10^{-14}$ | $-4.13045809 - 4.87142384i$ |
| 7 | **+0.33336324 − 0.18217042i** | $4.14 \times 10^{-14}$ | $-8.04725134 + 0.10007238i$ |
| 8 | **+0.33336324 + 0.18217042i** | $6.92 \times 10^{-14}$ | $+6.45071565 + 3.77328737i$ |
| 9 | **−0.54261232 − 0.00000000i** | $1.52 \times 10^{-14}$ | $-1.37030828 + 1.66944385i$ |
| 10 | **−0.08820357 − 0.00000000i** | $1.78 \times 10^{-13}$ | $-3.03924483 + 3.67246714i$ |
| 11 | $-0.43248417 + 0.71102419i$ | $6.22 \times 10^{-5}$ | $-0.00000000 - 0.00000000i$ |
| 12 | **−0.51507157 + 0.45900079i** | $8.56 \times 10^{-15}$ | $+1.97883460 + 0.79241891i$ |

Ten eigenvalues in the circle are given in bold

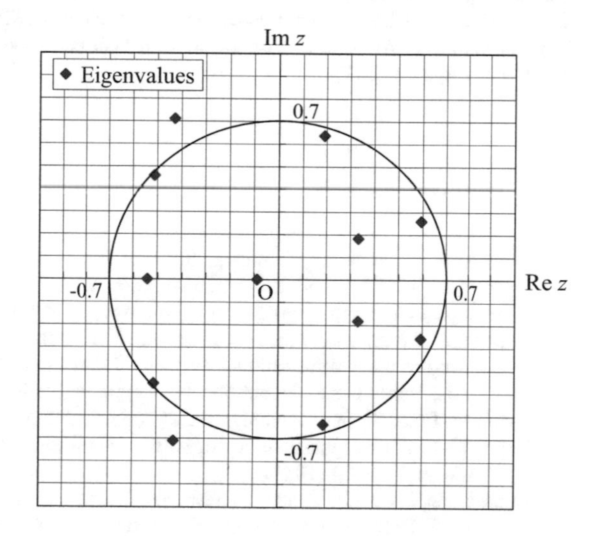


**Fig. 1** Distribution of the eigenvalues in Example 2

Among the twelve computed eigenvalues, the ten eigenvalues except for $z_2$ and $z_{11}$ are inside the circle and have large value of $\nu_i$. Accordingly, these are judged to be the wanted eigenvalues. This is confirmed by the fact that the corresponding residuals are all of order $10^{-13}$. Hence we can conclude that our algorithm again succeeded in finding all the wanted eigenvalues in this example. The computed eigenvalues in the complex plane are shown in Fig. 1.

# 5 Conclusion

In this paper, we presented an alternative derivation of the SS-type method for the nonlinear eigenvalue problem. We assumed that all the eigenvalues in the specified region are simple and considered contour integrals along infinitesimally small circles around the eigenvalues. This allowed us to use the analyticity of the eigenvalues and eigenvectors of a parametrized matrix $A(z)$, which is a well-known result in matrix perturbation theory, instead of the canonical forms of $A(z)$ described by the Smith form or the theorem of Keldysh. We believe this will provide an easily accessible approach to the theory of the nonlinear SS-type method.

**Acknowledgements** We thank Professor Masaaki Sugihara for valuable comments. We are also grateful to the anonymous referees, whose suggestions helped us much in improving the quality of this paper. Prof. Akira Imakura brought reference [8] to our attention. This study is supported in part by the Ministry of Education, Science, Sports and Culture, Grant-in-Aid for Scientific Research (Nos. 26286087, 15H02708, 15H02709, 16KT0016, 17H02828, 17K19966) and the Core Research for Evolutional Science and Technology (CREST) Program "Highly Productive, High Performance Application Frameworks for Post Petascale Computing" of Japan Science and Technology Agency (JST).

# References

1. Amako, T., Yamamoto, Y., Zhang, S.-L.: A large-grained parallel algorithm for nonlinear eigenvalue problems and its implementation using OmniRPC. In: Proceedings of IEEE International Conference on Cluster Computing, 2008, pp. 42–49. IEEE Press (2008)
2. Asakura, J., Sakurai, T., Tadano, H., Ikegami, T., Kimura, K.: A numerical method for nonlinear eigenvalue problems using contour integrals. JSIAM Lett. **1**, 52–55 (2009)
3. Betcke, T., Voss, H.: A Jacobi-Davidson-type projection method for nonlinear eigenvalue problems. Futur. Gener. Comput. Syst. **20**, 363–372 (2004)
4. Beyn, W.-J.: An integral method for solving nonlinear eigenvalue problems. Linear Algebra Appl. **436**, 3839–3863 (2012)
5. Gantmacher, F.R.: The Theory of Matrices. Chelsea, New York (1959)
6. Gohberg, I., Rodman, L.: Analytic matrix functions with prescribed local data. J. d'Analyse Mathématique **40**, 90–128 (1981)
7. Golub, G.H., Van Loan, C.F.: Matrix Computations, 3rd edn. Johns Hopkins University Press, Baltimore (1996)
8. Ikegami, T., Sakurai, T., Nagashima, U.: A filter diagonalization for generalized eigenvalue problems based on the Sakurai-Sugiura projection method. J. Comput. Appl. Math. **233**, 1927–1936 (2010)
9. Kato, T.: Perturbation Theory for Linear Operators, 2nd edn. Springer, Berlin (1976)
10. Kato, T.: A Short Introduction to Perturbation Theory for Linear Operators. Springer, New York (1982)
11. Keldysh, M.V.: On the characteristic values and characteristic functions of certain classes of non-self-adjoint equations. Doklady Akad. Nauk SSSR (N. S.) **77**, 11–14 (1951)
12. Keldysh, M.V.: The completeness of eigenfunctions of certain classes of nonselfadjoint linear operators. Uspehi Mat. Nauk **26**(4(160)), 15–41 (1971)
13. Neumaier, A.: Residual inverse iteration for the nonlinear eigenvalue problem. SIAM J. Numer. Anal. **22**, 914–923 (1985)

14. Ruhe, A.: Algorithms for the nonlinear eigenvalue problem. SIAM J. Numer. Anal. **10**, 674–689 (1973)
15. Saad, Y.: Numerical Methods for Large Eigenvalue Problems. Halsted Press, New York (1992)
16. Sakurai, T., Sugiura, H.: A projection method for generalized eigenvalue problems using numerical integration. J. Comput. Appl. Math. **159**, 119–128 (2003)
17. Sakurai, T., Tadano, H., Inadomi, Y., Nagashima, U.: A moment-based method for large-scale generalized eigenvalue problems. Appl. Numer. Anal. Comput. Math. **1**, 516–523 (2004)
18. Scott, D.S., Ward, R.C.: Solving symmetric-definite quadratic problems without factorization. SIAM J. Sci. Stat. Comput. **3**, 58–67 (1982)
19. Sloan, I.H., Joe, S.: Lattice Methods for Multiple Integration. Oxford University Press, New York (1994)
20. Trefethen, L.N., Weideman, J.A.C.: The exponentially convergent trapezoidal rule. SIAM Rev. **56**, 385–458 (2014)
21. Voss, H.: An Arnoldi method for nonlinear eigenvalue problems. BIT Numer. Math. **44**, 387–401 (2004)
22. Yokota, S., Sakurai, T.: A projection method for nonlinear eigenvalue problems using contour integrals. JSIAM Lett. **5**, 41–44 (2013)

# Fast Multipole Method as a Matrix-Free Hierarchical Low-Rank Approximation

**Rio Yokota, Huda Ibeid, and David Keyes**

**Abstract** There has been a large increase in the amount of work on hierarchical low-rank approximation methods, where the interest is shared by multiple communities that previously did not intersect. This objective of this article is two-fold; to provide a thorough review of the recent advancements in this field from both analytical and algebraic perspectives, and to present a comparative benchmark of two highly optimized implementations of contrasting methods for some simple yet representative test cases. The first half of this paper has the form of a survey paper, to achieve the former objective. We categorize the recent advances in this field from the perspective of compute-memory tradeoff, which has not been considered in much detail in this area. Benchmark tests reveal that there is a large difference in the memory consumption and performance between the different methods.

## 1 Introduction

The fast multipole method (FMM) was originally developed as an algorithm to bring down the $\mathcal{O}(N^2)$ complexity of the direct $N$-body problem to $\mathcal{O}(N)$ by approximating the hierarchically decomposed far field with multipole/local expansions. In its original form, the applicability of FMM is limited to problems that have a Green's function solution, for which the multipole/local expansions can be calculated analytically. Their function is also limited to matrix-vector multiplications, in contrast to the algebraic variants that can perform matrix-matrix multiplication and factorizations. However, these restrictions no longer apply to the FMM since the kernel independent FMM [103] does not require a Green's function, and inverse FMM [2] can be used as the inverse operator instead of the forward mat-vec. Therefore the FMM can be used for a wide range of scientific applications,

R. Yokota (✉)
Tokyo Institute of Technology, 2-12-1 O-okayama Meguro-ku, Tokyo, Japan
e-mail: rioyokota@gsic.titech.ac.jp

H. Ibeid • D. Keyes
King Abdullah University of Science and Technology, 4700 KAUST, Thuwal, Saudi Arabia
e-mail: huda.ibeid@kaust.edu.sa; david.keyes@kaust.edu.sa



T. Sakurai et al. (eds.), *Eigenvalue Problems: Algorithms, Software and Applications in Petascale Computing*, Lecture Notes in Computational Science and Engineering 117, https://doi.org/10.1007/978-3-319-62426-6_17

which can be broadly classified into elliptic partial differential equations (PDE) and kernel summation. Integral form of elliptic PDEs can be further categorized into boundary integrals for homogeneous problems, discrete volume integrals, and continuous volume integrals.

Scientific applications of FMM for boundary integrals include acoustics [59, 97], biomolecular electrostatics [105], electromagnetics [33, 42], fluid dynamics for Euler [96] and Stokes [88] flows, geomechanics [92], and seismology [22, 95]. Application areas of FMM for discrete volume integrals are astrophysics [14], Brownian dynamics [75], classical molecular dynamics [84], density functional theory [90], vortex dynamics [106], and force directed graph layout [107]. FMM for continuous volume integrals have been used to solve Schrödinger [108] and Stokes [79] equations. More generalized forms of FMM can be used as fast kernel summation for Bayesian inversion [3], Kalman filtering [74], Machine learning [49, 72], and radial basis function interpolation [54].

All of these applications have in common the key feature that they are global problems where the calculation at every location depends on the values everywhere else. Elliptic PDEs that represent a state of equilibrium, many iterations with global inner products for their solution, dense matrices in boundary integral problems, all-to-all interaction in $N$-body problems, and kernel summations with global support are all different manifestations of the same source of global data dependency. Due to this global data dependency, their concurrent execution on future computer architectures with heterogeneous and deep memory hierarchy is one of the main challenges of exascale computing. For global problems that require uniform resolution, FFT is often the method of choice, despite its suboptimal communication costs. The methods we describe here have an advantage for global problems that require non-uniform resolution. For such non-uniform global problems multigrid methods are known to do quite well. Whether the reduced synchronization and increased arithmetic intensity of the FMM will become advantageous compared to multigrid on future architectures is something that is yet to be determined.

Many of the original FMM researchers have now moved on to develop algebraic variants of FMM, such as $\mathscr{H}$-matrix [55], $\mathscr{H}^2$-matrix [57], hierarchically semi-separable (HSS) [26], hierarchically block-separable (HBS) [82], and hierarchically off-diagonal low-rank (HODLR) [1] matrices. The differences between these methods are concisely summarized by Ambikasaran and Darve [2]. These algebraic generalizations of the FMM can perform addition, multiplication, and even factorization of dense matrices with near linear complexity. This transition from analytic to algebraic did not happen suddenly, and semi-analytic variants were developed along the way [39, 103]. Optimization techniques for the FMM such as compressed translation operators and their precomputation, also fall somewhere between the analytic and algebraic extremes.

The spectrum that spans purely analytic and purely algebraic forms of these hierarchical low-rank approximation methods, represents the tradeoff between computation (Flops) and memory (Bytes). The purely analytic FMM is a matrix-free $\mathscr{H}^2$-matrix-vector product, and due to its matrix-free nature it has very high arithmetic intensity (Flop/Byte) [9]. On the other end we have the purely algebraic methods, which precompute and store the entire hierarchical matrix. This results in more storage and more data movement, both vertically and horizontally in the memory hierarchy. When the cost of data movement increases faster than arithmetic operations on future architectures, the methods that compute more to store/move less will become advantageous. Therefore, it is important to consider the whole spectrum of hierarchical low-rank approximation methods, and choose the appropriate method for a given pair of application and architecture.

There have been few attempts to quantitatively investigate the tradeoff between the analytic and algebraic hierarchical low-rank approximation methods. Previously, the applicability of the analytic variants were limited to problems with Green's functions, and could only be used for matrix-vector products but not to solve the matrix. With the advent of the kernel-independent FMM (KIFMM) [103] and inverse FMM (IFMM) [2], these restrictions no longer apply to the analytic variants. Furthermore, the common argument for using the algebraic variants because they can operate directly on the matrix without the need to pass geometric information is not very convincing. Major libraries like PETSc offer interfaces to insert one's own matrix free preconditioner as a function, and passing geometric information is something that users are willing to do if the result is increased performance. Therefore, there is no strong reason from the user's perspective to be monolithically inclined to use the algebraic variants. It is rather a matter of choosing the method with the right balance between its analytic (Flops) and algebraic (Bytes) features.

The topic of investigating the tradeoff between analytic and algebraic hierarchical low-rank approximation methods is too broad to cover in a page-constrained article. In the present work, we limit our investigation to the compute-memory tradeoff in a comparison between FMM and HSS for Laplace and Helmholtz kernels. We also investigate the use of FMM as a preconditioner for iterative solutions to the Laplace and Helmholtz problems with finite elements, for which we compare with geometric and algebraic multigrid methods.

## 2 Hierarchical Low-Rank Approximation: Analytic or Algebraic?

In this section we review the full spectrum of hierarchical low-rank approximations starting from the analytic side and proceeding to the algebraic side. The spectrum is depicted in Fig. 1, where various techniques like between the analytic and algebraic extremes. One can choose the appropriate method for a given architecture to achieve the best performance.

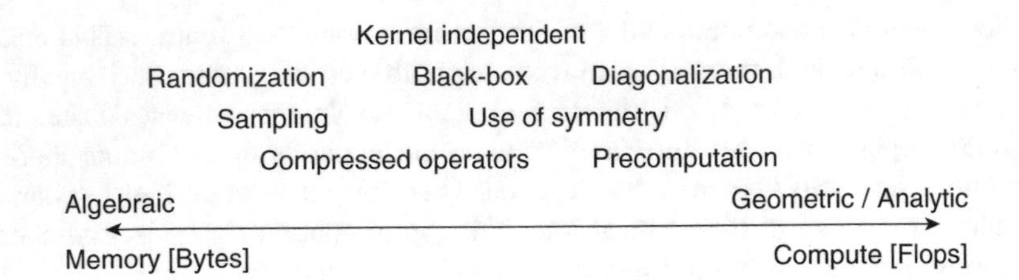


**Fig. 1** The compute-memory tradeoff between the analytic and algebraic hierarchical low-rank approximation methods. Various techniques lie between the analytic and algebraic extremes

## *2.1 Analytic Low-Rank Approximation*

On the analytic end of the spectrum, we have classical methods such as the Treecode [10], FMM [8, 50], and panel clustering methods [56]. These methods have extremely high arithmetic intensity (Flop/Byte) due to their matrix-free nature, and are compute-bound on most modern architectures. One important fact is that these are not brute force methods that do unnecessary Flops, but are (near) linear complexity methods that are only doing useful Flops, but they are still able to remain compute-bound. This is very different from achieving high Flops counts on dense matrix-matrix multiplication or LU decomposition that have $\mathscr{O}(N^3)$ complexity. The methods we describe in this section can approximate the same dense linear algebra calculation in $\mathscr{O}(N)$ or $\mathscr{O}(N \log N)$ time.

As an example of the absolute performance of the analytic variants, we refer to the Treecode implementation—`Bonsai`, which scales to the full node of Titan using 18,600 GPUs achieving 24.77 PFlops [14]. Bonsai's performance comes not only from its matrix-free nature, but also from domain specific optimizations for hardcoded quadrupoles and an assumption that all charges are positive. Therefore, this kind of performance cannot be transferred to other applications that require higher accuracy. However, viewing these methods as a preconditioner instead of a direct solver significantly reduces the accuracy requirements [5, 67].

## *2.2 Fast Translation Operators*

A large part of the calculation time of FMM is spent on the translation of multipole expansions to local expansions (or their equivalent charges). Therefore, much work has focused on developing fast translation operators to accelerate this part of the FMM. Rotation of spherical harmonics [94], Block FFT [37], Planewaves [51] are analytic options for fast translation operators.

These translation operators are applied to a pair of boxes in the FMM tree structure that satisfy a certain proximity threshold. This proximity is usually defined as the parent's neighbors' children that are non-neighbors. This produces a list of boxes that are far enough that the multipole/local expansion converges, but are close enough that the expansion does not converge for the their parents. Such an interaction list can contain up to $6^3 - 3^3 = 189$ source boxes for each target box. Out of these 189 boxes, the ones that are further from the target box can perform the translation operation using their parent box as the source without loss of accuracy. There are a few variants for these techniques that reduce the interaction list size such as the level-skip M2L method [93] and 8, 4, 2-box method [95]. There are also methods that use the dual tree traversal along with the multipole acceptance criterion to construct optimal interaction lists [35], which automates the process of finding the optimal interaction list size.

Another technique to accelerate the translation operators is the use of variable expansion order, as proposed in the very fast multipole method (VFMM) [87], Gaussian VFMM [21], optimal parameter FMM [29], and error controlled FMM [32]. There are two main reasons why spatially varying the expansion order in the translation operators is beneficial. One is because not all boxes in the interaction list are of equal distance, and the boxes that are further from each other can afford to use lower expansion order, while retaining the accuracy. The other reason is because some parts of the domain may have smaller values, and the contribution from that part can afford to use lower expansion order without sacrificing the overall accuracy.

The translation operators can be stored as matrices that operate on the vector of expansion coefficients. Therefore, singular value decomposition (SVD) can be used to compress this matrix [43] and BLAS can be used to maximize the cache utilization [40]. Some methods use a combination of these techniques like Chebychev with SVD [39] and planewave with adaptive cross approximation (ACA) and SVD [61]. The use of SVD is a systematic and optimal way of achieving what the variable expansion order techniques in the previous paragraph were trying to do manually. Precomputing these translation matrices and storing them is a typical optimization technique in many FMM implementations [78].

One important connection to make here is that these matrices for the translation operators are precisely what $\mathscr{H}^2$-matrices and *HSS* matrices store in the off-diagonal blocks after compression. One can think of FMM as a method that has the analytical form to generate these small matrices in the off-diagonal blocks, without relying on numerical low-rank approximation methods. To complete this analogy, we point out that the dense diagonal blocks in $\mathscr{H}^2$-matrices and *HSS* matrices are simply storing the direct operator (Green's function) in FMM. Noticing this equivalence leads to many possibilities of hybridization among the analytic and algebraic variants. Possibly the most profound is the following. Those that are familiar with FMM know that translation operators for boxes with the same relative positioning are identical. This suggests that many of the entries in the off-diagonal blocks of $\mathscr{H}^2$-matrices and *HSS* matrices are identical. For matrices that are generated from a mesh that has a regular structure even the diagonal blocks would be identical, which is what happens in FMMs for continuous volume integrals [78]. This leads to $\mathscr{O}(1)$ storage for the matrix entries at every level of the hierarchy, so the total storage cost of these hierarchical matrices could be reduced to $\mathscr{O}(\log N)$ if the identical entires are not stored redundantly. This aspect is currently underutilized in the algebraic variants, but seems obvious from the analytic side. By making use of the translational invariance and rotational symmetry of the interaction list one can reduce the amount of storage even further [31, 34, 91]. This also results in blocking techniques for better cache utilization.

## 2.3 Semi-analytical FMM

The methods described in the previous subsection all require the existence of an analytical form of the multipole/local translation operator, which is kernel dependent. There are a class of methods that remove this restriction by using equivalent charges instead of multipole expansions [7, 15, 77]. A well known implementation of this method is the kernel independent FMM (KIFMM) code [103]. There are also variants that use Chebychev polynomials [36], and a representative implementation of this is the Black-box FMM [39]. As the name of these codes suggest, these variants of the FMM have reduced requirements for the information that has to be provided by the user. The translation operators are kernel-independent, which frees the user from the most difficult task of having to provide an analytical form of the translation operators. For example, if one wants to calculate the Matérn function for covariance matrices, or multiquadrics for radial basis function interpolation, one simply needs to provide these functions and the location of the points and the FMM will handle the rest. It is important to note that these methods are not entirely kernel independent or black-box because the user still needs to provide the kernel dependent analytic form of the original equation they wish to calculate. Using the vocabulary of the algebraic variants, one could say that these analytical expressions for the hierarchical matrices are kernel independent only for the off-diagonal blocks, and for the diagonal blocks the analytical form is kernel dependent.

FMM for continuous volume integrals [38] also has important features when considering the analytic-algebraic tradeoff. The volume integrals are often combined with boundary integrals, as well [104]. One can think of these methods as an FMM that includes the discretization process [70]. Unlike the FMM for discrete particles, these methods have the ability to impose regular underlying geometry. This enables the use of precomputation of the direct interaction matrix in the analytic variants [78], and reduces the storage requirements of the dense diagonal blocks in the algebraic variants.

## 2.4 Algebraic Low-Rank Approximation

There are many variants of algebraic low-rank approximation methods. They can be categorized based on whether they are hierarchical, whether they use weak admissibility, or if the basis is nested, as shown in Table 1. For the definition of admissibility see [45]. Starting from the top, $\mathcal{H}$-matrices [12, 55] are hierarchical, usually use

**Table 1** Categorization of algebraic low-rank approximation methods

| Method | Hierarchical | Weak admissibility | Nested basis |
|---|---|---|---|
| $\mathscr{H}$-matrix [55] | Yes | Maybe | No |
| $\mathscr{H}^2$-matrix [57] | Yes | Maybe | Yes |
| HODLR [1] | Yes | Yes | No |
| HSS [26]/HBS [82] | Yes | Yes | Yes |
| BLR [4] | No | Yes | No |

standard or strong admissibility, and no nested basis. The analytic counterpart of the $\mathscr{H}$-matrix is the Treecode. The $\mathscr{H}^2$-matrices [16, 57] are also hierarchical and use standard or strong admissibility, but unlike $\mathscr{H}$-matrices use a nested basis. This brings the complexity down from $\mathscr{O}(NlogN)$ to $\mathscr{O}(N)$. The analytic counterpart of the $\mathscr{H}^2$-matrix is the FMM. The next three entries in Table 1 do not have analytic counterparts because analytic low-rank approximations do not converge under weak admissibility conditions. Hierarchical off-diagonal low-rank (HODLR) matrices [1, 6], are basically $\mathscr{H}$-matrices with weak admissibility conditions. Similarly, hierarchically semi-separable (HSS) [26, 101], and hierarchically block-separable (HBS) [82] matrices are $\mathscr{H}^2$-matrices with weak admissibility conditions. The block low-rank (BLR) matrices [4] are a non-hierarchical version of the HODLR, with just the bottom level. A summary of implementations and their characteristics are presented in [89].

For methods that do not have weak admissibility, it is common to use geometrical information to calculate the standard/strong admissibility condition. This dependence on the geometry of the algebraic variants is not ideal. There have been various proposals for algebraic clustering methods [46, 71, 85]. This problem requires even more advanced solutions for high dimension problems [80]. Stronger admissibility is also problem for parallelization since it results in more communication. There have been studies on how to partition hierarchical matrices on distributed memory [68]. There are also methods to reduce the amount of memory consumption during the construction of HSS matrices [73].

The categorization in Table 1 is for the hierarchical matrix structure, and any low-rank approximation method can be used with each of them during the compression phase. The singular value decomposition is the most naïve and expensive way to calculate a low-rank approximation. QR or LU decompositions can be used to find the numerical rank by using appropriate pivoting. Rank-revealing QR [24] has been proposed along with efficient pivoting strategies [25, 53, 64]. Rank-revealing LU [23] also requires efficient pivoting strategies [65, 66, 83]. Rank-revealing LU is typically faster than rank-revealing QR [86]. There are other methods like the pseudo-skeletal method [44] and adaptive cross approximation (ACA) [11, 13], which do not yield the optimal low-rank factorizations but have a much lower cost. ACA has a better pivoting strategy than pseudo-skeletal methods, but can still fail because of bad pivots [18]. The hybrid cross approximation (HCA) [17] has the same proven convergence as standard interpolation but also the same efficiency as ACA. Yet another class of low-rank approximation is the interpolative decomposition (ID) [28, 82], where a few of its columns are used to form a well-conditioned basis for the remaining columns. ID can be combined with randomized methods [76], which has much lower complexity. For a nice review on these randomized methods see [58].

# 3 Low-Rank Approximation for Factorization

## *3.1 Sparse Matrix Factorization*

Hierarchical low-rank approximation methods can be used as direct solvers with controllable accuracy. This makes them useful as preconditioners within a Krylov subspace method, which in turn reduces the accuracy requirements of the low-rank approximation. High accuracy and completely algebraic methods are demanding in terms of memory consumption and amount of communication, so they are unlikely to be the optimal choice unless they are the only solution to that problem.

There are two ways to use hierarchical low-rank approximations for factorization of a sparse matrix. The first way is to perform the LU decomposition on the sparse matrix, and use hierarchical low-rank approximations for the dense blocks that appear during the process [98, 100, 101]. The other way is to represent the sparse matrix with a hierarchical low-rank approximation and perform an LU decomposition on it [46–48]. The main difference is whether you view the base method as the nested dissection and the additional component as HLRA or vice versa. The former has the advantage of being able to leverage the existing highly optimized sparse direct solvers, whereas the latter has the advantage of handling both sparse and dense matrices with the same infrastructure.

There are various ways to minimize the fill-in and compress the dense blocks during factorization. These dense blocks (Schur complements) are an algebraic form of the Green's function [99], and have the same low-rank properties [27] stemming from the fact that some of the boundary points in the underlying geometry are distant from each other. Formulating a boundary integral equation is the analytical way of arriving to the same dense matrix. From an algebraic point of view, the sparse matrix for the volume turns into a dense matrix for the boundary, through the process of trying to minimize fill-in. Considering the minimization of fill-in and the compression of the dense matrices in separate phases leads to methods like HSS + multifrontal [98, 100, 101].

Ultimately, minimizing fill-in and minimizing off-diagonal rank should not be conflicting objectives. The former depends on the connectivity and the latter depends on the distance in the underlying geometry. In most applications, the closer points are connected (or interact) more densely, so reordering according to the distance should produce near optimal ordering for the connectivity as well. The same can be said about minimizing communication for the parallel implementation of these methods. Mapping the 3-D connectivity/distance to a 1-D locality in the memory space (or matrix column/row) is what we are ultimately trying to achieve.

### *3.2 Dense Matrix Factorization*

The methods in the previous subsection are direct solvers/preconditioners for sparse matrices. As we have mentioned, there is an analogy between minimizing fill-in in sparse matrices by looking at the connectivity, and minimizing the rank of off-diagonal blocks of dense matrices by looking at the distance. Using this analogy, the same concept as nested dissection for sparse matrices can be applied to dense matrices. This leads to methods like the recursive skeletonization [62], or hierarchical Poincare-Steklov (HPS) [41, 81]. HPS is like a bottom-up version of what nested dissection and recursive skeletonization do top-down. For high contrast coefficient problems, it makes sense to construct the domain dissection bottom-up, to align the bisectors with the coefficient jumps. There are also other methods that rely on a similar concept [19, 52, 69, 102]. Furthermore, since many of these methods use weak admissibility with growing ranks for 3-D problems, it is useful to have nested hierarchical decompositions, which is like a nested dimension reduction. In this respect, the recursive skeletonization has been extended to hierarchical interpolative factorization (HIF) [63], the HSS has been extended to HSS2D [99]. There is also a combination of HSS and Skeletonization [30]. There are methods that use this nested dimension reduction concept without the low-rank approximation [60] in the context of domain decomposition for incomplete LU factorization. One method that does not use weak admissibility is the inverse FMM [2], which makes it applicable to 3-D problems in $\mathscr{O}(N)$ without nested dimension reduction.

## 4 Experimental Results

### *4.1 FMM vs. HSS*

There have been few comparisons between the analytic and algebraic hierarchical low-rank approximation methods [20]. From a high performance computing perspective, the practical performance of highly optimized implementations of these various methods is of great interest. There have been many efforts to develop new methods in this area, which has resulted in a large amount of similar methods with different names without a clear overall picture of their relative performance on modern HPC architectures. The trend in architecture where arithmetic operations are becoming cheap compared to data movement, is something that must be considered carefully when predicting which method will perform better on computers of the future.

We acknowledge that the comparisons we present here are far from complete, and much more comparisons between all the different methods are needed in order to achieve our long term objective. The limitation actually comes from the lack of highly optimized implementations of these methods that are openly available to us at the moment.

In the present work we start by comparing exaFMM—a highly optimized implementation of FMM, with STRUMPACK—a highly optimized implementation of HSS. We select the 2D and 3D Laplace equation on uniform lattices as test cases. For HSS we directly construct the compressed matrix by calling the Green's function in the randomized low-rank approximation routine. We perform the matrix-vector multiplication using the FMM and HSS, and measure the time for the compression/precalculation and application of the matrix-vector multiplication. We also measure the peak memory consumption of both methods.

The elapsed time for the FMM and HSS for different problem sizes is shown in Fig. 2. In order to isolate the effect of the thread scalability of the two methods, these runs are performed on a single core of a 12-core Ivy Bridge (E5-2695 v2). For the 2D Laplace equation, the FMM shows some overhead for small $N$, but is about 3 orders of magnitude faster than HSS for larger problems. For the 3D Laplace equation, the FMM is about 2 orders of magnitude faster than HSS for smaller $N$, but HSS exhibits non-optimal behavior for large $N$ because the rank keeps growing.

The large difference in the computational time is actually coming from the heavy computation in the sampling phase and compression phase of the HSS. In Fig. 3,

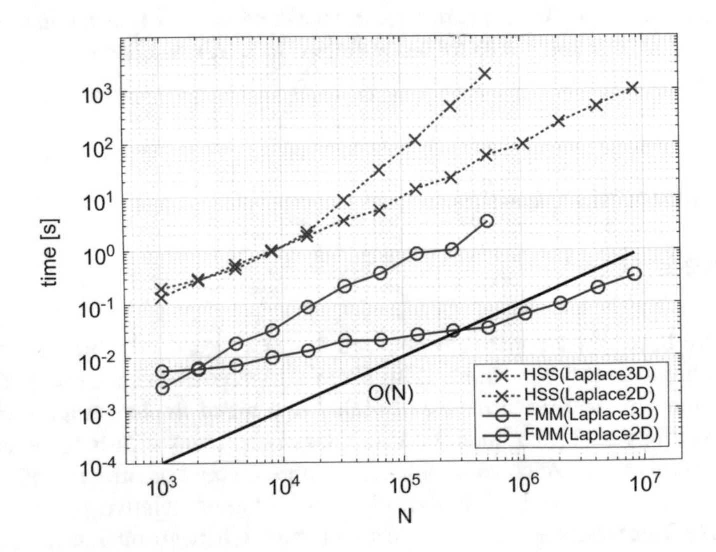


**Fig. 2** Elapsed time for the matrix-vector multiplication using FMM and HSS for different problem sizes

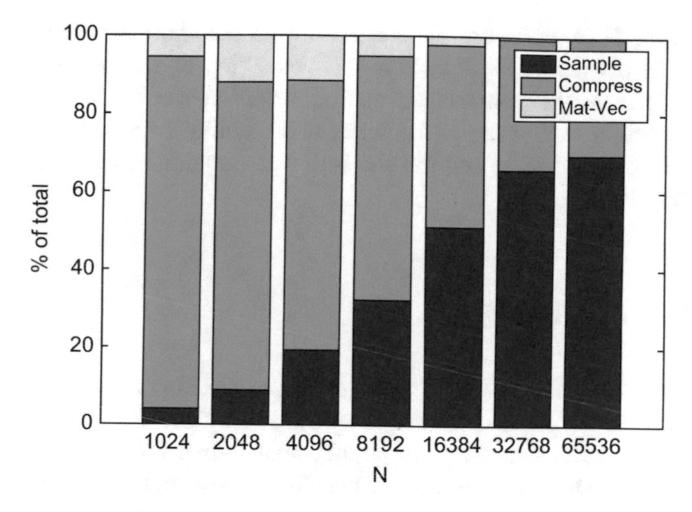


**Fig. 3** Percentage of the computation time of HSS for different problem sizes

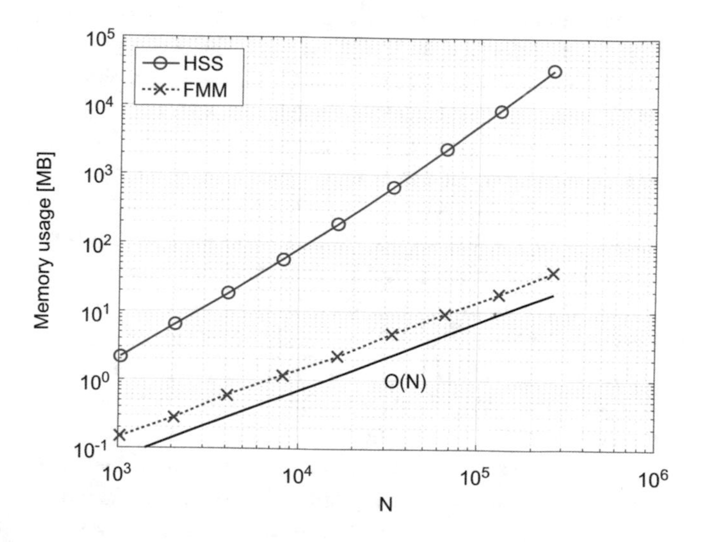


**Fig. 4** Peak memory usage of FMM and HSS for the 3D Laplace equation

we show the percentage of the computation time of HSS for different problem sizes $N$. "Sample" is the sampling time, "Compress" is the compression time, and "Mat-Vec" is the matrix-vector multiplication time. We can see that the sampling is taking longer and longer as the problem size increases. This is because the rank $k$ increases with the problem size $N$, and both sampling and compression time increase with the $k$ and $N$.

The peak memory usage of FMM and HSS is shown in Fig. 4 for the 3D Laplace equation. We see that the FMM has strictly $\mathcal{O}(N)$ storage requirements, but since the rank in the HSS grows for 3D kernels it does not show the ideal $\mathcal{O}(N \log N)$ behavior. The disadvantage of HSS is two-fold. First of all, its algebraic nature requires it to store the compressed matrix, where as the FMM is analytic and therefore matrix-free. Secondly, the weak admissibility causes the rank to grow for 3D problems, and with that the memory consumption grows at a suboptimal complexity.

## 4.2 FMM vs. Multigrid

If we are to use the FMM as a matrix-free $\mathcal{O}(N)$ preconditioner based on hierarchical low-rank approximation, the natural question to ask is "How does it compare against multigrid?", which is a much more popular matrix-free $\mathcal{O}(N)$ preconditioner for solving elliptic PDEs. We perform a benchmark test similar to the one in the previous subsection, for the Laplace equation and Helmholtz equation on a 3D cubic lattice $[-1, 1]^3$, but for this case we impose Dirichlet boundary conditions at the faces of the domain. The preconditioners are used inside a Krylov subspace solver. The runs were performed on Matlab using a finite element package IFISS. Our fast multipole preconditioner is compared with the incomplete Cholesky (IC) factorization with zero fill implemented in Matlab and the algebraic multigrid (AMG) and geometric multigrid (GMG) methods in IFISS. The FMM code is written in C and called as a MEX function.

The convergence rate of the FMM and Multigrid preconditioners for the Laplace equation is shown in Fig. 5, for a grid spacing of $h = 2^{-5}$. "AMG" is algebraic multigrid, "GMG" is geometric multigrid, "Inc Chol" is incomplete Cholesky. The $\epsilon$ value represents the accuracy of the FMM. We see that the FMM preconditioner has comparable convergence to the algebraic and geometric multigrid method. Even for a very low-accuracy FMM with $\epsilon = 10^{-2}$, the convergence rate is much better than the incomplete Cholesky. We refer to the work by Ibeid et al. [67] for more detailed comparisons between FMM and Multigrid.

A similar plot is shown for the Helmholtz equation with grid spacing of $h = 2^{-5}$ and wave number $\kappa = 7$ in Fig. 6. The nomenclature of the legend is identical to that of Fig. 5. In this case, we see a larger difference between the convergence rate of FMM and Multigrid. Even the FMM with the worst accuracy does better than the multigrid. We have also confirmed that the FMM preconditioner has a convergence rate that is independent of the problem size, up to moderate wave numbers of $\kappa$.

The strong scaling of FMM and AMG are shown in Fig. 7, which includes the setup phase and all iterations it took to converge. All calculations were performed on the TACC Stampede system without using the coprocessors. Stampede has 6400 nodes, each with two Xeon E5-2680 processors and one Intel Xeon Phi SE10P coprocessor and 32GB of memory. We used the Intel compiler (version 13.1.0.146) and configured PETSc with "`COPTFLAGS=-O3 FOPTFLAGS=-O3 -with-clanguage=cxx`

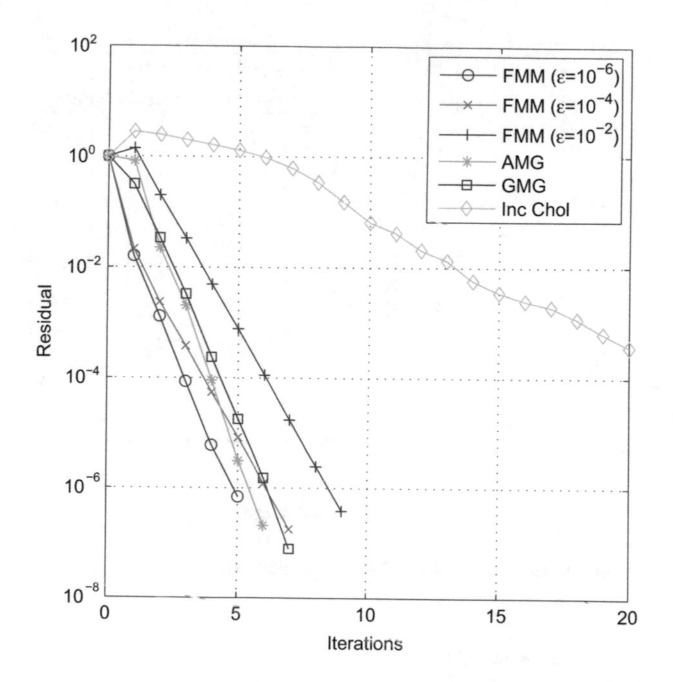


**Fig. 5** Convergence rate of the FMM and Multigrid preconditioners for the Laplace equation on a $[-1, 1]^3$ lattice with spacing $h = 2^{-5}$

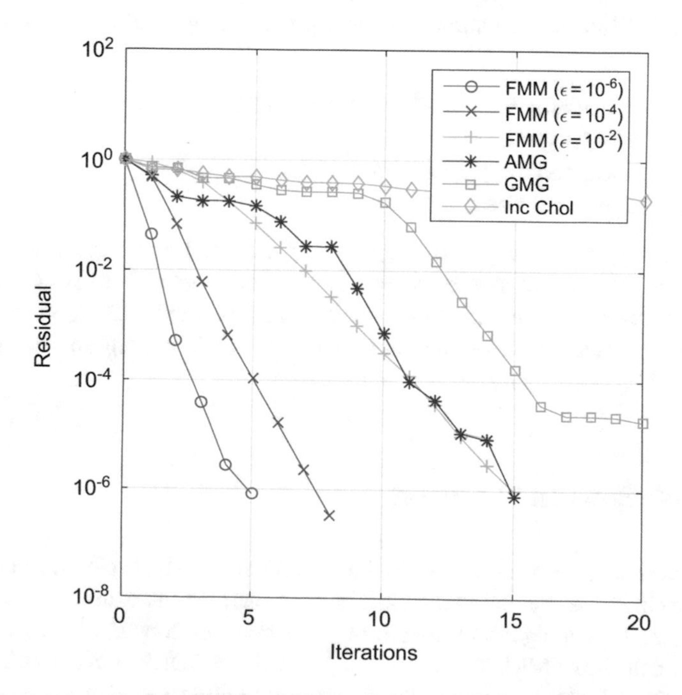


**Fig. 6** Convergence rate of the FMM and Multigrid preconditioners for the Helmholtz equation on a $[-1, 1]^3$ lattice with spacing $h = 2^{-5}$ and wave number $\kappa = 7$

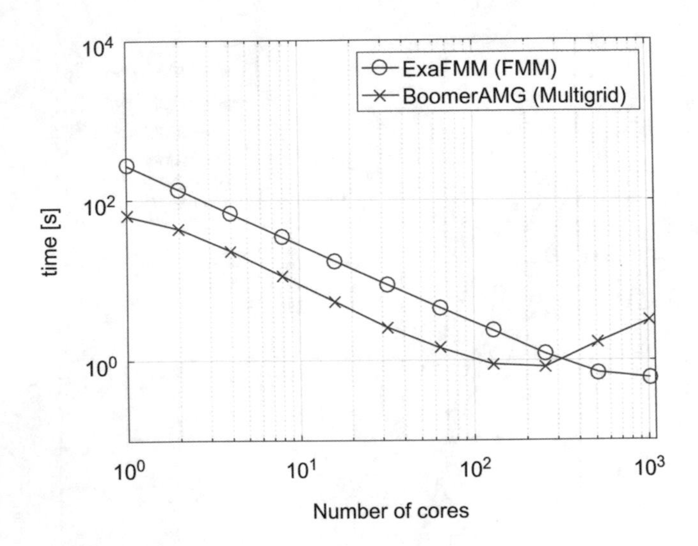


**Fig. 7** Strong scaling of the 2-D FMM and AMG preconditioners

```
-download-f-blas-lapack -download-hypre
-download-metis -download-parmetis
```

`-download-superlu_dist -with-debugging=0`". For BoomerAMG we compared different relaxation, coarsening, and interpolation methods and found that

```
"-pc_hypre_boomeramg_relax_type_all
backward-SOR/Jacobi
-pc_hypre_boomeramg_coarsen_type
modifiedRuge-Stueben
```

`-pc_hypre_bommeramg_interp_type classical`" gives the best performance. We use a grid size of $N = 4096^2$ and run from 1 to 1024 cores using up to 16 cores per node on Stampede. For this particular Poisson problem on this particular machine using this particular FMM code we see an advantage over BoomerAMG past 512 cores.

## 5 Conclusions and Outlook

We have shown the contrast between the analytical and algebraic hierarchical low-rank approximations, by reviewing the contributions over the years and placing them along the analytical-algebraic spectrum. The relation between Treecode, FMM, KIFMM, black-box FMM, $\mathcal{H}$-matrix, $\mathcal{H}^2$-matrix, HODLR, HSS, HBS, and BLR were explained from the perspective of compute-memory tradeoff. This birds-eye view of the entire hierarchical low-rank approximation landscape from analytical to algebraic, allows us to place ideas like precomputation of FMM translation matrices and relate that to storage reduction techniques for the algebraic variants.

Some important findings from this cross-disciplinary literature review are:

- Translational invariance of the FMM operators suggest that $\mathscr{H}^2$-matrices (and the like) have mostly duplicate entries, which many are redundantly storing at the moment.
- The analytical variants can now perform factorization and are kernel independent, so the decision to use the algebraic variants at the cost of consuming more memory should be made carefully.
- The kernel-independent variants of FMM can be used as a matrix-free $\mathscr{O}(N)$ compression technique.
- The use of SVD to compress the FMM translation matrices, makes the work on variable expansion order and its error optimized variants redundant.
- The hierarchical compression should not be applied directly to the inverse or factorizations of sparse matrices just because they fill-in. One must first try to minimize fill-in, and then compress only the dense blocks that cannot be avoided.

The comparison benchmarks between FMM and HSS are still preliminary tests for a very simple case. However, they clearly demonstrate the magnitude of the difference that lies between the various hierarchical low-rank approximation methods. The comparison between FMM and multigrid is also a very simple test case, but it reveals the previously unquantified convergence properties of low-accuracy FMM as a preconditioner. Of course, for such simple problems the FMM can give the exact solution in finite arithmetic and therefore solve the problem in a single iteration. The interesting point here is not the fact that it can be used as a preconditioner, but the practical performance of the low-accuracy FMM being significantly faster than the high accuracy FMM, even if it requires a few iterations.

There is much more that can be done if all of these complicated hierarchical low-rank approximation methods could somehow be made easier to code. We believe a modular view of these methods will help the developers though separation of concerns. Instead of everyone coding a slightly different version of the whole thing, we could each choose a module to focus on that fits our research interests, and contribute to a larger and more sustainable ecosystem. A few ideas to facilitate the transition to such a community effort are:

1. Create a common benchmark (mini app) for each of the modules.
2. Gradually propagate standards in the community, starting from the major codes.
3. Develop a common interface between the hierarchical structure and inner kernels.
4. Do not try to unify code, just have a standard with a common API (like MPI).

**Acknowledgements** We thank François-Henry Rouet, Pieter Ghysels, and Xiaoye, S. Li for providing the STRUMPACK interface for our comparisons between FMM and HSS. This work was supported by JSPS KAKENHI Grant-in-Aid for Research Activity Start-up Grant Number 15H06196. This publication was based on work supported in part by Award No KUK-C1-013-04, made by King Abdullah University of Science and Technology (KAUST). This work used the Extreme Science and Engineering Discovery Environment (XSEDE), which is supported by National Science Foundation grant number OCI-1053575.

## References

1. Ambikasaran, S., Darve, E.: An O(NlogN) fast direct solver for partial hierarchically semi-separable matrices. J. Sci. Comput. **57**, 477–501 (2013)
2. Ambikasaran, S., Darve, E.: The inverse fast multipole method. arXiv:1407.1572v1 (2014)
3. Ambikasaran, S., Li, J.-Y., Kitanidis, P.K., Darve, E.: Large-scale stochastic linear inversion using hierarchical matrices. Comput. Geosci. **17**(6), 913–927 (2013)
4. Amestoy, P., Ashcraft, C., Boiteau, O., Buttari, A., L'Excellent, J.-Y., Weisbecker, C.: Improving multifrontal methods by means of block low-rank representations. SIAM J. Sci. Comput. **37**(3), A1451–A1474 (2015)
5. Aminfar, A., Darve, E.: A fast, memory efficient and robust sparse preconditioner based on a multifrontal approach with applications to finite-element matrices. Int. J. Numer. Methods Eng. **107**, 520–540 (2016)
6. Aminfar, A., Ambikasaran, S., Darve, E.: A fast block low-rank dense solver with applications to finite-element matrices. J. Comput. Phys. **304**, 170–188 (2016)
7. Anderson, C.R.: An implementation of the fast multipole method without multipoles. SIAM J. Sci. Stat. Comput. **13**(4), 923–947 (1992)
8. Appel, A.W.: An efficient program for many-body simulation. SIAM J. Sci. Stat. Comput. **6**(1), 85–103 (1985)
9. Barba, L.A., Yokota, R.: How will the fast multipole method fare in the exascale era? SIAM News **46**(6), 1–3 (2013)
10. Barnes, J., Hut, P.: O(NlogN) force-calculation algorithm. Nature **324**, 446–449 (1986)
11. Bebendorf, M.: Approximation of boundary element matrices. Numer. Math. **86**, 565–589 (2000)
12. Bebendorf, M.: Hierarchical Matrices. Lecture Notes in Computational Science and Engineering, vol. 63. Springer, Berlin (2008)
13. Bebendorf, M., Rjasanow, S.: Adaptive low-rank approximation of collocation matrices. Computing **70**, 1–24 (2003)
14. Bédorf, J., Gaburov, E., Fujii, M.S., Nitadori, K., Ishiyama, T., Portegies Zwart, S.: 24.77 Pflops on a gravitational tree-code to simulate the milky way galaxy with 18600 GPUs. In: Proceedings of the 2014 ACM/IEEE International Conference for High Performance Computing, Networking, Storage and Analysis, pp. 1–12 (2014)
15. Berman, C.L.: Grid-multipole calculations. SIAM J. Sci. Comput. **16**(5), 1082–1091 (1995)
16. Börm, S.: Construction of data-sparse $h^2$-matrices by hierarchical compression. SIAM J. Sci. Comput. **31**(3), 1820–1839 (2009)
17. Börm, S., Grasedyck, L.: Hybrid cross approximation of integral operators. Numer. Math. **101**, 221–249 (2005)
18. Börm, S., Grasedyck, L., Hackbusch, W.: Introduction to hierarchical matrices with applications. Eng. Anal. Bound. Elem. **27**, 405–422 (2003)
19. Bremer, J.: A fast direct solver for the integral equations of scattering theory on planar curves with corners. J. Comput. Phys. **231**, 1879–1899 (2012)
20. Brunner, D., Junge, M., Rapp, P., Bebendorf, M., Gaul, L.: Comparison of the Fast Multipole Method with Hierarchical Matrices for the Helmholtz-BEM. Comput. Model. Eng. Sci. **58**(2), 131–160 (2010)
21. Burant, J.C., Strain, M.C., Scuseria, G.E., Frisch, M.J.: Analytic energy gradients for the Gaussian very fast multipole method (GvFMM). Chem. Phys. Lett. **248**, 43–49 (1996)
22. Chaillat, S., Bonnet, M., Semblat, J.-F.: A multi-level fast multipole BEM for 3-D elastodynamics in the frequency domain. Comput. Methods Appl. Mech. Eng. **197**, 4233–4249 (2008)

23. Chan, T.F.: On the existence and computation of LU-factorizations with small pivots. Math. Comput. **42**(166), 535–547 (1984)
24. Chan, T.F.: Rank revealing QR factorizations. Linear Algebra Appl. **88/89**, 67–82 (1987)
25. Chandrasekaran, S., Ipsen, I.C.F.: On rank-revealing factorizations. SIAM J. Matrix Anal. Appl. **15**(2), 592–622 (1994)
26. Chandrasekaran, S., Dewilde, P., Gu, M., Lyons, W., Pals, T.: A fast solver for HSS representations via sparse matrices. SIAM J. Matrix Anal. Appl. **29**(1), 67–81 (2006)
27. Chandrasekaran, S., Dewilde, P., Gu, M., Somasunderam, N.: On the numerical rank of the off-diagonal blocks of Schur complements of discretized elliptic PDEs. SIAM J. Matrix Anal. Appl. **31**(5), 2261–2290 (2010)
28. Cheng, H., Gimbutas, Z., Martinsson, P.G., Rokhlin, V.: On the compression of low rank matrices. SIAM J. Sci. Comput. **26**(4), 1389–1404 (2005)
29. Choi, C.H., Ruedenberg, K., Gordon, M.S.: New parallel optimal-parameter fast multipole method (OPFMM). J. Comput. Chem. **22**(13), 1484–1501 (2001)
30. Corona, E., Martinsson, P.G., Zorin, D.: An O(N) direct solver for integral equations on the plane. Appl. Comput. Harmon. Anal. **38**, 284–317 (2015)
31. Coulaud, O., Fortin, P., Roman, J.: High performance BLAS formulation of the multipole-to-local operator in the fast multipole method. J. Comput. Phys. **227**, 1836–1862 (2008)
32. Dachsel, H.: Corrected article: "an error-controlled fast multipole method". J. Chem. Phys. **132**, 119901 (2010)
33. Darve, E., Havé, P.: A fast multipole method for Maxwell equations stable at all frequencies. Philos. Trans. R. Soc. Lond. A **362**, 603–628 (2004)
34. Darve, E., Cecka, C., Takahashi, T.: The fast multipole method on parallel clusters, multicore processors, and graphics processing units. C.R. Mec. **339**, 185–193 (2011)
35. Dehnen, W.: A hierarchical O(N) force calculation algorithm. J. Comput Phys. **179**(1), 27–42 (2002)
36. Dutt, A., Gu, M., Rokhlin, V.: Fast algorithms for polynomial interpolation, integration, and differntiation. SIAM J. Numer. Anal. **33**(5), 1689–1711 (1996)
37. Elliott, W.D., Board, J.A.: Fast Fourier transform accelerated fast multipole algorithm. SIAM J. Sci. Comput. **17**(2), 398–415 (1996)
38. Ethridge, F., Greengard, L.: A new fast-multipole accelerated Poisson solver in two dimensions. SIAM J. Sci. Comput. **23**(3), 741–760 (2001)
39. Fong, W., Darve, E.: The black-box fast multipole method. J. Comput. Phys. **228**, 8712–8725 (2009)
40. Fortin, P.: Multipole-to-local operator in the fast multipole method: Comparison of FFT, rotations and BLAS improvements. Technical Report RR-5752, Rapports de recherche, et theses de l'Inria (2005)
41. Gillman, A., Barnett, A., Martinsson, P.G.: A spectrally accurate direct solution technique for frequency-domain scattering problems with variable media. BIT Numer. Math. **55**, 141–170 (2015)
42. Gimbutas, Z., Greengard, L.: Fast multi-particle scattering: a hybrid solver for the Maxwell equations in microstructured materials. J. Comput. Phys. **232**, 22–32 (2013)
43. Gimbutas, Z., Rokhlin, V.: A generalized fast multipole method for nonoscillatory kernels. SIAM J. Sci. Comput. **24**(3), 796–817 (2002)
44. Goreinov, S.A., Tyrtyshnikov, E.E., Zamarashkin, N.L.: A theory of pseudoskeleton approximations. Linear Algebra Appl. **261**(1–3), 1–21 (1997)
45. Grasedyck, L., Hackbusch, W.: Construction and arithmetics of H-matrices. Computing **70**, 295–334 (2003)
46. Grasedyck, L., Kriemann, R., Le Borne, S.: Parallel black box H-LU preconditioning for elliptic boundary value problems. Comput. Vis. Sci. **11**, 273–291 (2008)
47. Grasedyck, L., Hackbusch, W., Kriemann, R.: Performance of H-LU preconditioning for sparse matrices. Comput. Methods Appl. Math. **8**(4), 336–349 (2008)
48. Grasedyck, L., Kriemann, R., Le Borne, S.: Domain decomposition based H-LU preconditioning. Numer. Math. **112**, 565–600 (2009)

49. Gray, A.G., Moore, A.W.: N-body problems in statistical learning. In: Leen, T.K., Dietterich, T.G., Tresp, V. (eds.) Advances in Neural Information Processing Systems, vol. 13, pp. 521—527. MIT Press, Cambridge (2001)
50. Greengard, L., Rokhlin, V.: A fast algorithm for particle simulations. J. Comput. Phys. **73**(2), 325–348 (1987)
51. Greengard, L., Rokhlin, V.: A new version of the fast multipole method for the Laplace equation in three dimensions. Acta Numer. **6**, 229–269 (1997)
52. Greengard, L., Gueyffier, D., Martinsson, P.G., Rokhlin, V.: Fast direct solvers for integral equations in complex three dimensional domains. Acta Numer. **18**, 243–275 (2009)
53. Gu, M., Eisenstat, S.C.: Efficient algorithms for computing a strong rank-revealing QR factorization. SIAM J. Sci. Comput. **17**(4), 848–869 (1996)
54. Gumerov, N.A., Duraiswami, R.: Fast radial basis function interpolation via preconditioned Krylov iteration. SIAM J. Sci. Comput. **29**(5), 1876–1899 (2007)
55. Hackbusch, W.: A sparse matrix arithmetic based on H-matrices, part I: Introduction to H-matrices. Computing **62**, 89–108 (1999)
56. Hackbusch, W., Nowak, Z.P.: On the fast matrix multiplication in the boundary element method by panel clustering. Numer. Math. **54**, 463–491 (1989)
57. Hackbusch, W., Khoromskij, B., Sauter, S.A.: On $h^2$-matrices. In: Bungartz, H., Hoppe, R., Zenger, C. (eds.) Lectures on Applied Mathematics. Springer, Berlin (2000)
58. Halko, N., Martinsson, P.G., Tropp, J.A.: Finding structure with randomness: probabilistic algorithms for constructing approximate matrix decompositions. SIAM Rev. **53**(2), 217–288 (2011)
59. Hao, S., Martinsson, P.G., Young, P.: An efficient and highly accurate solver for multi-body acoustic scattering problems involving rotationally symmetric scatterers. Comput. Math. Appl. **69**, 304–318 (2015)
60. Hénon, P., Saad, Y.: A parallel multistage ILU factorization based on a hierarchical graph decomposition. SIAM J. Sci. Comput. **28**(6), 2266–2293 (2006)
61. Hesford, A.J., Waag, R.C.: Reduced-rank approximations to the far-field transform in the gridded fast multipole method. J. Comput. Phys. **230**, 3656–3667 (2011)
62. Ho, K.L., Greengard, L.: A fast direct solver for structured linear systems by recursive skeletonization. SIAM J. Sci. Comput. **34**(5), A2507–A2532 (2012)
63. Ho, K.L., Ying, L.: Hierarchical interpolative factorization for elliptic operators: Integral equations. arXiv:1307.2666 (2015)
64. Hong, Y.P., Pan, C.T.: Rank-revealing QR factorizations and the singular value decomposition. Math. Comput. **58**(197), 213–232 (1992)
65. Hwang, T.-M., Lin, W.-W., Yang, E.K.: Rank revealing LU factorizations. Linear Algebra Appl. **175**, 115–141 (1992)
66. Hwang, T.-M., Lin, W.-W., Pierce, D.: Improved bound for rank revealing LU factorizations. Linear Algebra Appl. **261**(1), 173–186 (1997)
67. Ibeid, H., Yokota, R., Pestana, J., Keyes, D.: Fast multipole preconditioners for sparse matrices arising from elliptic equations. arXiv:1308.3339 (2016)
68. Izadi, M.: Hierarchical Matrix Techniques on Massively Parallel Computers. Ph.D. thesis, Universitat Leipzig (2012)
69. Kong, W.Y., Bremer, J., Rokhlin, V.: An adaptive fast direct solver for boundary integral equations in two dimensions. Appl. Comput. Harmon. Anal. **31**, 346–369 (2011)
70. Langston, H., Greengard, L., Zorin, D.: A free-space adaptive FMM-based PDE solver in three dimensions. Commun. Appl. Math. Comput. Sci. **6**(1), 79–122 (2011)
71. Le Borne, S.: Multilevel hierarchical matrices. SIAM J. Matrix Anal. Appl. **28**(3), 871–889 (2006)
72. Lee, D., Vuduc, R., Gray, A.G.: A distributed kernel summation framework for general-dimension machine learning. In: Proceedings of the 2012 SIAM International Conference on Data Mining (2012)

73. Lessel, K., Hartman, M., Chandrasekaran, S.: A fast memory efficient construction algorithm for hierarchically semi-separable representations. http://scg.ece.ucsb.edu/publications/MemoryEfficientHSS.pdf (2015)
74. Li, J.-Y., Ambikasaran, S., Darve, E.F., Kitanidis, P.K.: A Kalman filter powered by $h^2$-matrices for quasi-continuous data assimilation problems. Water Resour. Res. **50**, 3734–3749 (2014)
75. Liang, Z., Gimbutas, Z., Greengard, L., Huang, J., Jiang, S.: A fast multipole method for the Rotne-Prager-Yamakawa tensor and its applications. J. Comput. Phys. **234**, 133–139 (2013)
76. Liberty, E., Woolfe, F., Martinsson, P.G., Rokhlin, V., Tygert, M.: Randomized algorithms for the low-rank approximation of matrices. Proc. Natl. Acad. Sci. U.S.A. **104**(51), 20167–20172 (2007)
77. Makino, J.: Yet another fast multipole method without multipoles – Pseudoparticle multipole method. J. Comput. Phys. **151**(2), 910–920 (1999)
78. Malhotra, D., Biros, G.: PVFMM: a parallel kernel independent FMM for particle and volume potentials. Commun. Comput. Phys. **18**(3), 808–830 (2015)
79. Malhotra, D., Gholami, A., Biros, G.: A volume integral equation stokes solver for problems with variable coefficients. In: Proceedings of the 2014 ACM/IEEE International Conference for High Performance Computing, Networking, Storage and Analysis, pp. 1–11 (2014)
80. March, W.B., Xiao, B., Biros, G.: ASKIT: approximate skeletonization kernel-independent treecode in high dimensions. SIAM J. Sci. Comput. **37**(2), A1089–A1110 (2015)
81. Martinsson, P.G.: The hierarchical Poincaré-Steklov (HPS) solver for elliptic PDEs: a tutorial. arXiv:1506.01308 (2015)
82. Martinsson, P.G., Rokhlin, V.: A fast direct solver for boundary integral equations in two dimensions. J. Comput. Phys. **205**, 1–23 (2005)
83. Miranian, L., Gu, M.: Strong rank revealing LU factorizations. Linear Algebra Appl. **367**, 1–16 (2003)
84. Ohno, Y., Yokota, R., Koyama, H., Morimoto, G., Hasegawa, A., Masumoto, G., Okimoto, N., Hirano, Y., Ibeid, H., Narumi, T., Taiji, M.: Petascale molecular dynamics simulation using the fast multipole method on k computer. Comput. Phys. Commun. **185**, 2575–2585 (2014)
85. Oliveira, S., Yang F.: An algebraic approach for H-matrix preconditioners. Computing **80**, 169–188 (2007)
86. Pan, C.T.: On the existence and computation of rank-revealing LU factorizations. Linear Algebra Appl. **316**, 199–222 (2000)
87. Petersen, H.G., Soelvason, D., Perram, J.W., Smith, E.R.: The very fast multipole method. J. Chem. Phys. **101**(10), 8870–8876 (1994)
88. Rahimian, A., Lashuk, I., Veerapaneni, K., Chandramowlishwaran, A., Malhotra, D., Moon, L., Sampath, R., Shringarpure, A., Vetter, J., Vuduc, R., Zorin, D., Biros, G.: Petascale direct numerical simulation of blood flow on 200k cores and heterogeneous architectures. In: Proceedings of the 2010 ACM/IEEE International Conference for High Performance Computing, Networking, Storage and Analysis, SC '10 (2010)
89. Rouet, F.-H., Li, X.-S., Ghysels, P., Napov, A.: A distributed-memory package for dense hierarchically semi-separable matrix computations using randomization. arXiv:1503.05464 (2015)
90. Shao, Y., White, C.A., Head-Gordon, M.: Efficient evaluation of the Coulomb force in density-functional theory calculations. J. Chem. Phys. **114**(15), 6572–6577 (2001)
91. Takahashi, T., Cecka, C., Fong, W., Darve, E.: Optimizing the multipole-to-local operator in the fast multipole method for graphical processing units. Int. J. Numer. Methods Eng. **89**, 105–133 (2012)
92. Verde, A., Ghassemi, A.: Fast multipole displacement discontinuity method (FM-DDM) for geomechanics reservoir simulations. Int. J. Numer. Anal. Methods Geomech. **39**(18), 1953–1974 (2015)
93. Wang, Y., Wang, Q., Deng, X., Xia, Z., Yan, J., Xu, H.: Graphics processing unit (GPU) accelerated fast multipole BEM with level-skip M2L for 3D elasticity problems. Adv. Eng. Softw. **82**, 105–118 (2015)

94. White, C.A., Head-Gordon, M.: Rotating around the quartic angular momentum barrier in fast multipole method calculations. J. Chem. Phys. **105**(12), 5061–5067 (1996)
95. Wilkes, D.R., Duncan, A.J.: A low frequency elastodynamic fast multipole boundary element method in three dimensions. Comput. Mech. **56**, 829–848 (2015)
96. Willis, D., Peraire, J., White, J.: FastAero – a precorrected FFT-fast multipole tree steady and unsteady potential flow solver. http://hdl.handle.net/1721.1/7378 (2005)
97. Wolf, W.R., Lele, S.K.: Aeroacoustic integrals accelerated by fast multipole method. AIAA J. **49**(7), 1466–1477 (2011)
98. Xia, J.: Randomized sparse direct solvers. SIAM J. Matrix Anal. Appl. **34**(1), 197–227 (2013)
99. Xia, J.: O(N) complexity randomized 3D direct solver with HSS2D structure. In: Proceedings of the Project Review, Geo-Mathematical Imaging Group, Purdue University, pp. 317–325 (2014)
100. Xia, J., Chandrasekaran, S., Gu, M., Li, X.S.: Superfast multifrontal method for large structured linear systems of equations. SIAM J. Matrix Anal. Appl. **31**(3), 1382–1411 (2009)
101. Xia, J., Chandrasekaran, S., Gu, M., Li, X.S.: Fast algorithms for hierarchically semiseperable matrices. Numer. Linear Algebra Appl. **17**, 953–976 (2010)
102. Yarvin, N., Rokhlin, V.: An improved fast multipole algorithm for potential fields on the line. SIAM J. Numer. Anal. **36**(2), 629–666 (1999)
103. Ying, L., Biros, G., Zorin, D.: A kernel-independent adaptive fast multipole algorithm in two and three dimensions. J. Comput. Phys. **196**(2), 591–626 (2004)
104. Ying, L., Biros, G., Zorin, D.: A high-order 3D boundary integral equation solver for elliptic PDEs in smooth domains. J. Comput. Phys. **219**, 247–275 (2006)
105. Yokota, R., Bardhan, J.P., Knepley, M.G., Barba, L.A., Hamada, T.: Biomolecular electrostatics using a fast multipole BEM on up to 512 GPUs and a billion unknowns. Comput. Phys. Commun. **182**, 1272–1283 (2011)
106. Yokota, R., Narumi, T., Yasuoka, K., Barba, L.A.: Petascale turbulence simulation using a highly parallel fast multipole method on GPUs. Comput. Phys. Commun. **184**, 445–455 (2013)
107. Yunis, E., Yokota, R., Ahmadia, A.: Scalable force directed graph layout algorithms using fast multipole methods. In: The 11th International Symposium on Parallel and Distributed Computing, Munich, June 2012
108. Zhao, Z., Kovvali, N., Lin, W., Ahn, C.-H., Couchman, L., Carin, L.: Volumetric fast multipole method for modeling Schrödinger's equation. J. Comput. Phys. **224**, 941–955 (2007)

# Recent Progress in Linear Response Eigenvalue Problems

**Zhaojun Bai and Ren-Cang Li**

**Abstract** Linear response eigenvalue problems arise from the calculation of excitation states of many-particle systems in computational materials science. In this paper, from the point of view of numerical linear algebra and matrix computations, we review the progress of linear response eigenvalue problems in theory and algorithms since 2012.

## 1 Introduction

The standard Linear Response Eigenvalue Problem (LREP) is the following eigenvalue problem

$$\begin{bmatrix} A & B \\ -B & -A \end{bmatrix} \begin{bmatrix} u \\ v \end{bmatrix} = \lambda \begin{bmatrix} u \\ v \end{bmatrix}, \tag{1}$$

where $A$ and $B$ are $n \times n$ real symmetric matrices such that the symmetric matrix $\begin{bmatrix} A & B \\ B & A \end{bmatrix}$ is positive definite. Such an eigenvalue problem arises from computing excitation states (energies) of physical systems in the study of collective motion of many particle systems, ranging from silicon nanoparticles and nanoscale materials to the analysis of interstellar clouds (see, for example, [12, 27, 33, 34, 38, 50] and references therein). In computational quantum chemistry and physics, the excitation states and absorption spectra for molecules or surface of solids are described by the *Random Phase Approximation (RPA)* or the *Bethe-Salpeter (BS) equation*. For

Z. Bai (✉)
Department of Computer Science and Department of Mathematics, University of California, Davis, CA 95616, USA
e-mail: bai@cs.ucdavis.edu

R.-C. Li
Department of Mathematics, University of Texas at Arlington, P.O. Box 19408, Arlington, TX 76019, USA
e-mail: rcli@uta.edu



T. Sakurai et al. (eds.), *Eigenvalue Problems: Algorithms, Software and Applications in Petascale Computing*, Lecture Notes in Computational Science and Engineering 117, https://doi.org/10.1007/978-3-319-62426-6_18

this reason, the LREP (1) is also called the RPA eigenvalue problem [17], or the BS eigenvalue problem [5, 6, 42]. There are immense recent interest in developing new theory, efficient numerical algorithms of the LREP (1) and the associated excitation response calculations of molecules for materials design in energy science [16, 28, 40, 41].

In this article, we survey recent progress in the LREP research from numerical linear algebra and matrix computations perspective. We focus on recent work since 2012. A survey of previous algorithmic work prior to 2012 can be found in [2, 51] and references therein. The rest of this paper is organized as follows. In Sect. 2, we survey the recent theoretical studies on the properties of the LREP and minimization principles. In Sect. 3, we briefly describe algorithmic advances for solving the LREP. In Sect. 4, we state recent results on perturbation and backward error analysis of the LREP. In Sect. 5, we remark on several related researches spawn from the LREP (1), including a generalized LREP.

## 2 Theory

Define the symmetric orthogonal matrix

$$J = \frac{1}{\sqrt{2}} \begin{bmatrix} I_n & I_n \\ I_n & -I_n \end{bmatrix}. \tag{2}$$

It can be verified that $J^{\mathrm{T}}J = J^2 = I_{2n}$ and

$$J^{\mathrm{T}} \begin{bmatrix} A & B \\ -B & -A \end{bmatrix} J = \begin{bmatrix} 0 & A-B \\ A+B & 0 \end{bmatrix}. \tag{3}$$

This means that the LREP (1) is orthogonally similar to

$$Hz := \begin{bmatrix} 0 & K \\ M & 0 \end{bmatrix} \begin{bmatrix} y \\ x \end{bmatrix} = \lambda \begin{bmatrix} y \\ x \end{bmatrix} =: \lambda z, \tag{4}$$

where $K = A - B$ and $M = A + B$. Both eigenvalue problems (1) and (4) have the same eigenvalues with corresponding eigenvectors related by

$$\begin{bmatrix} y \\ x \end{bmatrix} = J^{\mathrm{T}} \begin{bmatrix} u \\ v \end{bmatrix} \quad \text{and} \quad \begin{bmatrix} u \\ v \end{bmatrix} = J \begin{bmatrix} y \\ x \end{bmatrix}. \tag{5}$$

Furthermore, the positive definiteness of the matrix $\begin{bmatrix} A & B \\ B & A \end{bmatrix}$ is translated into that both $K$ and $M$ are positive definite since

$$J^{\mathrm{T}} \begin{bmatrix} A & B \\ B & A \end{bmatrix} J = \begin{bmatrix} A+B & 0 \\ 0 & A-B \end{bmatrix} = \begin{bmatrix} M & 0 \\ 0 & K \end{bmatrix}. \tag{6}$$

Because of the equivalence of the eigenvalue problems (1) and (4), we still refer to (4) as an LREP which will be one to be studied from now on, unless otherwise explicitly stated differently.

## 2.1 Basic Eigen-Properties

It is straightforward to verify that

$$H\begin{bmatrix} y \\ x \end{bmatrix} = \lambda \begin{bmatrix} y \\ x \end{bmatrix} \quad \Rightarrow \quad H\begin{bmatrix} y \\ -x \end{bmatrix} = -\lambda \begin{bmatrix} y \\ -x \end{bmatrix}. \tag{7}$$

This implies that the eigenvalues of $H$ come in pair $\{\lambda, -\lambda\}$ and their associated eigenvectors enjoy a simple relationship. In fact, as shown in [1], there exists a nonsingular $\Phi \in \mathbb{R}^{n\times n}$ such that

$$K = \Psi\Lambda^2\Psi^{\mathrm{T}} \quad \text{and} \quad M = \Phi\Phi^{\mathrm{T}}, \tag{8a}$$

where $\Lambda = \operatorname{diag}(\lambda_1, \lambda_2, \ldots, \lambda_n)$ and $\Psi = \Phi^{-\mathrm{T}}$. In particular

$$H\begin{bmatrix} \Psi\Lambda & \Psi\Lambda \\ \Phi & -\Phi \end{bmatrix} = \begin{bmatrix} \Psi\Lambda & \Psi\Lambda \\ \Phi & -\Phi \end{bmatrix}\begin{bmatrix} \Lambda & 0 \\ 0 & -\Lambda \end{bmatrix}. \tag{8b}$$

Thus $H$ is diagonalizable and has the eigen-decomposition (8b).

The notion of invariant subspace (*aka* eigenspace) is an important concept for the standard matrix eigenvalue problem not only in theory but also in numerical computation. In the context of LREP (4), with consideration of its eigen-properties as revealed by (7) and (8b), in [1, 2] we introduced a *pair of deflating subspaces* of $\{K, M\}$, by which we mean a pair $\{\mathscr{U}, \mathscr{V}\}$ of two $k$-dimensional subspaces $\mathscr{U} \subseteq \mathbb{R}^n$ and $\mathscr{V} \subseteq \mathbb{R}^n$ such that

$$K\mathscr{U} \subseteq \mathscr{V} \quad \text{and} \quad M\mathscr{V} \subseteq \mathscr{U}. \tag{9}$$

Let $U \in \mathbb{R}^{n\times k}$ and $V \in \mathbb{R}^{n\times k}$ be the basis matrices of $\mathscr{U}$ and $\mathscr{V}$, respectively. Then (9) can be restated as that there exist $K_{\mathrm{R}} \in \mathbb{R}^{k\times k}$ and $M_{\mathrm{R}} \in \mathbb{R}^{k\times k}$ such that

$$KU = VK_{\mathrm{R}} \quad \text{and} \quad MV = UM_{\mathrm{R}}, \tag{10}$$

and vice versa, or equivalently,

$$H\begin{bmatrix} V & 0 \\ 0 & U \end{bmatrix} = \begin{bmatrix} V & 0 \\ 0 & U \end{bmatrix} H_{\mathrm{R}} \quad \text{with} \quad H_{\mathrm{R}} := \begin{bmatrix} 0 & K_{\mathrm{R}} \\ M_{\mathrm{R}} & 0 \end{bmatrix},$$

i.e., $\mathscr{V} \oplus \mathscr{U}$ is an invariant subspace of $H$ [1, Theorem 2.4]. We call $\{U, V, K_{\mathrm{R}}, M_{\mathrm{R}}\}$ an *eigen-quaternary* of $\{K, M\}$ [57].

Given a pair of deflating subspaces $\{\mathscr{U}, \mathscr{V}\} = \{\mathscr{R}(U), \mathscr{R}(V)\}$, a part of the eigenpairs of $H$ can be obtained via solving the smaller eigenvalue problem [1, Theorem 2.5]. Specifically, if

$$H_{\mathrm{R}}\hat{z} := \begin{bmatrix} 0 & K_{\mathrm{R}} \\ M_{\mathrm{R}} & 0 \end{bmatrix} \begin{bmatrix} \hat{y} \\ \hat{x} \end{bmatrix} = \lambda \begin{bmatrix} \hat{y} \\ \hat{x} \end{bmatrix} =: \lambda\hat{z}, \tag{11}$$

then $(\lambda, \begin{bmatrix} V\hat{y} \\ U\hat{x} \end{bmatrix})$ is an eigenpair of $H$. The matrix $H_{\mathrm{R}}$ is the restriction of $H$ onto $\mathscr{V} \oplus \mathscr{U}$ with respect to the basis matrices $V \oplus U$. Moreover, the eigenvalues of $H_{\mathrm{R}}$ are uniquely determined by the pair of deflating subspaces $\{\mathscr{U}, \mathscr{V}\}$ [2].

There are infinitely many choices of $\{K_{\mathrm{R}}, M_{\mathrm{R}}\}$ in (10). The most important one introduced in [57] is the *Rayleigh quotient pair*, denoted by $\{K_{\mathrm{RQ}}, M_{\mathrm{RQ}}\}$, of the LREP (4) associated with $\{\mathscr{R}(U), \mathscr{R}(V)\}$:

$$K_{\mathrm{RQ}} := (U^{\mathrm{T}}V)^{-1}U^{\mathrm{T}}KU \quad \text{and} \quad M_{\mathrm{RQ}} := (V^{\mathrm{T}}U)^{-1}V^{\mathrm{T}}MV, \tag{12}$$

and accordingly,

$$H_{\mathrm{RQ}} = \begin{bmatrix} 0 & K_{\mathrm{RQ}} \\ M_{\mathrm{RQ}} & 0 \end{bmatrix}.$$

Note that $H_{\mathrm{RQ}}$ so defined is not of the LREP type because $K_{\mathrm{RQ}}$ and $M_{\mathrm{RQ}}$ are not symmetric unless $U^{\mathrm{T}}V = I_k$. To circumvent this, we factorize $W := U^{\mathrm{T}}V$ as $W = W_1^{\mathrm{T}}W_2$, where $W_i \in \mathbb{R}^{k\times k}$ are nonsingular, and define

$$H_{\mathrm{SR}} := \begin{bmatrix} 0 & W_1^{-\mathrm{T}}U^{\mathrm{T}}KUW_1^{-1} \\ W_2^{-\mathrm{T}}V^{\mathrm{T}}MVW_2^{-1} & 0 \end{bmatrix} = [W_2 \oplus W_1]H_{\mathrm{RQ}}[W_2 \oplus W_1]^{-1}. \tag{13}$$

Thus $H_{\mathrm{RQ}}$ is similar to $H_{\mathrm{SR}}$. The latter is of the LREP type and has played an important role in [1, 2] for the LREP, much the same role as played by the Rayleigh quotient matrix in the symmetric eigenvalue problem [36].

Up to this point, our discussion is under the assumption that $\{\mathscr{R}(U), \mathscr{R}(V)\}$ is a pair of deflating subspaces. But as far as the construction of $H_{\mathrm{RQ}}$ is concerned, this is not necessary, so long as $U^{\mathrm{T}}V$ is nonsingular. The same statement also goes for $H_{\mathrm{SR}}$. In fact, a key component in [2, 58] on eigenvalue approximations for the LREP is the use of the eigenvalues of $H_{\mathrm{SR}}$ to approximate part of the eigenvalues of $H$.

## 2.2 Thouless' Minimization Principle

Back to 1961, Thouless [49] showed that the smallest positive eigenvalue $\lambda_1$ of the LREP (1) admits the following minimization principle:

$$\lambda_1 = \min_{u,v} \rho_t(u, v), \tag{14}$$

where $\rho_t(u, v)$ is defined by

$$\rho_t(u, v) = \frac{\begin{bmatrix} u \\ v \end{bmatrix}^{\mathrm{T}} \begin{bmatrix} A & B \\ B & A \end{bmatrix} \begin{bmatrix} u \\ v \end{bmatrix}}{|u^{\mathrm{T}} u - v^{\mathrm{T}} v|}. \tag{15}$$

The minimization in (14) is taken among all vectors $u, v$ such that $u^{\mathrm{T}} u - v^{\mathrm{T}} v \neq 0$.

By the similarity transformation (3) and using the relationships in (5), we have

$$\rho_t(u, v) \equiv \rho(x, y) := \frac{x^{\mathrm{T}} K x + y^{\mathrm{T}} M y}{2|x^{\mathrm{T}} y|}, \tag{16}$$

and thus equivalently

$$\lambda_1 = \min_{x,y} \rho(x, y). \tag{17}$$

The minimization here is taken among all vectors $x, y$ such that $x^{\mathrm{T}} y \neq 0$ [53].

We will refer to both $\rho_t(u, v)$ and $\rho(x, y)$ as the *Thouless functionals* but in different forms. Although $\rho_t(u, v) \equiv \rho(x, y)$ under (5), in this paper we primarily work with $\rho(x, y)$ to state extensions of (17) and efficient numerical methods.

## 2.3 New Minimization Principles and Cauchy Interlacing Inequalities

In [1], we have systematically studied eigenvalue minimization principles for the LREP to mirror those for the standard symmetric eigenvalue problems [7, 36]. We proved the following subspace version of the minimization principle (14):

$$\sum_{i=1}^{k} \lambda_i = \frac{1}{2} \min_{U^{\mathrm{T}} V = I_k} \operatorname{trace}(U^{\mathrm{T}} K U + V^{\mathrm{T}} M V), \tag{18}$$

among all $U, V \in \mathbb{R}^{n \times k}$. Moreover if $\lambda_k < \lambda_{k+1}$, then for any $U$ and $V$ that attain the minimum, $\{\mathscr{R}(U), \mathscr{R}(V)\}$ is a pair of deflating subspaces of $\{K, M\}$ and the corresponding $H_{\mathrm{RQ}}$ has eigenvalues $\pm\lambda_i$ $(1 \leq i \leq k)$.

Equation (18) suggests that

$$\frac{1}{2}\,\mathrm{trace}(U^{\mathrm{T}}KU + V^{\mathrm{T}}MV) \quad \text{subject to } U^{\mathrm{T}}V = I_k \tag{19}$$

is a *proper subspace version* of the Thouless functional in the form of $\rho(\cdot,\cdot)$. Exploiting the close relation through (5) between the two different forms of the Thouless functionals $\rho_t(\cdot,\cdot)$ and $\rho(\cdot,\cdot)$, we see that

$$\frac{1}{2}\,\mathrm{trace}(\begin{bmatrix} U \\ V \end{bmatrix}^{\mathrm{T}} \begin{bmatrix} A & B \\ B & A \end{bmatrix} \begin{bmatrix} U \\ V \end{bmatrix}) \quad \text{subject to } U^{\mathrm{T}}U - V^{\mathrm{T}}V = 2I_k,\ U^{\mathrm{T}}V = V^{\mathrm{T}}U \tag{20}$$

is a *proper subspace version* of the Thouless functional in the form of $\rho_t(\cdot,\cdot)$. Also as a consequence of (18), we have

$$\sum_{i=1}^{k} \lambda_i = \frac{1}{2} \min_{\substack{U^{\mathrm{T}}U - V^{\mathrm{T}}V = 2I_k \\ U^{\mathrm{T}}V = V^{\mathrm{T}}U}} \mathrm{trace}(\begin{bmatrix} U \\ V \end{bmatrix}^{\mathrm{T}} \begin{bmatrix} A & B \\ B & A \end{bmatrix} \begin{bmatrix} U \\ V \end{bmatrix}) \tag{21}$$

among all $U, V \in \mathbb{R}^{n\times k}$.

In [1], we also derived the Cauchy-type interlacing inequalities. Specifically, let $U, V \in \mathbb{R}^{n\times k}$ such that $U^{\mathrm{T}}V$ is nonsingular, and denote by $\pm\mu_i$ $(1 \le i \le k)$ the eigenvalues of[1] $H_{\mathrm{RQ}}$, where $0 \le \mu_1 \le \cdots \le \mu_k$. Then

$$\lambda_i \le \mu_i \le \gamma\, \lambda_{i+n-k} \quad \text{for } 1 \le i \le k, \tag{22}$$

where $\gamma = \sqrt{\min\{\kappa(K), \kappa(M)\}}/\cos\angle(\mathscr{U}, \mathscr{V})$, $\mathscr{U} = \mathscr{R}(U)$ and $\mathscr{V} = \mathscr{R}(V)$. Furthermore, if $\lambda_k < \lambda_{k+1}$ and $\lambda_i = \mu_i$ for $1 \le i \le k$, then $\{\mathscr{U}, \mathscr{V}\}$ is a pair of deflating subspaces of $\{K, M\}$ corresponding to the eigenvalues $\pm\lambda_i$ $(1 \le i \le k)$ of $H$ when both $K$ and $M$ are definite.

## 2.4 Bounds on Eigenvalue Approximations

Let $U$, $V \in \mathbb{R}^{n\times k}$ and $U^{\mathrm{T}}V = I_k$. $\{\mathscr{R}(U), \mathscr{R}(V)\}$ is a pair of approximate deflating subspaces intended to approximate $\{\mathscr{R}(\Phi_1), \mathscr{R}(\Psi_1)\}$, where $\Phi_1 = \Phi_{(:,1:k)}$ and $\Psi_1 = \Psi_{(:,1:k)}$. Construct $H_{\mathrm{SR}}$ as in (13). We see $H_{\mathrm{SR}} = H_{\mathrm{RQ}}$ since $U^{\mathrm{T}}V = I_k$. Denote the eigenvalues of $H_{\mathrm{SR}}$ by

$$-\mu_k \le \cdots \le -\mu_1 \le \mu_1 \le \cdots \le \mu_k.$$

[1] In [1], it was stated in terms of the eigenvalues of $H_{\mathrm{SR}}$ which is similar to $H_{\mathrm{RQ}}$ and thus both have the same eigenvalues.

We are interested in bounding

1. the errors in $\mu_i$ as approximations to $\lambda_i$ in terms of the error in $\{\mathscr{R}(U), \mathscr{R}(V)\}$ as an approximation to $\{\mathscr{R}(\Phi_1), \mathscr{R}(\Psi_1)\}$, and conversely
2. the error in $\{\mathscr{R}(U), \mathscr{R}(V)\}$ as an approximation to $\{\mathscr{R}(\Phi_1), \mathscr{R}(\Psi_1)\}$ in terms of the errors in $\mu_i$ as approximations to $\lambda_i$.

To these goals, define

$$\delta_k := \sum_{i=1}^{k}(\mu_i^2 - \lambda_i^2). \tag{23}$$

We know $0 < \lambda_i \le \mu_i$ by (22); so $\delta_k$ defines an error measurement in all $\mu_i$ as approximations to $\lambda_i$ for $1 \le i \le k$. Suppose $\lambda_k < \lambda_{k+1}$. It is proved in [58] that

$$(\lambda_{k+1}^2 - \lambda_k^2)\|\sin\Theta_{M^{-1}}(U,\Phi_1)\|_{\mathrm{F}}^2 \le \delta_k \le \sum_{i=1}^{k}\lambda_i^2 \cdot \tan^2\theta_{M^{-1}}(U,MV) + \frac{\lambda_n^2 - \lambda_1^2}{\cos^2\theta_{M^{-1}}(U,MV)}\|\sin\Theta_{M^{-1}}(U,\Phi_1)\|_{\mathrm{F}}^2, \tag{24a}$$

$$(\lambda_{k+1}^2 - \lambda_k^2)\|\sin\Theta_{K^{-1}}(V,\Psi_1)\|_{\mathrm{F}}^2 \le \delta_k \le \sum_{i=1}^{k}\lambda_i^2 \cdot \tan^2\theta_{K^{-1}}(V,KU) + \frac{\lambda_n^2 - \lambda_1^2}{\cos^2\theta_{K^{-1}}(V,KU)}\|\sin\Theta_{K^{-1}}(V,\Psi_1)\|_{\mathrm{F}}^2, \tag{24b}$$

where $\Theta_{M^{-1}}(U,\Phi_1)$ is the diagonal matrix of the canonical angles between subspaces $\mathscr{R}(U)$ and $\mathscr{R}(\Phi)$ in the $M^{-1}$-inner product, the largest of which is denoted by $\theta_{M^{-1}}(U,\Phi_1)$, and similarly for $\theta_{M^{-1}}(U,MV)$, $\Theta_{K^{-1}}(V,\Psi_1)$, and $\theta_{K^{-1}}(V,KU)$ (see, e.g., [58] for precise definitions). As a result,

$$\|\sin\Theta_{M^{-1}}(U,\Phi_1)\|_{\mathrm{F}} \le \sqrt{\frac{\delta_k}{\lambda_{k+1}^2 - \lambda_k^2}}, \tag{25a}$$

$$\|\sin\Theta_{K^{-1}}(V,\Psi_1)\|_{\mathrm{F}} \le \sqrt{\frac{\delta_k}{\lambda_{k+1}^2 - \lambda_k^2}}. \tag{25b}$$

The inequalities in (24) address item 1 above, while item 2 is answered by these in (25).

# 3 Numerical Algorithms

In [2], we reviewed a list of algorithms for solving the small dense and large sparse LREPs up to 2012. In the recent work [42] for solving dense complex and real LREP, authors established the equivalence between the eigenvalue problem and real Hamiltonian eigenvalue problem. Consequently, a structure preserving algorithm is proposed and implemented using ScaLAPACK [10] on distributed memory computer systems. In this section, we will review recently proposed algorithms for solving large sparse LREPs.

## *3.1 Deflation*

Whether already known or computed eigenpairs can be effectively deflated away to avoid being recomputed is crucial to numerical efficiency in the process of computing more eigenpairs while avoiding the known ones. In [4], we developed a shifting deflation technique by a low-rank update to either $K$ or $M$ and thus the resulting $K$ or $M$ performs at about comparable cost as the original $K$ or $M$ when it comes to do matrix-vector multiplication operations. This deflation strategy is made possible by the following result.

Let $\mathbb{J} = \{i_j \,:\, 1 \le j \le k\} \subset \{1, 2, \dots, n\}$, and let $V \in \mathbb{R}^{n\times k}$ with $\operatorname{rank}(V) = k$ satisfying $\mathscr{R}(V) = \mathscr{R}(\Psi_{(:,\mathbb{J})})$, or equivalently $V = \Psi_{(:,\mathbb{J})} Q$ for some nonsingular $Q \in \mathbb{R}^{k\times k}$. Let $\xi > 0$, and define

$$\underline{H} = \begin{bmatrix} 0 & \underline{K} \\ M & 0 \end{bmatrix} := \begin{bmatrix} 0 & K + \xi VV^{\mathrm{T}} \\ M & 0 \end{bmatrix}. \tag{26}$$

Then $H$ and $\underline{H}$ share the same eigenvalues $\pm\lambda_i$ for $i \notin \mathbb{J}$ and the corresponding eigenvectors, and the rest of eigenvalues of $\underline{H}$ are the square roots of the eigenvalues of $\Lambda_1^2 + \xi QQ^{\mathrm{T}}$, where $\Lambda_1 = \operatorname{diag}(\lambda_{i_1}, \dots, \lambda_{i_k})$. There is a version of this result for updating $M$ only, too.

## *3.2 CG Type Algorithms*

One of the most important numerical implications of the eigenvalue minimization principles such as the ones presented in Sect. 2.2 is the possibility of using optimization approaches such as the steepest descent (SD) method, conjugate gradient (CG) type methods, and their improvements. A key component in these approaches is the line search. But in our case, it turns out that the 4D search is a more natural approach to take. Consider the Thouless functional $\rho(x, y)$. Given a search direction $\begin{bmatrix} q \\ p \end{bmatrix}$ from the current position $\begin{bmatrix} y \\ x \end{bmatrix}$, the basic idea of the line search [27, 29] is to look for the best possible scalar argument $t$ to minimize $\rho$:

$$\min_t \rho(x + tp, y + tq) \tag{27}$$

on the line $\left\{ \begin{bmatrix} y \\ x \end{bmatrix} + t \begin{bmatrix} q \\ p \end{bmatrix} : t \in \mathbb{R} \right\}$. While (27) does have an explicit solution through calculus, it is cumbersome. Another related search idea is the so-called *dual-channel* search [13] through solving the minimization problem

$$\min_{s,t} \rho(x + sp, y + tq), \tag{28}$$

where the search directions $p$ and $q$ are selected as the partial gradients $\nabla_x \rho$ and $\nabla_y \rho$ to be given in (31). The minimization problem (28) is then solved iteratively by freezing one of $s$ and $t$ and minimizing the functional $\rho$ over the other in an alternative manner.

In [2] we proposed to look for four scalars $\alpha$, $\beta$, $s$, and $t$ for the minimization problem

$$\inf_{\alpha,\beta,s,t} \rho(\alpha x + sp, \beta y + tq) = \min_{u \in \mathscr{R}(U),\, v \in \mathscr{R}(V)} \rho(u, v), \tag{29}$$

where $U = [x, p]$ and $V = [y, q]$. This no longer performs a line search (27) but a 4-*dimensional subspace search* (4D *search* for short) within the 4-*dimensional subspace*:

$$\left\{ \begin{bmatrix} \beta y + tq \\ \alpha x + sp \end{bmatrix} \quad \text{for all scalars } \alpha,\ \beta,\ s, \text{ and } t \right\}. \tag{30}$$

There are several advantages of this 4D search over the line search (27) and dual-channel search (28): (1) the right-hand side of (29) can be solved by the LREP for the $4 \times 4$ $H_{\mathrm{SR}}$ constructed with $U = [x, p]$ and $V = [y, q]$, provided $U^{\mathrm{T}} V$ is nonsingular; (2) the 4D search yields a better approximation because of the larger search subspace; (3) most importantly, it paves the way for a block version to simultaneously approximate several interested eigenpairs.

The partial gradients of the Thouless functional $\rho(x, y)$ with respect to $x$ and $y$ will be needed for various minimization approaches. Let $x$ and $y$ be perturbed to $x + p$ and $y + q$, respectively, where $p$ and $q$ are assumed to be small in magnitude. Assuming $x^{\mathrm{T}} y \neq 0$, up to the first order in $p$ and $q$, we have [2]

$$\rho(x + p, y + q) = \rho(x, y) + \frac{1}{x^{\mathrm{T}} y} p^{\mathrm{T}} [Kx - \rho(x, y)\, y] + \frac{1}{x^{\mathrm{T}} y} q^{\mathrm{T}} [My - \rho(x, y)\, x]$$

to give the partial gradients of $\rho(x, y)$ with respect to $x$ and $y$

$$\nabla_x \rho = \frac{1}{x^{\mathrm{T}} y} [Kx - \rho(x, y)\, y], \quad \nabla_y \rho = \frac{1}{x^{\mathrm{T}} y} [My - \rho(x, y)\, x]. \tag{31}$$

With the partial gradients (31) and the 4D-search, extensions of the SD method and nonlinear CG method for the LREP are straightforward. But more efficient approaches lie in their block versions. In [39], a block 4D SD algorithm is presented and validated for excitation energies calculations of simple molecules in time-dependent density functional theory. Most recently, borrowing many proven techniques in the symmetric eigenvalue problem such as LOBPCG [19] and augmented projection subspace approaches [15, 18, 23, 37, 55], we developed an *extended locally optimal block preconditioned 4-D CG algorithm* (ELOBP4dCG) in [4]. The key idea for its iterative step is as follows. Consider the eigenvalue problem for

$$\underline{\boldsymbol{A}} - \lambda \boldsymbol{B} \equiv \begin{bmatrix} M & 0 \\ 0 & \underline{K} \end{bmatrix} - \lambda \begin{bmatrix} 0 & I \\ I & 0 \end{bmatrix} \tag{32}$$

which is equivalent to the LREP for $\underline{H}$ in (26). This is a positive semidefinite pencil in the sense that $\boldsymbol{A} - \lambda_0 \boldsymbol{B} \succeq 0$ for $\lambda_0 = 0$ [25, 26]. Now at the beginning of the $(i+1)$st iterative step, we have approximate eigenvectors

$$z_j^{(i)} := \begin{bmatrix} y_j^{(i)} \\ x_j^{(i)} \end{bmatrix}, \quad z_j^{(i-1)} := \begin{bmatrix} y_j^{(i-1)} \\ x_j^{(i-1)} \end{bmatrix} \quad \text{for } 1 \le j \le n_b,$$

where $n_b$ is the block size, the superscripts $(i-1)$ and $(i)$ indicate that they are for the $(i-1)$st and $i$th iterative steps, respectively. We then compute a basis matrix $\begin{bmatrix} V_1 \\ U_1 \end{bmatrix}$ of

$$\bigcup_{j=1}^{n_b} \mathscr{K}_m(\Pi[\underline{\boldsymbol{A}} - \rho(x_j^{(i)}, y_j^{(i)})\boldsymbol{B}], z_j^{(i)}), \tag{33}$$

where $\Pi$ is some preconditioner such as $\underline{\boldsymbol{A}}^{-1}$ and $\mathscr{K}_m(\Pi[\underline{\boldsymbol{A}} - \rho(x_j^{(i)}, y_j^{(i)})\boldsymbol{B}], z_j^{(i)})$ is the $m$th Krylov subspace, and then compute two basis matrices $V$ and $U$ for the subspaces

$$\mathscr{V} = \mathscr{R}(V_1) + \operatorname{span}\{y_j^{(i-1)}, \text{ for } 1 \le j \le n_b\}, \tag{34a}$$

$$\mathscr{U} = \mathscr{R}(U_1) + \operatorname{span}\{x_j^{(i-1)}, \text{ for } 1 \le j \le n_b\}, \tag{34b}$$

respectively, and finally solve the projected eigenvalue problem for

$$\begin{bmatrix} U & 0 \\ 0 & V \end{bmatrix}^{\mathrm{T}} (H - \lambda I) \begin{bmatrix} V & 0 \\ 0 & U \end{bmatrix} = \begin{bmatrix} 0 & U^{\mathrm{T}}KU \\ V^{\mathrm{T}}MV & 0 \end{bmatrix} - \lambda \begin{bmatrix} U^{\mathrm{T}}V & \\ & V^{\mathrm{T}}U \end{bmatrix} \tag{35}$$

to construct new approximations $z_j^{(i+1)}$ for $1 \le j \le n_b$. When $m = 2$ in (33), it gives the LOBP4dCG of [1].

As an illustrative example to display the convergence behavior of ELOBP4dCG, Fig. 1, first presented in [4], shows iterative history plots of LOBP4dCG and ELOBP4dCG on an LREP arising from a time-dependent density-functional theory simulation of a $Na_2$ sodium in QUANTUM EXPRESSO [39]. At each iteration $i$, there are 4 normalized residuals $\|\underline{H}z - \mu z\|_1/((\|\underline{H}\|_1 + \mu)\|z\|_1)$ which move down as $i$ goes. As soon as one reaches $10^{-8}$, the corresponding eigenpair $(\mu, z)$ is deflated and locked away, and a new residual shows up at the top. We see dramatic reductions in the numbers of iterations required in going from from $m = 2$ to $m = 3$, and

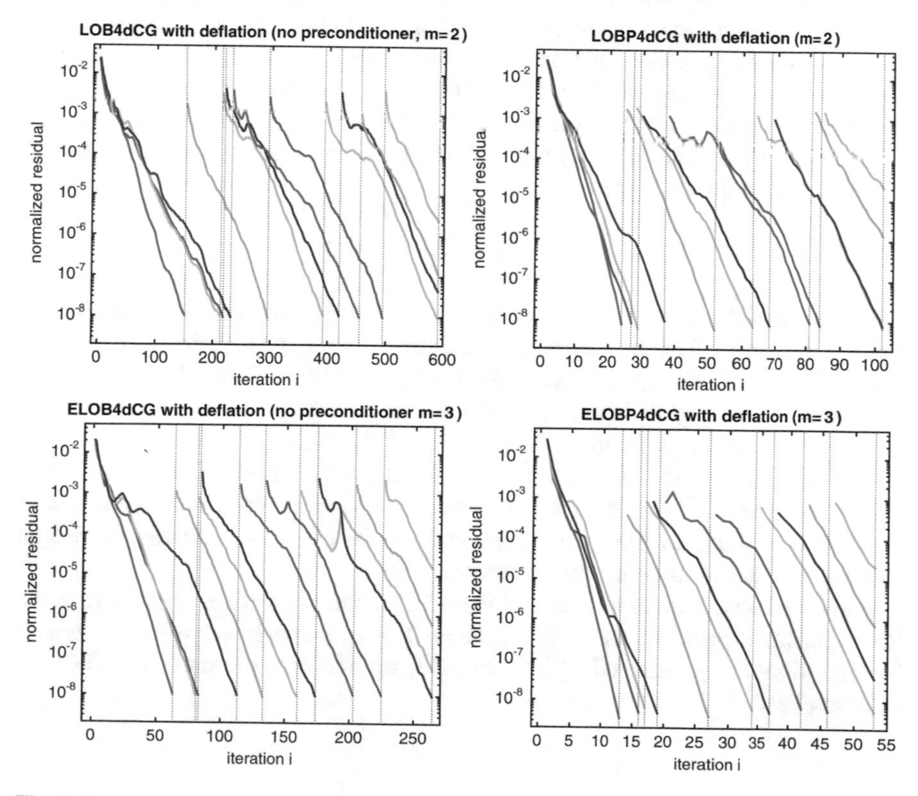


**Fig. 1** Top row: convergence of LOB4dCG (i.e., $m = 2$) without preconditioning (left) and with deflation (right). Bottom row: convergence of extended LOB4dCG (ELOB4dCG) with $m = 3$ without preconditioning (left) and with deflation (right)

in going from "without preconditioning" to "with preconditioning". The powers of using a preconditioner and extending the searching subspace are in display prominently. More detail can be found in [4].

### 3.3 Other Methods

There is a natural extension of Lanczos method based on the following decompositions. Given $0 \neq v_0 \in \mathbb{R}^n$ and $0 \neq u_0 \in \mathbb{R}^n$ such that $Mv_0 = u_0$, there exist nonsingular $U, V \in \mathbb{R}^{n \times n}$ such that $Ve_1 = \alpha v_0$ and $Ue_1 = \beta u_0$ for some $\alpha, \beta \in \mathbb{R}$, and

$$U^{\mathrm{T}}KU = T, \quad V^{\mathrm{T}}MV = D, \tag{36}$$

where $T$ is tridiagonal, $D$ is diagonal and $U^{\mathrm{T}}V = I_n$. Partially realizing (36) leads to the first Lanczos process in [46]. A similar Lanczos process is also studied in [11] for estimating absorption spectrum with the linear response time-dependent density functional theory. There is an early work by Tsiper [52, 53] on a Lanczos-type process to reduce both $K$ and $M$ to tridiagonal. Generically, Tsiper's Lanczos process converges at only half the speed of the Lanczos process based on (36).

Recently, Xu and Zhong [56] proposed a Golub-Kahan-Lanczos type process that partially realize the factorizations:

$$KX = YG, \quad MY = XG^{\mathrm{T}},$$

where $G$ is bidiagonal, $X^{\mathrm{T}}KX = I_n$ and $Y^{\mathrm{T}}MY = I_n$. The basic idea is to use the singular values of the partially realized $G$ to approximate some positive eigenvalues of $H$. Numerical results there suggest that the Golub-Kahan-Lanczos process performs slightly better than the Lanczos process based on (36).

The equations in (8a) implies $KM = \Phi\Lambda^2\Phi^{-1}$. Noticing $\lambda_i^2$ for $1 \leq i \leq k$ lie in low end of the spectrum of $KM$, in [48] the authors devised a block Chebyshev-Davidson approach to build subspaces through suppress components of vectors in the direction of eigenvectors associated with $\lambda_i^2$ for $i > k + 1$. Numerical results there show that the approach can work quite well.

Most recently, structurally inverse-based iterative solvers for very large scale BS eigenvalue problem using the reduced basis approach via low-rank tensor factorizations are presented in [5, 6]. In [21], an indefinite variant of LOBPCG is also proposed.

# 4 Perturbation and Error Analysis

First we consider the perturbation of the LREP (4). Recall the eigen-decompositions in (8), and let

$$Z = \begin{bmatrix} \Psi\Lambda^{1/2} & \Psi\Lambda^{1/2} \\ -\Phi\Lambda^{-1/2} & \Phi\Lambda^{-1/2} \end{bmatrix}. \tag{37}$$

Suppose $H$ is perturbed to $\widetilde{H}$ with correspondingly positive definite $\widetilde{K}$ and $\widetilde{M}$. The same decompositions as in (8) for $\widetilde{H}$ exist. Adopt the same notations for the perturbed LREP for $\widetilde{H}$ as those for $H$ except with a *tilde* on each symbol. It was proved in [57] that

$$\max_{1\le i\le n} |\widetilde{\lambda}_i - \lambda_i| \le \|Z\|_2\|\widetilde{Z}\|_2 \max\{\|\widetilde{M} - M\|_2, \|\widetilde{K} - K\|_2\}, \tag{38a}$$

$$\sqrt{\sum_{i=1}^{n} |\widetilde{\lambda}_i - \lambda_i|^2} \le \frac{1}{\sqrt{2}}\|Z\|_2\|\widetilde{Z}\|_2 \sqrt{\|\widetilde{M} - M\|_{\mathrm{F}}^2 + \|\widetilde{K} - K\|_{\mathrm{F}}^2}. \tag{38b}$$

These inequalities involve the norms $\|Z\|_2$ and $\|\widetilde{Z}\|_2$ which are not known *a priori*. But they can be bounded in terms of the norms of $K$, $M$, their inverses, and bounds on $\lambda_1$ and $\lambda_n$.

Previously in Sect. 2.1, we note that for an exact pair $\{\mathscr{U}, \mathscr{V}\}$ of deflating subspaces we have (10). In particular, $KU = VK_{\mathrm{RQ}}$ and $MV = UM_{\mathrm{RQ}}$, where $U \in \mathbb{R}^{n\times k}$ and $V \in \mathbb{R}^{n\times k}$ are the basis matrices for $\mathscr{U}$ and $\mathscr{V}$, respectively. When $\{\mathscr{U}, \mathscr{V}\}$ is only an approximate pair, it would be interesting to seek backward perturbations $\Delta K$ and $\Delta M$ to $K$ and $M$, respectively, such that

$$(K + \Delta K)U = VK_{\mathrm{RQ}} \quad \text{and} \quad (M + \Delta M)V = UM_{\mathrm{RQ}}. \tag{39}$$

In the other word, $\{\mathscr{U}, \mathscr{V}\}$ is an exact pair for $\{K + \Delta K, M + \Delta M\}$. Since $K$ and $M$ are symmetric, we further restrict $\Delta K$ and $\Delta M$ to be symmetric, too. The first and foremost question is, naturally, if such perturbations $\Delta K$ and $\Delta M$ exist, i.e., if the set

$$\mathbb{B} := \big\{(\Delta K, \Delta M) \,:\, \Delta K^{\mathrm{T}} = \Delta K,\ \Delta M^{\mathrm{T}} = \Delta M \in \mathbb{R}^{n\times n} \text{ satisfying (39)}\big\}, \tag{40}$$

is not empty. Indeed $\mathbb{B} \ne \varnothing$ [57]. Next we are interested in knowing

$$\eta(U, V) := \min_{(\Delta K, \Delta M)\in\mathbb{B}} (\|\Delta K\| + \|\Delta M\|), \tag{41}$$

where $\|\cdot\|$ is some matrix norm. Without loss of generality, we assume $U^{\mathrm{T}}U = V^{\mathrm{T}}V = I_k$. It is obtained in [57] that

$$\eta_{\mathrm{F}}(U,V) = \sqrt{2\|\mathscr{R}_K(K_{\mathrm{RQ}})\|_{\mathrm{F}}^2 - \|U^{\mathrm{T}}\mathscr{R}_K(K_{\mathrm{RQ}})\|_{\mathrm{F}}^2} + \sqrt{2\|\mathscr{R}_M(M_{\mathrm{RQ}})\|_{\mathrm{F}}^2 - \|V^{\mathrm{T}}\mathscr{R}_M(M_{\mathrm{RQ}})\|_{\mathrm{F}}^2}, \tag{42}$$

$$\eta_2(U,V) = \|\mathscr{R}_K(K_{\mathrm{RQ}})\|_2 + \|\mathscr{R}_M(M_{\mathrm{RQ}})\|_2, \tag{43}$$

where $\eta_{\mathrm{F}}$ and $\eta_2$ are the ones of (41) with the Frobenius and spectral norms, respectively, and $\mathscr{R}_K(K_{\mathrm{RQ}}) := KU - VK_{\mathrm{RQ}}$ and $\mathscr{R}_M(M_{\mathrm{RQ}}) := MV - UM_{\mathrm{RQ}}$. An immediate consequence of such backward error analysis is bounds on approximation errors by the eigenvalues of $H_{\mathrm{RQ}}$ to some of those of $H$.

There are a couple of recent work [47, 54] on the perturbation of partitioned LREP. Let $K$ and $M$ be partitioned as

$$K = \begin{array}{c} \\ {\scriptstyle k} \\ {\scriptstyle n-k} \end{array}\!\!\begin{array}{c} \begin{array}{cc} {\scriptstyle k} & {\scriptstyle n-k} \end{array} \\ \begin{bmatrix} K_1 & K_{21}^{\mathrm{T}} \\ K_{21} & K_2 \end{bmatrix} \end{array} \quad \text{and} \quad M = \begin{array}{c} \\ {\scriptstyle k} \\ {\scriptstyle n-k} \end{array}\!\!\begin{array}{c} \begin{array}{cc} {\scriptstyle k} & {\scriptstyle n-k} \end{array} \\ \begin{bmatrix} M_1 & M_{21}^{\mathrm{T}} \\ M_{21} & M_2 \end{bmatrix} \end{array}. \tag{44}$$

If $K_{21} = M_{21} = 0$, then $\{\mathscr{U}_0, \mathscr{V}_0\}$ is a pair of deflating subspaces, where $\mathscr{U}_0 = \mathscr{V}_0 = \mathscr{R}(\begin{bmatrix} I_k \\ 0 \end{bmatrix})$. But what if $K_{21} \neq 0$ and/or $M_{21} \neq 0$ but tiny in magnitude? Then $\{\mathscr{U}_0, \mathscr{V}_0\}$ can only be regarded as a pair of *approximate* deflating subspaces, and likely there would exist an *exact* pair $\{\widetilde{\mathscr{U}}, \widetilde{\mathscr{V}}\}$ of deflating subspaces nearby. Specifically, we may seek

$$\widetilde{\mathscr{U}} = \mathscr{R}(\widetilde{U}),\ \widetilde{\mathscr{V}} = \mathscr{R}(\widetilde{V}) \quad \text{with} \quad \widetilde{U} = \begin{bmatrix} I_k \\ P \end{bmatrix},\quad \widetilde{V} = \begin{bmatrix} I_k \\ Q \end{bmatrix}$$

for some $P$ and $Q$. It resembles the well-known Stewart's perturbation analysis for the standard and generalized eigenvalue problems [43–45]. The study along this line for the LREP has been recently conducted in [54].

Alternatively, if $K_{21} = M_{21} = 0$ in (44), then $\mathrm{eig}(H) = \mathrm{eig}(H_1) \cup \mathrm{eig}(H_2)$, where $H_i = \begin{bmatrix} 0 & K_i \\ M_i & 0 \end{bmatrix}$ for $i = 1, 2$, and $\mathrm{eig}(H)$ is the set of eigenvalues of $H$ and similarly for $\mathrm{eig}(H_i)$. Again what if $K_{21} \neq 0$ and/or $M_{21} \neq 0$ but tiny in magnitude? They may be treated as tiny perturbations. It would be interesting to know the effect on the eigenvalues from resetting them to 0, as conceivably to decouple $H$ into two smaller LREPs. It is shown that such an action brings changes to the eigenvalues of $H$ at most proportional to $\|K_{21}\|_2^2 + \|M_{21}\|_2^2$ and reciprocally proportional to the gaps between $\mathrm{eig}(H_1)$ and $\mathrm{eig}(H_2)$ [47].

## 5 Concluding Remarks

Throughout, we have focused on recent studies of the standard LREP (4) with the assumption that $K$ and $M$ are real and symmetric as deduced from the original LREP (1). There are several directions to expand these studies by relaxing the assumption on $K$ and $M$ and, for that matter, accordingly on $A$ and $B$.

An immediate expansion is to allow $K$ and $M$ to be complex but Hermitian and still positive definite. All surveyed results with a minor modification (by changing all transposes to conjugate transposes) hold. Most of the theoretical results in Sects. 2.2 and 2.3 are still valid when only one of $K$ and $M$ is positive and the other is semidefinite, after changing "min" in (17) and (14) to "inf".

Although often $K$ and $M$ are definite, there are cases that one of them is indefinite while the other is still definite [35]. In such cases, all theoretical results in Sects. 2.2–2.4 no longer hold. But some of the numerical methods mentioned in Sect. 3.3, namely, the Lanczos type methods in [46] and the Chebyshev-Davidson approach [48], still work. Recently in [24], a symmetric structure-preserving $\Gamma$QR algorithm is developed for LREPs in the form of (1) without any definiteness assumption.

The following generalized linear response eigenvalue problem (GLREP) [14, 32, 33]

$$\begin{bmatrix} A & B \\ -B & -A \end{bmatrix} \begin{bmatrix} u \\ v \end{bmatrix} = \lambda \begin{bmatrix} \Sigma & \Delta \\ \Delta & \Sigma \end{bmatrix} \begin{bmatrix} u \\ v \end{bmatrix} \tag{45}$$

was studied in [3], where $A$ and $B$ are the same as the ones in (1), and and $\Sigma$ and $\Delta$ are also $n \times n$ with $\Sigma$ being symmetric while $\Delta$ skew-symmetric (i.e., $\Delta^{\mathrm{T}} = -\Delta$) such that $\begin{bmatrix} \Sigma & \Delta \\ \Delta & \Sigma \end{bmatrix}$ is nonsingular. Performing the same orthogonal similarity transformation, we can transform GLREP (45) equivalently to

$$\begin{bmatrix} 0 & K \\ M & 0 \end{bmatrix} \begin{bmatrix} y \\ x \end{bmatrix} = \lambda \begin{bmatrix} E_+ & 0 \\ 0 & E_- \end{bmatrix} \begin{bmatrix} y \\ x \end{bmatrix}, \tag{46}$$

where $E_+^{\mathrm{T}} = E_-$ is nonsingular. Many results parallel to what we surveyed so far for the LREP (4) are obtained in [3].

Both (4) and (46) are equivalent to the generalized eigenvalue problem for

$$\boldsymbol{A} - \lambda \boldsymbol{B} \text{ with } \boldsymbol{A} = \begin{bmatrix} M & 0 \\ 0 & K \end{bmatrix}, \ \boldsymbol{B} = \begin{bmatrix} 0 & I_n \\ I_n & 0 \end{bmatrix} \text{ or } \begin{bmatrix} 0 & E_- \\ E_+ & 0 \end{bmatrix}.$$

Since $\boldsymbol{A} - 0 \cdot \boldsymbol{B} = \boldsymbol{A}$ is positive definite, $\boldsymbol{A} - \lambda \boldsymbol{B}$ falls into the category of the so-called *positive semi-definite matrix pencils* (*positive definite* if both $K$ and $M$ are positive definite). Numerous eigenvalue min-max principles, as generalizations of the classical ones, are obtained in [8, 9, 20, 22, 30, 31] and, more recently, [25, 26].

**Acknowledgements** Bai is supported in part by NSF grants DMS-1522697 and CCF-1527091, and Li is supported in part by NSF grants DMS-1317330 and CCF-1527104, and NSFC grant 11428104.

## References

1. Bai, Z., Li, R.-C.: Minimization principles for the linear response eigenvalue problem, I: theory. SIAM J. Matrix Anal. Appl. **33**(4), 1075–1100 (2012)
2. Bai, Z., Li, R.-C.: Minimization principle for linear response eigenvalue problem, II: Computation. SIAM J. Matrix Anal. Appl. **34**(2), 392–416 (2013)
3. Bai, Z., Li, R.-C.: Minimization principles and computation for the generalized linear response eigenvalue problem. BIT Numer. Math. **54**(1), 31–54 (2014)
4. Bai, Z., Li, R.-C., Lin, W.-W.: Linear response eigenvalue problem solved by extended locally optimal preconditioned conjugate gradient methods. Sci. China Math. **59**(8), 1443–1460 (2016)
5. Benner, P., Khoromskaia, V., Khoromskij, B.N.: A reduced basis approach for calculation of the Bethe-Salpeter excitation energies using low-rank tensor factorization. Technical Report, `arXiv:1505.02696v1` (2015)
6. Benner, P., Dolgov, S., Khoromskaia, V., Khoromskij, B.N.: Fast iterative solution of the Bethe-Salpeter eigenvalue problem using low-rank and QTT tensor approximation. Technical Report, `arXiv:1602.02646v1` (2016)
7. Bhatia, R.: Matrix Analysis. Springer, New York (1996)
8. Binding, P., Ye, Q.: Variational principles for indefinite eigenvalue problems. Linear Algebra Appl. **218**, 251–262 (1995)
9. Binding, P., Najman, B., Ye, Q.: A variational principle for eigenvalues of pencils of Hermitian matrices. Integr. Equ. Oper. Theory **35**, 398–422 (1999)
10. Blackford, L., Choi, J., Cleary, A., D'Azevedo, E., Demmel, J., Dhillon, I., Dongarra, J., Hammarling, S., Henry, G., Petitet, A., Stanley, K., Walker, D., Whaley, R.: ScaLAPACK Users' Guide. SIAM, Philadelphia (1997)
11. Brabec, J., Lin, L., Shao, M., Govind, N., Yang, C., Saad, Y., Ng, E.: Efficient algorithm for estimating the absorption spectrum within linear response TDDFT. J. Chem. Theory Comput. **11**(11), 5197–5208 (2015)
12. Casida, M.E.: Time-dependent density-functional response theory for molecules. In: Chong, D.P. (ed.) Recent advances in Density Functional Methods, pp. 155–189. World Scientific, Singapore (1995)
13. Challacombe, M.: Linear scaling solution of the time-dependent self-consistent-field equations. Computation **2**, 1–11 (2014)
14. Flaschka, U., Lin, W.-W., Wu, J.-L.: A KQZ algorithm for solving linear-response eigenvalue equations. Linear Algebra Appl. **165**, 93–123 (1992)
15. Golub, G., Ye, Q.: An inverse free preconditioned Krylov subspace methods for symmetric eigenvalue problems. SIAM J. Sci. Comput. **24**, 312–334 (2002)
16. Grüning, M., Marini, A., Gonze, X.: Exciton-plasmon states in nanoscale materials: breakdown of the Tamm-Dancoff approximation. Nano Lett. **9**, 2820–2824 (2009)
17. Grüning, M., Marini, A., Gonze, X.: Implementation and testing of Lanczos-based algorithms for random-phase approximation eigenproblems. Comput. Mater. Sci. **50**(7), 2148–2156 (2011)
18. Imakura, A., Du, L., Sakurai, T.: Error bounds of Rayleigh-Ritz type contour integral-based eigensolver for solving generalized eigenvalue problems. Numer. Algorithms **71**, 103–120 (2016)
19. Knyazev, A.V.: Toward the optimal preconditioned eigensolver: locally optimal block preconditioned conjugate gradient method. SIAM J. Sci. Comput. **23**(2), 517–541 (2001)

20. Kovač-Striko, J., Veselić, K.: Trace minimization and definiteness of symmetric pencils. Linear Algebra Appl. **216**, 139–158 (1995)
21. Kressner, D., Pandur, M.M., Shao, M.: An indefinite variant of LOBPCG for definite matrix pencils. Numer. Algorithms **66**, 681–703 (2014)
22. Lancaster, P., Ye, Q.: Variational properties and Rayleigh quotient algorithms for symmetric matrix pencils. Oper. Theory Adv. Appl. **40**, 247–278 (1989)
23. Li, R.-C.: Rayleigh quotient based optimization methods for eigenvalue problems. In: Bai, Z., Gao, W., Su, Y. (eds.) Matrix Functions and Matrix Equations. Series in Contemporary Applied Mathematics, vol. 19, pp. 76–108. World Scientific, Singapore (2015)
24. Li, T., Li, R.-C., Lin, W.-W.: A symmetric structure-preserving gamma-qr algorithm for linear response eigenvalue problems. Technical Report 2016-02, Department of Mathematics, University of Texas at Arlington (2016). Available at http://www.uta.edu/math/preprint/
25. Liang, X., Li, R.-C.: Extensions of Wielandt's min-max principles for positive semi-definite pencils. Linear Multilinear Algebra **62**(8), 1032–1048 (2014)
26. Liang, X., Li, R.-C., Bai, Z.: Trace minimization principles for positive semi-definite pencils. Linear Algebra Appl. **438**, 3085–3106 (2013)
27. Lucero, M.J., Niklasson, A.M.N., Tretiak, S., Challacombe, M.: Molecular-orbital-free algorithm for excited states in time-dependent perturbation theory. J. Chem. Phys. **129**(6), 064114 (2008)
28. Lusk, M.T., Mattsson, A.E.: High-performance computing for materials design to advance energy science. MRS Bull. **36**, 169–174 (2011)
29. Muta, A., Iwata, J.-I., Hashimoto, Y., Yabana, K.: Solving the RPA eigenvalue equation in real-space. Progress Theor. Phys. **108**(6), 1065–1076 (2002)
30. Najman, B., Ye, Q.: A minimax characterization of eigenvalues of Hermitian pencils. Linear Algebra Appl. **144**, 217–230 (1991)
31. Najman, B., Ye, Q.: A minimax characterization of eigenvalues of Hermitian pencils II. Linear Algebra Appl. **191**, 183–197 (1993)
32. Olsen, J., Jørgensen, P.: Linear and nonlinear response functions for an exact state and for an MCSCF state. J. Chem. Phys. **82**(7), 3235–3264 (1985)
33. Olsen, J., Jensen, H.J.A., Jørgensen, P.: Solution of the large matrix equations which occur in response theory. J. Comput. Phys. **74**(2), 265–282 (1988)
34. Onida, G., Reining, L., Rubio, A.: Electronic excitations: density-functional versus many-body Green's function approaches. Rev. Mod. Phys **74**(2), 601–659 (2002)
35. Papakonstantinou, P.: Reduction of the RPA eigenvalue problem and a generalized Cholesky decomposition for real-symmetric matrices. Europhys. Lett. **78**(1), 12001 (2007)
36. Parlett, B.N.: The Symmetric Eigenvalue Problem. SIAM, Philadelphia (1998)
37. Quillen, P., Ye, Q.: A block inverse-free preconditioned Krylov subspace method for symmetric generalized eigenvalue problems. J. Comput. Appl. Math. **233**(5), 1298–1313 (2010)
38. Ring, P., Schuck, P.: The Nuclear Many-Body Problem. Springer, New York (1980)
39. Rocca, D., Bai, Z., Li, R.-C., Galli, G.: A block variational procedure for the iterative diagonalization of non-Hermitian random-phase approximation matrices. J. Chem. Phys. **136**, 034111 (2012)
40. Rocca, D., Lu, D., Galli, G.: Ab initio calculations of optical absorpation spectra: solution of the Bethe-Salpeter equation within density matrix perturbation theory. J. Chem. Phys. **133**(16), 164109 (2010)
41. Saad, Y., Chelikowsky, J.R., Shontz, S.M.: Numerical methods for electronic structure calculations of materials. SIAM Rev. **52**, 3–54 (2010)
42. Shao, M., da Jornada, F.H., Yang, C., Deslippe, J., Louie, S.G.: Structure preserving parallel algorithms for solving the Bethe-Salpeter eigenvalue problem. Linear Algebra Appl. **488**, 148–167 (2016)
43. Stewart, G.W.: Error bounds for approximate invariant subspaces of closed linear operators. SIAM J. Numer. Anal. **8**, 796–808 (1971)
44. Stewart, G.W.: On the sensitivity of the eigenvalue problem $Ax = \lambda Bx$. SIAM J. Numer. Anal. **4**, 669–686 (1972)

45. Stewart, G.W.: Error and perturbation bounds for subspaces associated with certain eigenvalue problems. SIAM Rev. **15**, 727–764 (1973)
46. Teng, Z., Li, R.-C.: Convergence analysis of Lanczos-type methods for the linear response eigenvalue problem. J. Comput. Appl. Math. **247**, 17–33 (2013)
47. Teng, Z., Lu, L., Li, R.-C.: Perturbation of partitioned linear response eigenvalue problems. Electron. Trans. Numer. Anal. **44**, 624–638 (2015)
48. Teng, Z., Zhou, Y., Li, R.-C.: A block Chebyshev-Davidson method for linear response eigenvalue problems. Adv. Comput. Math. (2016). `link.springer.com/article/10.1007/s10444-016-9455-2`
49. Thouless, D.J.: Vibrational states of nuclei in the random phase approximation. Nucl. Phys. **22**(1), 78–95 (1961)
50. Thouless, D.J.: The Quantum Mechanics of Many-Body Systems. Academic, New York (1972)
51. Tretiak, S., Isborn, C.M., Niklasson, A.M.N., Challacombe, M.: Representation independent algorithms for molecular response calculations in time-dependent self-consistent field theories. J. Chem. Phys. **130**(5), 054111 (2009)
52. Tsiper, E.V.: Variational procedure and generalized Lanczos recursion for small-amplitude classical oscillations. J. Exp. Theor. Phys. Lett. **70**(11), 751–755 (1999)
53. Tsiper, E.V.: A classical mechanics technique for quantum linear response. J. Phys. B: At. Mol. Opt. Phys. **34**(12), L401–L407 (2001)
54. Wang, W.-G., Zhang, L.-H., Li, R.-C.: Error bounds for approximate deflating subspaces for linear response eigenvalue problems. Technical Report 2016-01, Department of Mathematics, University of Texas at Arlington (2016). Available at http://www.uta.edu/math/preprint/
55. Wen, Z., Zhang, Y.: Block algorithms with augmented Rayleigh-Ritz projections for large-scale eigenpair computation. Technical Report, arxiv: 1507.06078 (2015)
56. Xu, H., Zhong, H.: Weighted Golub-Kahan-Lanczos algorithms and applications. Department of Mathematics, University of Kansas, Lawrence, KS, January (2016)
57. Zhang, L.-H., Lin, W.-W., Li, R.-C.: Backward perturbation analysis and residual-based error bounds for the linear response eigenvalue problem. BIT Numer. Math. **55**(3), 869–896 (2015)
58. Zhang, L.-H., Xue, J., Li, R.-C.: Rayleigh–Ritz approximation for the linear response eigenvalue problem. SIAM J. Matrix Anal. Appl. **35**(2), 765–782 (2014)

# *Editorial Policy*

1. Volumes in the following three categories will be published in LNCSE:

i) Research monographs
ii) Tutorials
iii) Conference proceedings

Those considering a book which might be suitable for the series are strongly advised to contact the publisher or the series editors at an early stage.

2. Categories i) and ii). Tutorials are lecture notes typically arising via summer schools or similar events, which are used to teach graduate students. These categories will be emphasized by Lecture Notes in Computational Science and Engineering. **Submissions by interdisciplinary teams of authors are encouraged.** The goal is to report new developments – quickly, informally, and in a way that will make them accessible to non-specialists. In the evaluation of submissions timeliness of the work is an important criterion. Texts should be well-rounded, well-written and reasonably self-contained. In most cases the work will contain results of others as well as those of the author(s). In each case the author(s) should provide sufficient motivation, examples, and applications. In this respect, Ph.D. theses will usually be deemed unsuitable for the Lecture Notes series. Proposals for volumes in these categories should be submitted either to one of the series editors or to Springer-Verlag, Heidelberg, and will be refereed. A provisional judgement on the acceptability of a project can be based on partial information about the work: a detailed outline describing the contents of each chapter, the estimated length, a bibliography, and one or two sample chapters – or a first draft. A final decision whether to accept will rest on an evaluation of the completed work which should include

- at least 100 pages of text;
- a table of contents;
- an informative introduction perhaps with some historical remarks which should be accessible to readers unfamiliar with the topic treated;
- a subject index.

3. Category iii). Conference proceedings will be considered for publication provided that they are both of exceptional interest and devoted to a single topic. One (or more) expert participants will act as the scientific editor(s) of the volume. They select the papers which are suitable for inclusion and have them individually refereed as for a journal. Papers not closely related to the central topic are to be excluded. Organizers should contact the Editor for CSE at Springer at the planning stage, see *Addresses* below.

In exceptional cases some other multi-author-volumes may be considered in this category.

4. Only works in English will be considered. For evaluation purposes, manuscripts may be submitted in print or electronic form, in the latter case, preferably as pdf- or zipped ps-files. Authors are requested to use the LaTeX style files available from Springer at http://www.springer.com/gp/authors-editors/book-authors-editors/manuscript-preparation/5636 (Click on LaTeX Template → monographs or contributed books).

For categories ii) and iii) we strongly recommend that all contributions in a volume be written in the same LaTeX version, preferably LaTeX2e. Electronic material can be included if appropriate. Please contact the publisher.

Careful preparation of the manuscripts will help keep production time short besides ensuring satisfactory appearance of the finished book in print and online.

5. The following terms and conditions hold. Categories i), ii) and iii):

Authors receive 50 free copies of their book. No royalty is paid.
Volume editors receive a total of 50 free copies of their volume to be shared with authors, but no royalties.

Authors and volume editors are entitled to a discount of 33.3 % on the price of Springer books purchased for their personal use, if ordering directly from Springer.

6. Springer secures the copyright for each volume.

Addresses:

Timothy J. Barth
NASA Ames Research Center
NAS Division
Moffett Field, CA 94035, USA
barth@nas.nasa.gov

Michael Griebel
Institut für Numerische Simulation
der Universität Bonn
Wegelerstr. 6
53115 Bonn, Germany
griebel@ins.uni-bonn.de

David E. Keyes
Mathematical and Computer Sciences
and Engineering
King Abdullah University of Science
and Technology
P.O. Box 55455
Jeddah 21534, Saudi Arabia
david.keyes@kaust.edu.sa

and

Department of Applied Physics
and Applied Mathematics
Columbia University
500 W. 120 th Street
New York, NY 10027, USA
kd2112@columbia.edu

Risto M. Nieminen
Department of Applied Physics
Aalto University School of Science
and Technology
00076 Aalto, Finland
risto.nieminen@aalto.fi

Dirk Roose
Department of Computer Science
Katholieke Universiteit Leuven
Celestijnenlaan 200A
3001 Leuven-Heverlee, Belgium
dirk.roose@cs.kuleuven.be

Tamar Schlick
Department of Chemistry
and Courant Institute
of Mathematical Sciences
New York University
251 Mercer Street
New York, NY 10012, USA
schlick@nyu.edu

Editor for Computational Science
and Engineering at Springer:
Martin Peters
Springer-Verlag
Mathematics Editorial IV
Tiergartenstrasse 17
69121 Heidelberg, Germany
martin.peters@springer.com

# Lecture Notes in Computational Science and Engineering

1. D. Funaro, *Spectral Elements for Transport-Dominated Equations.*
2. H.P. Langtangen, *Computational Partial Differential Equations.* Numerical Methods and Diffpack Programming.
3. W. Hackbusch, G. Wittum (eds.), *Multigrid Methods V.*
4. P. Deuflhard, J. Hermans, B. Leimkuhler, A.E. Mark, S. Reich, R.D. Skeel (eds.), *Computational Molecular Dynamics: Challenges, Methods, Ideas.*
5. D. Kröner, M. Ohlberger, C. Rohde (eds.), *An Introduction to Recent Developments in Theory and Numerics for Conservation Laws.*
6. S. Turek, *Efficient Solvers for Incompressible Flow Problems.* An Algorithmic and Computational Approach.
7. R. von Schwerin, ***M**ulti **B**ody **S**ystem **SIM**ulation.* Numerical Methods, Algorithms, and Software.
8. H.-J. Bungartz, F. Durst, C. Zenger (eds.), *High Performance Scientific and Engineering Computing.*
9. T.J. Barth, H. Deconinck (eds.), *High-Order Methods for Computational Physics.*
10. H.P. Langtangen, A.M. Bruaset, E. Quak (eds.), *Advances in Software Tools for Scientific Computing.*
11. B. Cockburn, G.E. Karniadakis, C.-W. Shu (eds.), *Discontinuous Galerkin Methods.* Theory, Computation and Applications.
12. U. van Rienen, *Numerical Methods in Computational Electrodynamics.* Linear Systems in Practical Applications.
13. B. Engquist, L. Johnsson, M. Hammill, F. Short (eds.), *Simulation and Visualization on the Grid.*
14. E. Dick, K. Riemslagh, J. Vierendeels (eds.), *Multigrid Methods VI.*
15. A. Frommer, T. Lippert, B. Medeke, K. Schilling (eds.), *Numerical Challenges in Lattice Quantum Chromodynamics.*
16. J. Lang, *Adaptive Multilevel Solution of Nonlinear Parabolic PDE Systems.* Theory, Algorithm, and Applications.
17. B.I. Wohlmuth, *Discretization Methods and Iterative Solvers Based on Domain Decomposition.*
18. U. van Rienen, M. Günther, D. Hecht (eds.), *Scientific Computing in Electrical Engineering.*
19. I. Babuška, P.G. Ciarlet, T. Miyoshi (eds.), *Mathematical Modeling and Numerical Simulation in Continuum Mechanics.*
20. T.J. Barth, T. Chan, R. Haimes (eds.), *Multiscale and Multiresolution Methods.* Theory and Applications.
21. M. Breuer, F. Durst, C. Zenger (eds.), *High Performance Scientific and Engineering Computing.*
22. K. Urban, *Wavelets in Numerical Simulation.* Problem Adapted Construction and Applications.
23. L.F. Pavarino, A. Toselli (eds.), *Recent Developments in Domain Decomposition Methods.*

24. T. Schlick, H.H. Gan (eds.), *Computational Methods for Macromolecules: Challenges and Applications.*

25. T.J. Barth, H. Deconinck (eds.), *Error Estimation and Adaptive Discretization Methods in Computational Fluid Dynamics.*

26. M. Griebel, M.A. Schweitzer (eds.), *Meshfree Methods for Partial Differential Equations.*

27. S. Müller, *Adaptive Multiscale Schemes for Conservation Laws.*

28. C. Carstensen, S. Funken, W. Hackbusch, R.H.W. Hoppe, P. Monk (eds.), *Computational Electromagnetics.*

29. M.A. Schweitzer, *A Parallel Multilevel Partition of Unity Method for Elliptic Partial Differential Equations.*

30. T. Biegler, O. Ghattas, M. Heinkenschloss, B. van Bloemen Waanders (eds.), *Large-Scale PDE-Constrained Optimization.*

31. M. Ainsworth, P. Davies, D. Duncan, P. Martin, B. Rynne (eds.), *Topics in Computational Wave Propagation.* Direct and Inverse Problems.

32. H. Emmerich, B. Nestler, M. Schreckenberg (eds.), *Interface and Transport Dynamics.* Computational Modelling.

33. H.P. Langtangen, A. Tveito (eds.), *Advanced Topics in Computational Partial Differential Equations.* Numerical Methods and Diffpack Programming.

34. V. John, *Large Eddy Simulation of Turbulent Incompressible Flows.* Analytical and Numerical Results for a Class of LES Models.

35. E. Bänsch (ed.), *Challenges in Scientific Computing - CISC 2002.*

36. B.N. Khoromskij, G. Wittum, *Numerical Solution of Elliptic Differential Equations by Reduction to the Interface.*

37. A. Iskc, *Multiresolution Methods in Scattered Data Modelling.*

38. S.-I. Niculescu, K. Gu (eds.), *Advances in Time-Delay Systems.*

39. S. Attinger, P. Koumoutsakos (eds.), *Multiscale Modelling and Simulation.*

40. R. Kornhuber, R. Hoppe, J. Périaux, O. Pironneau, O. Wildlund, J. Xu (eds.), *Domain Decomposition Methods in Science and Engineering.*

41. T. Plewa, T. Linde, V.G. Weirs (eds.), *Adaptive Mesh Refinement – Theory and Applications.*

42. A. Schmidt, K.G. Siebert, *Design of Adaptive Finite Element Software.* The Finite Element Toolbox ALBERTA.

43. M. Griebel, M.A. Schweitzer (eds.), *Meshfree Methods for Partial Differential Equations II.*

44. B. Engquist, P. Lötstedt, O. Runborg (eds.), *Multiscale Methods in Science and Engineering.*

45. P. Benner, V. Mehrmann, D.C. Sorensen (eds.), *Dimension Reduction of Large-Scale Systems.*

46. D. Kressner, *Numerical Methods for General and Structured Eigenvalue Problems.*

47. A. Boriçi, A. Frommer, B. Joó, A. Kennedy, B. Pendleton (eds.), *QCD and Numerical Analysis III.*

48. F. Graziani (ed.), *Computational Methods in Transport.*

49. B. Leimkuhler, C. Chipot, R. Elber, A. Laaksonen, A. Mark, T. Schlick, C. Schütte, R. Skeel (eds.), *New Algorithms for Macromolecular Simulation.*

50. M. Bücker, G. Corliss, P. Hovland, U. Naumann, B. Norris (eds.), *Automatic Differentiation: Applications, Theory, and Implementations.*

51. A.M. Bruaset, A. Tveito (eds.), *Numerical Solution of Partial Differential Equations on Parallel Computers.*

52. K.H. Hoffmann, A. Meyer (eds.), *Parallel Algorithms and Cluster Computing.*

53. H.-J. Bungartz, M. Schäfer (eds.), *Fluid-Structure Interaction.*

54. J. Behrens, *Adaptive Atmospheric Modeling.*

55. O. Widlund, D. Keyes (eds.), *Domain Decomposition Methods in Science and Engineering XVI.*

56. S. Kassinos, C. Langer, G. Iaccarino, P. Moin (eds.), *Complex Effects in Large Eddy Simulations.*

57. M. Griebel, M.A Schweitzer (eds.), *Meshfree Methods for Partial Differential Equations III.*

58. A.N. Gorban, B. Kégl, D.C. Wunsch, A. Zinovyev (eds.), *Principal Manifolds for Data Visualization and Dimension Reduction.*

59. H. Ammari (ed.), *Modeling and Computations in Electromagnetics: A Volume Dedicated to Jean-Claude Nédélec.*

60. U. Langer, M. Discacciati, D. Keyes, O. Widlund, W. Zulehner (eds.), *Domain Decomposition Methods in Science and Engineering XVII.*

61. T. Mathew, *Domain Decomposition Methods for the Numerical Solution of Partial Differential Equations.*

62. F. Graziani (ed.), *Computational Methods in Transport: Verification and Validation.*

63. M. Bebendorf, *Hierarchical Matrices.* A Means to Efficiently Solve Elliptic Boundary Value Problems.

64. C.H. Bischof, H.M. Bücker, P. Hovland, U. Naumann, J. Utke (eds.), *Advances in Automatic Differentiation.*

65. M. Griebel, M.A. Schweitzer (eds.), *Meshfree Methods for Partial Differential Equations IV.*

66. B. Engquist, P. Lötstedt, O. Runborg (eds.), *Multiscale Modeling and Simulation in Science.*

67. I.H. Tuncer, Ü. Gülcat, D.R. Emerson, K. Matsuno (eds.), *Parallel Computational Fluid Dynamics 2007.*

68. S. Yip, T. Diaz de la Rubia (eds.), *Scientific Modeling and Simulations.*

69. A. Hegarty, N. Kopteva, E. O'Riordan, M. Stynes (eds.), *BAIL* 2008 – *Boundary and Interior Layers.*

70. M. Bercovier, M.J. Gander, R. Kornhuber, O. Widlund (eds.), *Domain Decomposition Methods in Science and Engineering XVIII.*

71. B. Koren, C. Vuik (eds.), *Advanced Computational Methods in Science and Engineering.*

72. M. Peters (ed.), *Computational Fluid Dynamics for Sport Simulation.*

73. H.-J. Bungartz, M. Mehl, M. Schäfer (eds.), *Fluid Structure Interaction II - Modelling, Simulation, Optimization.*

74. D. Tromeur-Dervout, G. Brenner, D.R. Emerson, J. Erhel (eds.), *Parallel Computational Fluid Dynamics 2008.*

75. A.N. Gorban, D. Roose (eds.), *Coping with Complexity: Model Reduction and Data Analysis.*

76. J.S. Hesthaven, E.M. Rønquist (eds.), *Spectral and High Order Methods for Partial Differential Equations.*

77. M. Holtz, *Sparse Grid Quadrature in High Dimensions with Applications in Finance and Insurance.*

78. Y. Huang, R. Kornhuber, O.Widlund, J. Xu (eds.), *Domain Decomposition Methods in Science and Engineering XIX.*

79. M. Griebel, M.A. Schweitzer (eds.), *Meshfree Methods for Partial Differential Equations V.*

80. P.H. Lauritzen, C. Jablonowski, M.A. Taylor, R.D. Nair (eds.), *Numerical Techniques for Global Atmospheric Models.*

81. C. Clavero, J.L. Gracia, F.J. Lisbona (eds.), *BAIL 2010 – Boundary and Interior Layers, Computational and Asymptotic Methods.*

82. B. Engquist, O. Runborg, Y.R. Tsai (eds.), *Numerical Analysis and Multiscale Computations.*

83. I.G. Graham, T.Y. Hou, O. Lakkis, R. Scheichl (eds.), *Numerical Analysis of Multiscale Problems.*

84. A. Logg, K.-A. Mardal, G. Wells (eds.), *Automated Solution of Differential Equations by the Finite Element Method.*

85. J. Blowey, M. Jensen (eds.), *Frontiers in Numerical Analysis - Durham 2010.*

86. O. Kolditz, U.-J. Gorke, H. Shao, W. Wang (eds.), *Thermo-Hydro-Mechanical-Chemical Processes in Fractured Porous Media - Benchmarks and Examples.*

87. S. Forth, P. Hovland, E. Phipps, J. Utke, A. Walther (eds.), *Recent Advances in Algorithmic Differentiation.*

88. J. Garcke, M. Griebel (eds.), *Sparse Grids and Applications.*

89. M. Griebel, M.A. Schweitzer (eds.), *Meshfree Methods for Partial Differential Equations VI.*

90. C. Pechstein, *Finite and Boundary Element Tearing and Interconnecting Solvers for Multiscale Problems.*

91. R. Bank, M. Holst, O. Widlund, J. Xu (eds.), *Domain Decomposition Methods in Science and Engineering XX.*

92. H. Bijl, D. Lucor, S. Mishra, C. Schwab (eds.), *Uncertainty Quantification in Computational Fluid Dynamics.*

93. M. Bader, H.-J. Bungartz, T. Weinzierl (eds.), *Advanced Computing.*

94. M. Ehrhardt, T. Koprucki (eds.), *Advanced Mathematical Models and Numerical Techniques for Multi-Band Effective Mass Approximations.*

95. M. Azaïez, H. El Fekih, J.S. Hesthaven (eds.), *Spectral and High Order Methods for Partial Differential Equations ICOSAHOM 2012.*

96. F. Graziani, M.P. Desjarlais, R. Redmer, S.B. Trickey (eds.), *Frontiers and Challenges in Warm Dense Matter.*

97. J. Garcke, D. Pflüger (eds.), *Sparse Grids and Applications – Munich 2012.*

98. J. Erhel, M. Gander, L. Halpern, G. Pichot, T. Sassi, O. Widlund (eds.), *Domain Decomposition Methods in Science and Engineering XXI.*

99. R. Abgrall, H. Beaugendre, P.M. Congedo, C. Dobrzynski, V. Perrier, M. Ricchiuto (eds.), *High Order Nonlinear Numerical Methods for Evolutionary PDEs - HONOM 2013.*

100. M. Griebel, M.A. Schweitzer (eds.), *Meshfree Methods for Partial Differential Equations VII.*

101. R. Hoppe (ed.), *Optimization with PDE Constraints - OPTPDE 2014.*

102. S. Dahlke, W. Dahmen, M. Griebel, W. Hackbusch, K. Ritter, R. Schneider, C. Schwab, H. Yserentant (eds.), *Extraction of Quantifiable Information from Complex Systems.*

103. A. Abdulle, S. Deparis, D. Kressner, F. Nobile, M. Picasso (eds.), *Numerical Mathematics and Advanced Applications - ENUMATH 2013.*

104. T. Dickopf, M.J. Gander, L. Halpern, R. Krause, L.F. Pavarino (eds.), *Domain Decomposition Methods in Science and Engineering XXII.*

105. M. Mehl, M. Bischoff, M. Schäfer (eds.), *Recent Trends in Computational Engineering - CE2014.* Optimization, Uncertainty, Parallel Algorithms, Coupled and Complex Problems.

106. R.M. Kirby, M. Berzins, J.S. Hesthaven (eds.), *Spectral and High Order Methods for Partial Differential Equations - ICOSAHOM'14.*

107. B. Jüttler, B. Simeon (eds.), *Isogeometric Analysis and Applications 2014.*

108. P. Knobloch (ed.), *Boundary and Interior Layers, Computational and Asymptotic Methods – BAIL 2014.*

109. J. Garcke, D. Pflüger (eds.), *Sparse Grids and Applications – Stuttgart 2014.*

110. H. P. Langtangen, *Finite Difference Computing with Exponential Decay Models.*

111. A. Tveito, G.T. Lines, *Computing Characterizations of Drugs for Ion Channels and Receptors Using Markov Models.*

112. B. Karazösen, M. Manguoğlu, M. Tezer-Sezgin, S. Göktepe, Ö. Uğur (eds.), *Numerical Mathematics and Advanced Applications - ENUMATH 2015.*

113. H.-J. Bungartz, P. Neumann, W.E. Nagel (eds.), *Software for Exascale Computing - SPPEXA 2013-2015.*

114. G.R. Barrenechea, F. Brezzi, A. Cangiani, E.H. Georgoulis (eds.), *Building Bridges: Connections and Challenges in Modern Approaches to Numerical Partial Differential Equations.*

115. M. Griebel, M.A. Schweitzer (eds.), *Meshfree Methods for Partial Differential Equations VIII.*

116. C.-O. Lee, X.-C. Cai, D.E. Keyes, H.H. Kim, A. Klawonn, E.-J. Park, O.B. Widlund (eds.), *Domain Decomposition Methods in Science and Engineering XXIII.*

117. T. Sakurai, S.-L. Zhang, T. Imamura, Y. Yamamoto, Y. Kuramashi, T. Hoshi (eds.), *Eigenvalue Problems: Algorithms, Software and Applications in Petascale Computing.* EPASA 2015, Tsukuba, Japan, September 2015.

118. T. Richter (ed.), *Fluid-structure Interactions.* Models, Analysis and Finite Elements.

119. M.L. Bittencourt, N.A. Dumont, J.S. Hesthaven (eds.), *Spectral and High Order Methods for Partial Differential Equations ICOSAHOM 2016.* Selected Papers from the ICOSAHOM Conference, June 27-July 1, 2016, Rio de Janeiro, Brazil.

120. Z. Huang, M. Stynes, Z. Zhang (eds.), *Boundary and Interior Layers, Computational and Asymptotic Methods BAIL 2016.*

121. S.P.A. Bordas, E.N. Burman, M.G. Larson, M.A. Olshanskii (eds.), *Geometrically Unfitted Finite Element Methods and Applications.* Proceedings of the UCL Workshop 2016.

*For further information on these books please have a look at our mathematics catalogue at the following URL:* `www.springer.com/series/3527`

# Monographs in Computational Science and Engineering

1. J. Sundnes, G.T. Lines, X. Cai, B.F. Nielsen, K.-A. Mardal, A. Tveito, *Computing the Electrical Activity in the Heart.*

*For further information on this book, please have a look at our mathematics catalogue at the following URL:* www.springer.com/series/7417

# Texts in Computational Science and Engineering

1. H. P. Langtangen, *Computational Partial Differential Equations.* Numerical Methods and Diffpack Programming. 2nd Edition
2. A. Quarteroni, F. Saleri, P. Gervasio, *Scientific Computing with MATLAB and Octave.* 4th Edition
3. H. P. Langtangen, *Python Scripting for Computational Science.* 3rd Edition
4. H. Gardner, G. Manduchi, *Design Patterns for e-Science.*
5. M. Griebel, S. Knapek, G. Zumbusch, *Numerical Simulation in Molecular Dynamics.*
6. H. P. Langtangen, *A Primer on Scientific Programming with Python.* 5th Edition
7. A. Tveito, H. P. Langtangen, B. F. Nielsen, X. Cai, *Elements of Scientific Computing.*
8. B. Gustafsson, *Fundamentals of Scientific Computing.*
9. M. Bader, *Space-Filling Curves.*
10. M. Larson, F. Bengzon, *The Finite Element Method: Theory, Implementation and Applications.*
11. W. Gander, M. Gander, F. Kwok, *Scientific Computing: An Introduction using Maple and MATLAB.*
12. P. Deuflhard, S. Röblitz, *A Guide to Numerical Modelling in Systems Biology.*
13. M. H. Holmes, *Introduction to Scientific Computing and Data Analysis.*
14. S. Linge, H. P. Langtangen, *Programming for Computations* - A Gentle Introduction to Numerical Simulations with MATLAB/Octave.
15. S. Linge, H. P. Langtangen, *Programming for Computations* - A Gentle Introduction to Numerical Simulations with Python.
16. H.P. Langtangen, S. Linge, *Finite Difference Computing with PDEs* - A Modern Software Approach.
17. B. Gustafsson, *Scientific Computing from a Historical Perspective.*
18. J. A. Trangenstein, *Scientific Computing - Vol. I.* Linear and Nonlinear Equations.

19. J. A. Trangenstein, *Scientific Computing - Vol. II.* Eigenvalues and Optimization.

20. J. A. Trangenstein, *Scientific Computing - Vol. III.* Approximation and Integration.

*For further information on these books please have a look at our mathematics catalogue at the following URL:* `www.springer.com/series/5151`

Printed in the United States
By Bookmasters